INTO THE ICE

ALSO BY MARK SYNNOTT

The Impossible Climb

The Third Pole

INTO THE ICE

The Northwest Passage, the *Polar Sun*, and a 175-Year-Old Mystery

MARK
SYNNOTT

DUTTON

DUTTON
An imprint of Penguin Random House LLC
1745 Broadway, New York, NY 10019
penguinrandomhouse.com

BOOK DESIGN BY LAURA K. CORLESS

Maps by Daniel P. Huffman
Art on title page and pages 1, 171, and 269 © Renan Ozturk;
art on page 111 © Getty/Wikimedia Commons

LIBRARY OF CONGRESS CATALOGING-IN-PUBLICATION DATA
Names: Synnott, Mark, author.
Title: Into the ice : the Northwest Passage, the Polar Sun,
and a 175-year-old mystery / Mark Synnott.
Other titles: Northwest Passage, the Polar Sun,
and a one-hundred-seventy-five-year-old mystery
Description: First edition. | New York : Dutton, an imprint of Penguin Random House LLC, [2025] | Includes bibliographical references and index.
Identifiers: LCCN 2024059864 | ISBN 9780593471524 (hardcover) |
ISBN 9780593471531 (ebook)
Subjects: LCSH: Synnott, Mark—Voyages and travels. | Sailing—Northwest Passage. |
John Franklin Arctic Expedition (1845-1851) | Franklin, John, 1786-1847. | Erebus (Ship) |
Terror (Ship) | Northwest Passage—Discovery and exploration—British. |
Arctic regions—Discovery and exploration—British.
Classification: LCC G640 .S96 2025 | DDC 910.9163/27—dc23/eng/20250213
LC record available at https://lccn.loc.gov/2024059864

Printed in the United States of America
1st Printing

The authorized representative in the EU for product safety and compliance is Penguin Random House Ireland, Morrison Chambers, 32 Nassau Street, Dublin D02 YH68, Ireland, https://eu-contact.penguin.ie.

To Hampton,

who I would follow

to the ends of the earth

CONTENTS

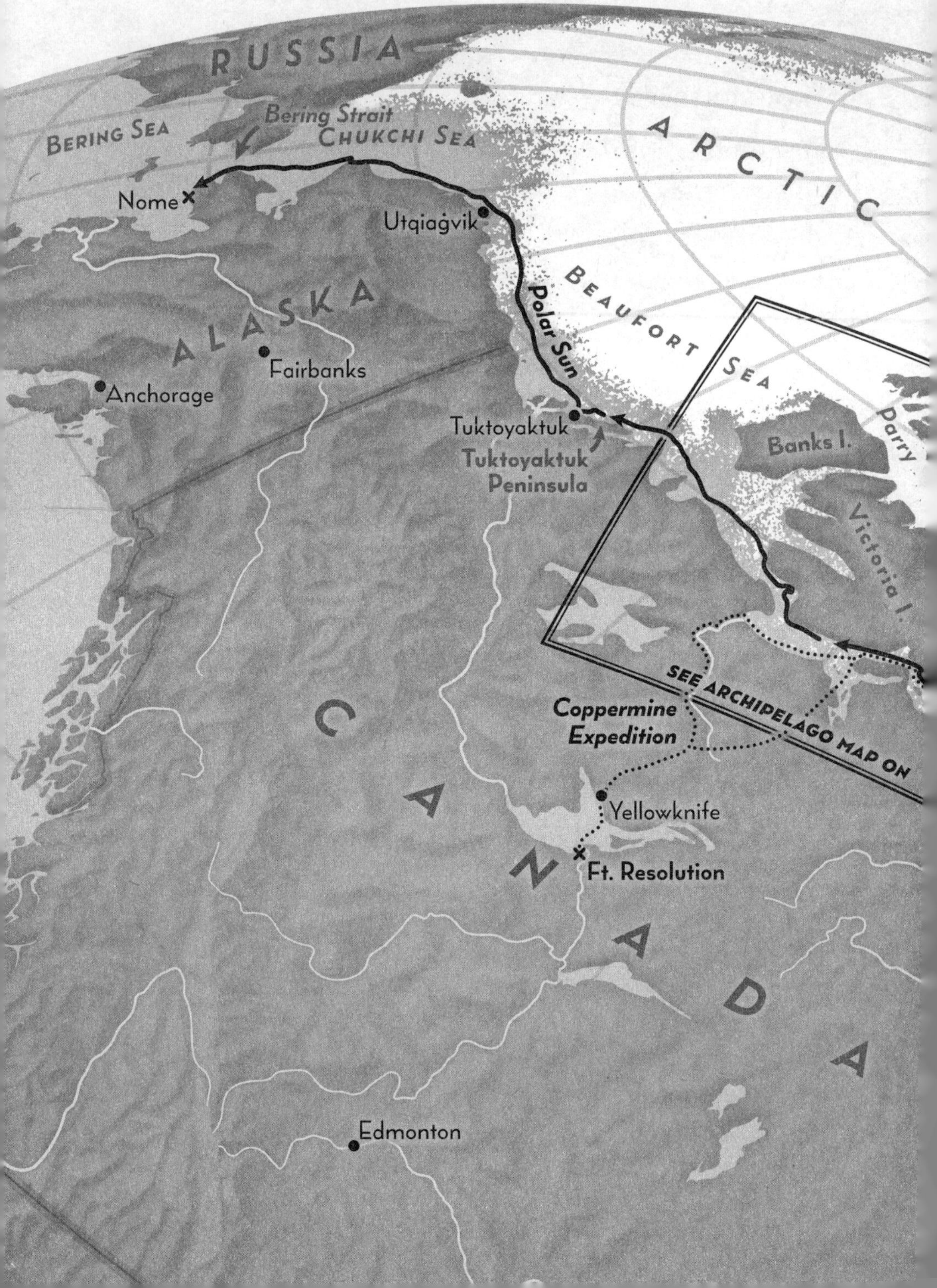

RUSSIA
Bering Strait
BERING SEA
CHUKCHI SEA
ARCTIC
Nome
Utqiaġvik
ALASKA
BEAUFORT SEA
Polar Sun
Fairbanks
Anchorage
Tuktoyaktuk
Tuktoyaktuk Peninsula
Banks I.
Parry
Victoria I.
SEE ARCHIPELAGO MAP ON
Coppermine Expedition
CANADA
Yellowknife
Ft. Resolution
Edmonton

North Pole
OCEAN
Ellesmere I.
Channel
Beechey Island
Devon I.
Bellot Strait
Franklin route
Arctic Bay
GREENLAND
BAFFIN BAY
James Ross Strait
Pond Inlet
Kangiqtualuk Uqquqti (Sam Ford Fjord)
Baffin Island
Polar Sun Spire
Polar Sun
FOLLOWING PAGES
Davis Strait
Nuuk
HUDSON BAY

Beaufort Sea
M'Clure Strait
Melville Island
Banks Island
Viscount Melville Sound
Prince of Wales Strait
Victoria Island
Amundsen Gulf
Dolphin and Union Strait
Cambridge Bay
Polar Sun
Kugluktuk
Coronation Gulf

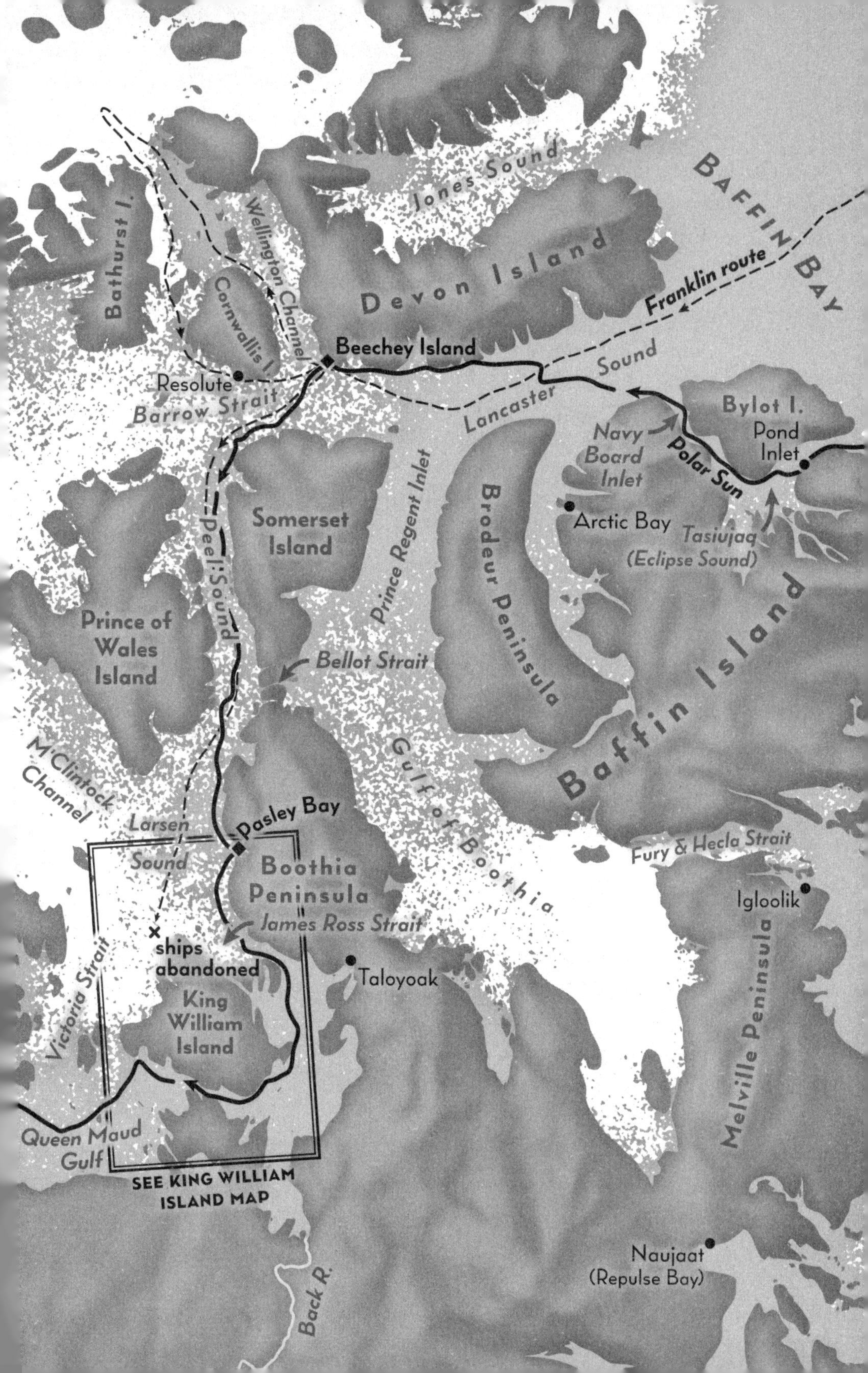

Jones Sound
BAFFIN BAY
Devon Island
Franklin route
Bathurst I.
Wellington Channel
Cornwallis I.
Beechey Island
Resolute
Barrow Strait
Lancaster
Sound
Bylot I.
Pond Inlet
Navy Board Inlet
Polar Sun
Arctic Bay
Tasiujaq
(Eclipse Sound)
Somerset Island
Peel Sound
Prince Regent Inlet
Brodeur Peninsula
Prince of Wales Island
Bellot Strait
Baffin Island
M'Clintock Channel
Gulf of Boothia
Larsen Sound
Pasley Bay
Boothia Peninsula
Fury & Hecla Strait
Igloolik
James Ross Strait
ships abandoned
Taloyoak
Victoria Strait
King William Island
Melville Peninsula
Queen Maud Gulf
SEE KING WILLIAM ISLAND MAP
Naujaat
(Repulse Bay)
Back R.

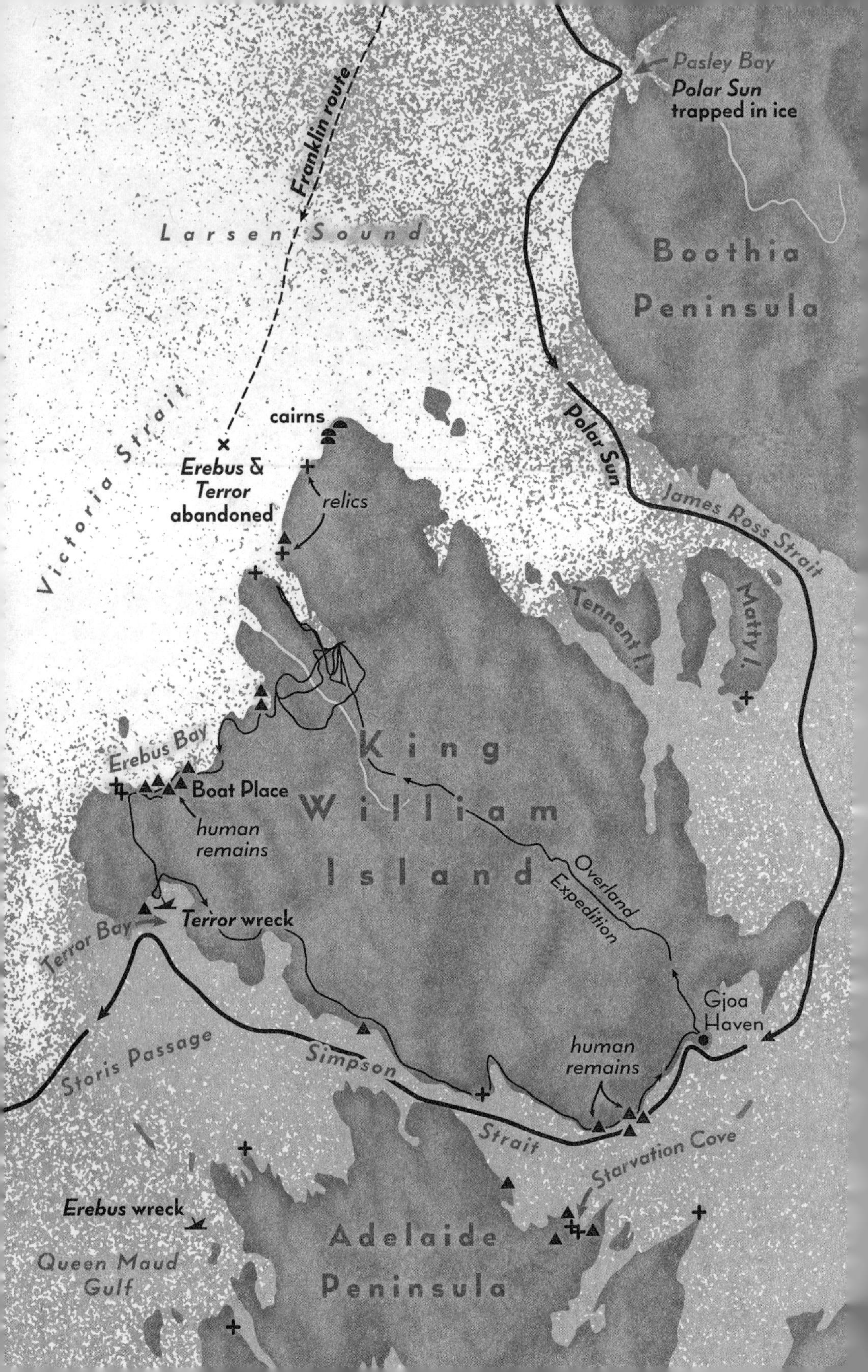

Franklin route
Pasley Bay
Polar Sun trapped in ice
Larsen Sound
Boothia Peninsula
Victoria Strait
cairns
Erebus & *Terror* abandoned
relics
Polar Sun
James Ross Strait
Tennent I.
Matty I.
Erebus Bay
Boat Place
human remains
King William Island
Overland Expedition
Terror wreck
Terror Bay
Gjoa Haven
Storis Passage
Simpson Strait
human remains
Starvation Cove
Erebus wreck
Adelaide Peninsula
Queen Maud Gulf

Introduction

In some ways, this story begins with a name and a long-nurtured secret dream that one day I'd own an oceangoing sailboat, and when I did, I would christen it *Polar Sun*. I never told a single person of this plan, and I'm sure it would have surprised most people who knew me because I was a guy whose life had always revolved around mountains, not the sea.

Ever since I was a young man growing up in New Hampshire amidst the storm-lashed White Mountains, I've centered my existence on outdoor adventure, and I suppose I sort of stumbled into the realization that it is out in the high and wild places in this world that I've always felt the closest to whomever it is that I really am. This never-ending search for my true self has by now led me on more than three dozen international climbing expeditions to places like the Arctic, Patagonia, the Himalayas, the Sahara, and the Amazon jungle. I've been extraordinarily lucky, and along the way I figured out how to turn my passion into a job. In addition to working as a mountain guide and running a climbing school, I've been a member of the North Face Global Athlete Team for more than half my life, and I routinely share stories about my adventures on the stage and in magazines like *Climbing*, *Outside*, and *National Geographic*. More recently, I've written two nonfiction books, *The Impossible Climb* and *The Third Pole*.

But when I look back over all the mountains I have climbed, there is one that stands like a sacred totem of my identity as a climber: an Arctic rock tower that rises nearly a mile from the ocean in a remote fjord on the east coast of Canada's Baffin Island. It's called the Polar Sun Spire, and on my second attempt to climb it in the summer of 1996, Jeff Chapman, Warren Hollinger, and I reached the summit via its north face after thirty-nine days on the wall.

Our route followed a crack system that arced through a massive overhanging amphitheater of dark gray gneiss, with rock quality that varied from diamond hard to a pumice-like choss that we could dig through with our fingers. The cliff was so steep and our line so traversing that we soon passed a point of no return—the only way off was to climb up and over and go down the back. Thus committed, we kept pushing upward, risking huge falls in the searing cold and slowly turning into madmen as we lived on the vertical and slept in a hanging cot called a portaledge. Ours was a custom-designed double-decker, with a tight two-person upper bunk and a hammock below that we nicknamed "Little Rico."

Since the sun never set, we went on an Arctic schedule, climbing for twenty-four hours, then resting for twelve. We had launched with food for thirty days, and when we'd been on the wall for a month and still hadn't reached the summit, we cut rations to one Snickers bar and a single Lipton noodle pack that we'd split three ways every thirty-six hours. For the last ten days of the climb, we went to bed hungry, which, combined with the other stresses of living on the side of a cliff in a cramped space with two other people, often brought out the worst in our characters. "Your leg is on my side," I'd say to Jeff, who'd take a long draw on his hand-rolled cigarette, then silently exhale the smoke in my face. Warren was an EMT and he would share random medical facts with me just to see how I would react. He knew that I was a worrier—and that we were weeks away from the nearest hospital. "Man would it suck if you got appendicitis up here," he said one day as he leered over the side of the top bunk down into Little Rico.

Maybe it was the intensity of the hardship, but when the pendulum swung the other way, as it always did, we found magic up on the Polar Sun Spire—moments of sublime beauty when the sun bathed our portaledge in auburn rays at two in the morning; a profound, pulsing silence broken only by the susurrus whispering of the wind; and pure joy that would sometimes erupt unbidden with cries of exultation heard by no one but us as we dangled high above the frozen ocean. In the Arctic, one can't help but feel one's place in the cosmos: tiny, insignificant, yet finely woven into the fabric that connects all things. It's a feeling I'd never had before, and after bathing in it for so long on that wall, it became a part of my soul forevermore.

It wasn't until many years later that I became a sailor and learned that at sea the essential elements are similar to those found on a big wall. After years of living in cramped portaledges and suffering through storms in the mountains, I realized that life on a sailboat, where one constantly battles with self-doubt and heavy weather, felt strangely familiar—as did the transcendent moments that always punctuated the misery.

And so it seemed fitting, maybe even predestined, that my totem mountain would become the namesake of the boat on which I would one day sail back to the land of the midnight sun.

First, though, I had to find my way from the mountains to the sea, and looking back, I'm not sure that would have happened if not for a chance encounter. It all started with a book called *The Bounty*, a magnificent telling of one of the greatest stories in the history of seafaring. Halfway through Caroline Alexander's account of the 1789 mutiny aboard HMS *Bounty*, I turned the page and found myself spellbound by a circa 1825 painting by Frederick William Beechey of the *Bounty* mutineers rowing one of their longboats across Pitcairn Island's Bounty Bay.

Pitcairn, I learned, is a 1.5-square-mile rock sticking out of the middle of the South Pacific Ocean. It's the most remote inhabited landmass on Earth, and it also has the distinction of being the place where Fletcher Christian attempted to create a utopian society after leading his mutiny against the tyrannical captain William Bligh.

I might have just kept flipping through that photo insert, but in the background of the painting, majestic rock spires rose from the ocean into a misty sky. I, of course, focused immediately on the cliffs and wondered if anyone had ever climbed them. Pitcairn is too small and rugged for an airstrip and too remote for any kind of ferry service, so the only reliable way to get there is via a private yacht. It didn't take me long to hatch a plan to charter a sailboat and lead the first-ever exploratory rock-climbing expedition to Pitcairn, and with financial backing from the North Face and National Geographic, I eventually found a sixty-six-foot sailboat based in French Polynesia called *Picasso*, which was available for private charter. And so, in the spring of 2005, I set off for Pitcairn with three friends, including a budding young filmmaker named Jimmy Chin.

When I first stepped aboard *Picasso*, Captain Mike led me to the tiki-torch-lit cockpit, where a tray of hors d'oeuvres awaited. We sat down, and he showed me how I could reach between my legs to a cooler filled with cold beer. As we clinked bottles, I breathed in the scent of the tropical flowers wafting over from the verdant shore. *I could get used to boat life,* I thought, feeling at ease in the world.

But a few days later, when *Picasso* exited the coral atoll surrounding the Gambier Islands and we sailed out into the deep ocean that encircles the globe along the Tropic of Capricorn, the world as I knew it dropped out from beneath my feet. Before we set off, Captain Mike had told me in his thick Aussie drawl that seasickness was all about "mind over matter, mate." *I've got a strong mind,* I remember thinking. *I should be fine.* But I probably should have factored in that I had somehow made it well into my thirties without ever stepping onto a sailboat.

By staying up in the cockpit and staring at the horizon, I managed to hold it together for several hours as *Picasso* took off on a wild rodeo ride through the waves. But later that night, when I ducked below deck to find my bunk, the boat lurched in an odd way and my stomach did a sudden somersault. That strength of mind immediately vacated my person, and I threw up for hours, eventually dry heaving so hard that I pulled a muscle in my neck. When the sun finally came up the next day, I straggled back to the cockpit, where Captain Mike was tethered to the helm with a soggy lit cigarette dangling from his lips as waves of frothy water swept across the deck. When he saw my pathetic green visage, he cracked a half smile and, without touching his cigarette, advised me that it was "time to man up, dawg."

A few hours later, I talked Captain Mike into turning back, and we spent the next ten days waiting for another weather window in a lagoon off the island of Mangareva. When we set off for Pitcairn the second time, I was wearing a scopolamine seasickness patch behind my ear. That night, I stood watch alone. The breeze was fair, and the sails were full. Captain Mike had prepped me enough that I knew how to read the instruments and keep us roughly on course, so I turned off the autopilot and took the wheel. As *Picasso* glided east, a trail of bioluminescence glowed green in the wake behind us. Ahead, a full moon lit a corridor across the sea and the waves scintillated in a light so heavenly, it felt as if Neptune himself were drawing me down the path to enlightenment. At that moment, it washed over me that I had only ever felt such tranquility before when I was deep in the mountains.

We eventually made it to Pitcairn, where we found that those majestic cliffs in the painting were composed of compressed volcanic ash that crumbled in our hands like overbaked brownies. We didn't climb anything of significance, but it was still the best trip of my life. By the time I got home to New Hampshire, I was hell-bent on getting a boat and learning how to sail—an enthusiasm that I quickly learned was not shared by my first wife.

We already have enough going on and we can't afford a boat, she reasoned. *The kids need you at home.* I'd heard similar arguments before when it came to the next adventure or the next mountain on my list. But then a deal fell into my lap that I couldn't refuse, and a few months later, we were the proud new owners of a twenty-seven-foot sailboat called *Capella*. The price was only one dollar, but it would take some time before I realized why.

|||

Years later, after sailing *Capella* up and down the coast of Maine, including several multiday excursions, I decided it was my turn to sell her to a friend for one dollar and invest in a more ocean-worthy vessel in which I could start venturing offshore. But I was going through a divorce and it was argued (not unreasonably) that it was an inopportune time to make a major life purchase. On the other hand, I wasn't sure what my financial status would be when I came out the other side, and I wanted to lock down a new sailboat while I still had the means.

And that's how, in the summer of 2013, I found myself sailing *Capella*'s upgrade—*Camelot*, a thirty-two-foot ketch with two masts—up the coast of southern Maine. I had bought the boat in New London, Connecticut, and was delivering her with the help of two friends to her new home in Portland. I like to think it was fate that the mainsail tore apart a few miles off a small village called Cape Porpoise, because it happened to be where a woman named

Hampton Kew was temporarily staying with her parents. A couple weeks earlier, I'd taken her up a climb on a cliff called Whitehorse Ledge in North Conway, New Hampshire. And I'd fallen hard for her.

When I called Hampton to say that we had an issue and needed a port of refuge, the Kews, who are lifelong sailors, directed us to a mooring at the head of Cape Porpoise Harbor. After tying up, we rowed our dinghy to a dock where Hampton and her dad were waiting. I still didn't know Hampton very well, but that night, over drinks on the Kews' back patio, I told her about my dream of one day sailing to the South Pacific. It wasn't something that was going to happen quickly, as I still had three kids to raise and my youngest wouldn't graduate high school for another ten years. For the next decade, my plan was to focus on parenting and becoming a competent mariner. Hampton, it turned out, was a newly minted U.S. Coast Guard–licensed captain, and she had a similar dream to sail around the world.

Hampton and I were married in Jackson, New Hampshire, in February 2015. A year later Tommy came into our lives, and that summer, before the little guy could even crawl, we rigged his car seat to hang overhead in *Camelot*'s cabin, counterbalanced with a milk jug on a line run through a pulley to dampen the swing. We sailed as a family for months that year, and again the year after, always following the unwritten rule that no one would ever speak of the possibility that Tommy wouldn't take to the sea like a natural-born mariner.

It was Hampton who suggested that we upgrade to our forever boat sooner rather than later. *Camelot* was capable of sailing across an ocean, but she was tired and badly in need of upgrades that would cost more than she was worth. And at thirty-two feet, she just didn't have enough room for the whole family, which includes my three children—Will, Matt, and Lilla—from my first marriage. Hampton had grown up sailing on her father's fifty-two-foot sailboat, so she knew how much more enjoyable cruising could be with elbow room. Why not get the round-the-world boat now, she reasoned, so that we'd have a few years to learn its idiosyncrasies before setting off on our big voyage?

We sold *Camelot* and a small house I had built in Downeast Maine to fund the purchase of a forty-seven-foot sailboat built in 1982 called a Stevens 47. In the *Used Boat Notebook*, considered the last word on the subject, author John Kretschmer included her as one of "ten great used boats to sail around the world" and called her "a proven offshore thoroughbred." *Hurrah*, as she was then named, had been sitting unused for a few years in Annapolis, Maryland, and she looked worn out. Many of her systems were old and outdated, her sails tatty and weather-beaten, her paint peeling. But during the prepurchase survey, an old mariner who'd spent time in the Taiwan boatyard where she'd been built told me that he'd surveyed thousands of boats, and while *Hurrah* might look a bit "like a stinker," she was actually "a diamond in the rough." She was cutter rigged, which meant she had two headsails. The smaller one—the staysail, which sat closer to the mast—was a storm sail built for heavy weather. The forward sail was the everyday workhorse called the jib or genoa. The cockpit had a full canvas enclosure and was situated in the center of the boat, opening up the aft end for a master suite down below. She would need a lot of TLC to get up to snuff, but she could sleep eight in three cabins and was designed to cross oceans.

When all of my work dried up during the first few months of the pandemic, Hampton and I stocked up on beans and rice and moved aboard *Hurrah* with Tommy. We stayed away from towns, gunkholing along the coast of Maine, getting to know our new vessel and debating whether we dared incur Neptune's wrath by changing the name. By now, Hampton knew about my long-held dream to name a boat after my totem mountain on Baffin Island. *Hurrah* was my third sailboat, but unlike the first two, I knew that she was "the one."

So after much discussion and a bit of cajoling, I took a scraper to the weather-beaten decals on the topsides and we rechristened her *Polar Sun* by smashing a bottle of champagne wrapped in cellophane over the bow.

Later that summer we were moored in Boothbay Harbor on a perfect late-July evening. I sat alone up in the cockpit watching the sun set as Hampton read Tommy a bedtime story below. The boat swung in a light breeze, and as I stared out toward the mouth of the bay, listening to Hampton's voice, I kept circling around a question that had begun to crowd out all other thoughts in my mind: *Where to next?* By this point, I had poured myself into sailing for fifteen years, and my apprenticeship was close to complete. In my first book, *The Impossible Climb*, I'd written about my old friend and climbing partner Alex Honnold and his 2017 free solo ascent of El Capitan in Yosemite. When people asked for his autograph, he often appended it with the words "Go Big!" And now, I decided, it was time for me to find my own El Capitan.

On that fated second date, Hampton and I had hatched a dream to circumnavigate the globe under sail. But that was still a few years out, and I wanted to take *Polar Sun* on a real shakedown voyage first. The most obvious objective was sitting right in front of me—a transatlantic. In recent months, I'd been researching the pros and cons of the two main routes to Europe from New England—north via Newfoundland to Ireland, which was the shortest but thorniest path; or the warm route that led south to Bermuda, the Azores, and the Mediterranean. Crossing the Atlantic is a major endeavor, of course, because unless you're planning to park yourself in Europe (which isn't even possible due to visa restrictions), you have to come back. The round trip is known as an Atlantic Circle, and boats from New England typically head east in May or June, spend the summer in

Europe, then come back across via the Canary Islands and the Caribbean in the late fall. From there, of course, I'd still have a long trip back to Maine.

One of the most important lessons I'd learned from climbing is that objectives matter, and there is probably nothing more elemental than making sure that one is always following one's heart—and not the expectations of others. In fact, listening to my heart was exactly why I was currently contemplating an ocean crossing instead of climbing another mountain. And so I asked myself that night: *If I could go anywhere in the world in this boat, where would that be?*

My father, who was born in Fall River, Massachusetts, in 1929, at the beginning of the Great Depression, had dropped out of high school at age fourteen to work at a factory where they made iron boxes. He'd eventually gotten his GED, gone to college at night, and then somehow found his way into the Advanced Management Program at Harvard Business School. After that, he worked his ass off in Boston for the rest of his career as a banker—a job he did not love. He'd always told me that he was paying his dues so that he could enjoy life when he retired at age sixty-five. But he also often said that he expected to live until he was eighty. I assume he must have done the math because I know I did. Fifteen years sure didn't seem like enough of a payout to me. If my father is proof of anything, it's that we should all envision a high number for ourselves because, sure enough, he died six weeks shy of his eighty-first birthday in 2010. I was with him in the hospital in Florida right before he passed, and he kept telling me that he needed to get the hell out of there so he could get back out on the golf course.

My mom soldiered on without him, but before long the years caught up with her too. In her case, it wasn't kidney disease; it was Alzheimer's and macular degeneration. As her mental facilities and eyesight waned, she lost one beloved aspect of her life after another. The first to go was her weekly bridge game, after one of her friends politely informed her that she couldn't hack it anymore. She became increasingly self-conscious and soon withdrew from her social life altogether. Eventually she wasn't doing much aside from sitting in her chair staring at the wall. And when I visited, she didn't recognize me.

I'm not a fatalistic person, but if your parents die or lose their minds in their eighties, then odds are you probably will too. *I've got thirty years, at the most,* I remember thinking. It sounds like a lot of time, but it's really not. To size it up, all I had to do was look back three decades, to when I was in my early twenties. I wouldn't trade those years for anything, but maybe because they'd been so good, they had flown by.

So what do you REALLY want to do with the time that you have left? I asked myself as *Polar Sun* rocked in the gentle swell of Boothbay Harbor. I wanted to spend as much of it as possible with the two human beings sleeping below and with my three other children. And I know that for a lot of people that's probably all they want—to spend time with their families. But if I'm honest with myself and with you, the truth is that I'm someone who has always needed more than that. I need epic adventure and exploration in my life—like the kind I've found on Polar Sun Spire, high in the Himalaya, and deep in the Amazon jungle—and without it, I know that I'll never be truly happy. So as Tommy and Hampton snoozed below, I looked even deeper within, and that's when it hit me, with a conviction that I'm not sure I can adequately explain, that the compass needle of my heart was pointing north—not east. If I had my druthers and I could sail *Polar Sun* anywhere in the world, the place I wanted to go was back to the Arctic.

Inspired by this sudden revelation, I sprang out of my seat and slipped below, realizing that I had just the book with which to fan this flame. Before moving aboard that summer, I'd filled a small duffel with seafaring titles that I'd grabbed off the shelf at home. I quietly reached into the teak cabinet by my bunk and grabbed a thick tome I had first read in college: *The Arctic Grail: The Quest for the North West Passage and the North Pole, 1818–1909* by Pierre Berton. Back in the cockpit, as I thumbed through the 672 dog-eared pages by headlamp, I found some old notes I'd written years before in the margins of a map series that showed Arctic geography across different decades of the nineteenth century. The blank spaces slowly began to disappear.

|||

Rereading Berton's book reminded me of something I had long forgotten: that the history of early European exploration in the Arctic was intimately entwined with a popular theory called the "Open Polar Sea." Most European explorers in the nineteenth century believed, with little in the way of supporting evidence, that the North Pole and its environs were ice-free. The origin of this myth can be traced back to a map drawn by the Flemish cartographer Gerardus Mercator in 1569. It was called *Nova et Aucta Orbis Terrae Descriptio ad Usum Navigantium Emendate Accommodata*, Latin for "new and more complete representation of the terrestrial globe properly adapted for use in navigation." This was the map on which Mercator first shared his famed "projection," which took the curved latitudinal lines you'd find on a globe and stretched them out so that sailors could use them to plot "great circle" courses across the oceans. To make his projection work, Mercator had to distort the high latitudes, making them look far larger than they are. This is why, even today, we all think that Greenland is three times larger than it really is.

According to the math used to create the Mercator projection, the top (and bottom) of the world would approach infinity in size. Rather than distort his map even further, Mercator created an insert with a top-down view of the Arctic. But in the sixteenth century,

Europeans knew almost nothing about the high latitudes. Inuit and their predecessors had lived and thrived north of the Arctic Circle for thousands of years in what would later become parts of Alaska, Canada, and Greenland, as had the Sami in northern Finland. But at this point, there had effectively been no exchange between these cultures and those in Europe and Asia.

With little to go on in terms of direct knowledge of the Arctic, Mercator turned to some sketchy sources to fill in the gaps. Most notable among these was a book called *Inventio Fortunata* or *Fortunate Discovery*. The author of this mysterious work is unknown, but Mercator believed that it had been written by an English friar in the fourteenth century. By the time Mercator created his famous map, the friar's original text had been lost for over a century, and some questioned whether it ever even existed. In it, the author purportedly claimed to have traveled from Norway to the top of the world, where he discovered a magnetic island encircled by a whirlpool that sucked the frothing ocean into the center of the Earth. Mercator put this island, which he called *Rupes Nigra et Altissima* (the "Black and Very High Rock"), directly on top of the North Pole. Surrounding this thirty-three-mile-wide lodestone, he drew four large islands separated by powerful rivers that poured into the whirlpool, where they disappeared into "the bowels of the earth." And as if to punctuate his map with even more fantastical details, Mercator added a few dragons and a band of pygmies "whose length is four feet" living on one of the polar islands.

The idea of an ice-free North Pole, as ludicrous as it might seem now, persisted for centuries, and by the early 1800s, it was widely accepted as scientific fact. It was Mercator and his overly active imagination that may have been responsible for this widely held fallacy, but out of all the believers in the Open Polar Sea, perhaps the most influential was a man named John Barrow, who, in 1804 at the age of thirty-nine, was appointed as the second secretary to the British

Admiralty. Barrow—whom Berton describes as a "moon-faced man with short cropped hair, bristling black brows, and the tenacious temperament of a bull terrier"—had risen to the highest echelons of the British navy despite humble beginnings.

The precocious son of a tanner from a tiny village in north Lancashire, Barrow left school at age thirteen to found a Sunday school for the poor. After a stint as a clerk at an iron foundry in Liverpool, he worked briefly as a Greenland whaler and then as a mathematics teacher in Greenwich, where he tutored the son of Sir George Staunton, the attorney general of Grenada. It was Barrow's first brush with Britain's upper crust, and never one to miss an opportunity for advancement, he impressed Sir George enough to get himself recommended to Lord Macartney, who had just been appointed as the first British ambassador to China.

Macartney, who had famously written in 1773 that the British ruled a "vast empire, on which the sun never sets," hired Barrow as his household comptroller. During his years at the embassy in Peking (now Beijing), Barrow became fluent in three different dialects of Chinese while also studying Asian literature and science. Macartney promoted Barrow to be his private secretary and then brought him to South Africa, where he helped broker a peace treaty between the Boers and the Zulus before serving as Cape Town's auditor-general. When Barrow returned to England in 1804, Lord Melville, another patron, appointed him to run the Admiralty.

At the time, Great Britain was embroiled in the Napoleonic Wars, and it was Barrow who sent Horatio Nelson off to the Battle of Trafalgar, in which he was mortally wounded—but not before securing a victory that would cement British naval supremacy for the next century. In the years that followed, under Barrow's administration, the British Royal Navy wiped out the last of the French fleet in 1811, fought the United States to a draw in the War of 1812, and then finally defeated Napoleon at the Battle of Waterloo in 1815. According to some, it was Barrow who suggested St. Helena as the location for Napoleon's exile.

Through the course of his forty-year career at the Admiralty, Barrow would oversee the navy's engagement in five wars, but it was the peace that ensued after Waterloo that would forever alter the history of Arctic exploration. As naval ships were decommissioned and officers placed on half pay, Barrow saw an opportunity to put these resources back to work in the name of exploration and conquest. At the time, there were two last great geographical problems that had yet to be solved: the North Pole and the Northwest Passage. And who better to solve them, thought Barrow, than the world's most powerful navy?

England and other European nations had been dreaming about a shortcut to the Far East by heading northwest—instead of the long and dangerous rambles around the great horns of South America and Africa—since the late fifteenth century, when Pope Alexander VI had divided the discovered world between Spain and Portugal. The Treaty of Tordesillas created a line of demarcation that ran north-south down a longitudinal meridian 370 leagues west of the Cape Verde Islands in the Atlantic Ocean. Everything to the west of this line would belong to Spain, and everything to the east, Portugal. It was a time in history when the great European powers were battling one another in a sort of cold war for control of the world's trade markets and natural resources. As Spain, Portugal, France, and Great Britain jockeyed endlessly to expand their empires, they justified their subjugation of millions of indigenous people by telling anyone who would listen that as Christians it was their duty to spread civilization to the "rude and savage tribes of the world," as one British explorer later put it.

Spain, with its extensive holdings in Central and South America, dominated world trade for centuries, but when Great Britain established colonies in India, North America, and the West Indies early in the seventeenth century, it broke the Spanish stranglehold. The British had never accepted the Treaty of Tordesillas, and even if they had, it covered only known lands. All the blanks on the map, including most of the Arctic, were still fair game.

Great Britain had passed a law in 1745 offering a prize of £20,000 (equivalent to about $2.5 million in 2025) to any subject who discovered "a Northern Passage for Vessels by Sea, between the Atlantic and Pacific Oceans, and also to such as shall first approach to the Northern Pole." The reward was revised in 1776 and then again in 1818, with incremental milestones that Barrow hoped would lure more expeditions to throw their hats into the ring. The first to sail into the Arctic from the Atlantic and reach 110 degrees west longitude would claim £5,000 and, for every 20 degrees farther west, another £5,000 would be added (these points are roughly equivalent to the locations today of the Arctic hamlets of Cambridge Bay and Tuktoyaktuk) until someone claimed the full prize by reaching the Bering Strait—or 89 degrees north—which apparently was considered close enough to the North Pole to call it good.

As part of his sales pitch to the Admiralty, Barrow espoused the many scientific discoveries that such expeditions would contribute to humankind—filling in blanks on the map, cataloging new species, mapping ocean currents, and, of course, conducting ethnographic studies of the Arctic's indigenous inhabitants. But the real prize was the glory and nationalistic pride to be had once this geographic first was claimed for Mother England. As Barrow would later put it, if the conquest of the Northwest Passage "were left to be performed by some other power, England, by her neglect of it . . . would be laughed at by all the world for having hesitated to cross the threshold."

And so, John Barrow—lured by a now centuries-old idea that an earthly paradise awaited anyone who could penetrate beyond 80 degrees north latitude, never mind the prize and the potential financial gain—began launching one expedition after another into the Arctic.

In the early nineteenth century, no one in the British Royal Navy had any experience with ice. The only sailors who did were the Greenland whalers, but Barrow and other captains in the navy couldn't bear to take advice from fishermen, whom they looked down upon as an inferior class. As Barrow pushed naval officers like David Buchan, John Ross, William Edward Parry, and John Franklin into the unknown kingdoms of the far north, the fabled Open Polar Sea always seemed to lie just out of reach. Instead of frolicking pygmies and coconut-filled palm trees, they found a sea covered in ice and frozen fog, and a realm where their compasses spun uselessly in circles.

Many of these expeditions ended ignominiously, but none so spectacularly as the one led by Sir John Franklin, who set off from Greenhithe, England, on May 19, 1845, to discover the Northwest Passage with 128 men aboard two state-of-the-art ships, HMS *Erebus* and HMS *Terror*. Both ships had been heavily reinforced with extra oak planking and heavy steel plates bolted to their bows to help plow through the ice. But unbeknownst to anyone back in England at the time, *Erebus* and *Terror* became trapped in pack ice in September 1846 near King William Island, stranding Franklin and his men deep in the central Arctic. Not a single one of them made it out, and no detailed written account of their ordeal has ever been found. This void in the historical record, collectively known as "the Franklin mystery," has led to more than 175 years of speculation and spawned generations of devoted "Franklinites" obsessed with piecing together the story.

Some of the only details we know for sure—much of which comes from a single scrap of paper known as the Victory Point Record—are that Franklin died on June 11, 1847, and the following spring, a group of 105 survivors attempted an overland escape from the ships. They marched south across the ice and tundra, dragging their longboats (which the ship's carpenters had significantly modified and lightened) in hopes of finding open water. The large group eventually fragmented into smaller bands, and some might have eventually doubled back and returned to the ships. But one by one, every single sailor must have succumbed to a variety of maladies including, we can assume, starvation, tuberculosis, scurvy, and trench foot.

According to the scant archaeological evidence that has been dug up over the centuries, the story of Franklin's expedition is one of cannibalism and chaotic disintegration of order. But it's also, possibly, a story of survival. Tantalizing clues point toward at least one small band that may have survived for years. Others hint that more evidence—Franklin's tomb and the logbooks of *Erebus* and *Terror*—is still out there somewhere.

We know the tragic stories of other doomed polar explorers like Robert Falcon Scott because their diaries were later found, but nearly every shred of the Franklin expedition's recorded history has been lost to the winds of time. But what if these documents could be located? The frigid Arctic climate is well-known for preserving bodies and artifacts, and I figured there was more than a slim chance that Franklin's records could be hiding amongst the rocks and ice of King William Island. *What story would those pages tell?* I wondered.

And so, on a balmy summer evening in Maine, inspired by a classic book and a dream to sail back into the halcyon days of my youth, I began to wonder: *What if . . . ?* What if I didn't just sail to the Arctic, but I kept going into the mazelike network of straits and bays that make up the Northwest Passage, and I didn't stop until I emerged on the other side of the continent, off the coast of Alaska? Wouldn't this deliver *Polar Sun* to the Pacific, where I wanted to get her anyway? Along the way, I could follow in the wake of *Erebus* and *Terror*, anchor in the same harbors, see what Franklin and his men saw—just 175-odd years later. And who knew? Maybe if I fully immersed myself into the Franklin mystery, I might discover what really happened to him and his men.

But could I sail a forty-year-old fiberglass boat from Maine to Alaska—a voyage of some seven thousand miles—and live to tell my own tale? It was a question that would soon all but consume me, in the same way I knew it had done to the European explorers who had ventured into these same waters long ago when this part of the world was still a blank on their maps. Something in the human psyche has always found the unknown irresistible. And whatever that thing is, it was now drawing me north, where it would either make me stronger or beat the spirit of adventure out of me once and for all.

I knew, of course, that for a voyage like this to be possible, A LOT of improbable things would have to fall into place. This is the story of how, against all odds, they did.

PART 1

CROSSING THE THRESHOLD

CHAPTER 1

Peanut Butter and Jelly

Date: Summer 2020
Position: New England

I suppose I could be forgiven if I had taken a hard look at myself in the mirror the morning after that manic night in Boothbay Harbor and then let the Northwest Passage scheme float back out into the snowy dreamworld from whence it had come. But in the days that followed, the idea stayed lodged like a splinter in my brain. And instead of creating a laundry list of excuses why I couldn't do it—too much time away from home, too dangerous, too far, too cold—I asked myself one simple question: Was it even *possible* to sail a boat with a plastic hull through the Northwest Passage?

When I popped that question into Google, it spat forth an article from *Yachting World* magazine by the legendary circumnavigator and author Jimmy Cornell. It was a story about his 2015 east-west transit of the Northwest Passage, which Cornell described as "the Everest of sailing." I knew that roughly six thousand people had stood on the top of the world, but how many had done the Northwest Passage? More googling turned up a list maintained by the Scott Polar Research Institute at the University of Cambridge. In the summer of 2020, it included a total of 313 transits. Of these, sixty-one had done the passage more than once, with the record going to *Kapitan Khlebnikov*, a Russian icebreaker that had eighteen transits. What I was most interested in, of course, was the number of smaller sailboats that had made it through. I counted 136 sailing vessels of sixty feet or less that had completed the passage through the end of the 2019 season. If each of these vessels had a crew of four, this adds up to about 550 people—one for every ten who have stood atop Mount Everest.

To meet the criteria for inclusion on the list, a ship needs to cross the Arctic Circle twice—first in Davis Strait on the east side, and then in Chukchi Sea on the west, or vice versa if going the other way; but an exception is made for vessels beginning their voyages north of the Arctic Circle on the west coast of Greenland. Measured in this way, the total distance of a Northwest Passage transit is only about three thousand miles, but since most of these voyages begin and end far to the south of the Arctic Circle, the total distance is usually much greater.

According to the institute, there are actually seven different routes that weave between the 36,500 islands of the Canadian Arctic Archipelago, but only two are passable with any regularity. The rest are almost always chock-full of ice. The standard route is also the first, opened by the Norwegian explorer Roald Amundsen and his six-person crew between 1903 and 1906 aboard *Gjøa*, a seventy-foot wooden fishing vessel. Traveling east to west, Amundsen spent three winters in the Northwest Passage, including two in a small central Arctic harbor on the south coast of King William Island that now bears the name of his ship. Next came Henry Larsen, another Norwegian, who took command of a Royal Canadian Mounted Police (RCMP) motor schooner in 1940 and completed the passage from west to east in two years. In 1943, Larsen turned around and went the other way, becoming the first to transit the Northwest Passage in a single season. Over the next three decades, a handful of U.S. and

Canadian Coast Guard icebreakers, buoy tenders, and research vessels made it through. But it wasn't until 1977 that *Williwaw*, a custom-built forty-foot steel ketch skippered by Willy de Roos of the Netherlands, became the first private sailing yacht to complete the Northwest Passage.

The turning point came thirty years later in 2007, when the passage, or least the most passable route, was entirely ice-free for the first time in recorded history. Taking advantage of this unprecedented event, several small boats snuck through in a single season. In the years since, the massive floes of ice have never returned to historical levels and more boats have successfully made the transit than in all the previous years combined.

The list doesn't include hull type, but after more research, I discovered that at least a handful of the small vessels had fiberglass hulls, like *Polar Sun*'s. Most notable of these was a voyage undertaken by a thirty-year-old Rhode Island man named Matt Rutherford, who set off from Annapolis, Maryland, in June 2011 aboard a forty-year-old, twenty-seven-foot Albin Vega called *St. Brendan*. Rutherford sailed back into Annapolis 309 days later, after having circumnavigated the Americas by way of the Northwest Passage and Cape Horn—a twenty-seven-thousand-mile voyage that he completed single-handed and, amazingly, without stopping. His motto was "*fortitudine vincimus*" ("by endurance we conquer"), which he borrowed from his hero, Ernest Shackleton.

My friend Eli had taken part in a Northwest Passage attempt in 2014. It was a bad year for ice, and they didn't make it through. They did, however, get to Erebus and Terror Bay at Beechey Island, where the Franklin expedition had spent their first winter. And they did so in a Baba 40, a classic fiberglass sailboat that was built at the same boatyard in Taiwan where *Polar Sun* was born. Eli put me in touch with the skipper, a retired NASA mechanical engineer named Sam Lowry, who sails out of Rockport, Maine. I owe a lot to Sam because not only did he refrain from laughing me out of the water; he sent me his logbook, which contained a literal treasure trove of data detailing his ninety-day voyage in three-hour increments, including latitude/longitude, water temperature, track (heading), wind speed, boat speed, barometric pressure, sail plan, battery levels, engine hours, and more. Sitting down with his log and Google Earth, I was able to see exactly where he had been and when and roughly what the conditions had been like. Later, Sam would gift me all his paper charts from the Northwest Passage (worth thousands of dollars), which he hand-delivered to *Polar Sun*.

According to Sam, a metal hull, either steel or aluminum, was definitely the preferred choice for sailing amongst ice, but he didn't consider a fiberglass boat to be a deal-breaker. His boat, *Lillian B*, had collided with a few ice floes in Dundas Harbour on Devon Island, but nothing that had done more than scrape off a bit of paint. He warned me, though, that I'd need to be super careful not to let the boat get trapped in the ice, especially in heavy seas, because the surging floes could crush a fiberglass hull like a beer can. And if your ship goes down in conditions like this, it's not like you can climb out onto the ice or launch your life raft.

My research indicated that more and more small sailboats were attempting the Northwest Passage every season, mainly due to the unfortunate reality that Arctic sea ice is melting at an unprecedented rate. For the first time in history, it *was* possible to sail a fiberglass boat into the Northwest Passage. And while I found this encouraging, I took little joy in it.

People might debate exactly what is causing our cryosphere to melt, but no one can deny that it's happening. According to data collected by the National Snow and Ice Data Center, the average temperature in the Arctic has risen by 5.5 degrees Fahrenheit over the past fifty years—an increase three times greater than in any other region on Earth. As a result of these higher temperatures, Arctic sea ice has decreased by 13 percent per decade since 1979. What is particularly worrying, if you care about things like sea level rise and polar bears, is the decline in multiyear ice, which grows thicker and rougher than new, first-year ice and provides important habitat for marine mammals. At the time of this writing, multiyear ice, which covered a third of the Arctic Ocean in 1985, has declined by 95 percent. The lowest Arctic-sea-ice extent in more than forty years of satellite tracking was in 2012, with 2020 a close second. In either of those years, you could have cruised right through the Northwest Passage without coming in contact with a single speck of ice.

I knew, of course, how messed up it was that in order to sail a fiberglass boat through the Northwest Passage, I would invariably be hoping for the very conditions that are destroying the essence of what makes the Arctic such a singular place. A good year for a Northwest Passage voyage is a bad year for our planet. But I also knew that if I was going to see the Arctic and record and share what I saw there before all the ice disappeared, there was no time to lose.

It was time, then, to see if there was anyone out there who might want to do something like this with me. The story as to why there was one name, and one name only, at the top of my short list of potential partners is a long and convoluted one that begins in the mid-nineties when Ben Zartman and I lived in adjacent caves in Yosemite National Park.

I had recently graduated from college with a degree in philosophy. But none of the jobs recommended by the career-counseling office were even remotely appealing, so I chose to live in Disneyland for climbers and collect cans for my living. I'd been part of the dirtbag-climbing scene in "the ditch," as we called it in those days, for a few years when Ben showed up one day from Ohio, having chosen the rock-climbing crucible of Yosemite Valley over college. Back then, before he got his nickname "Ben Wah," which was later appended with "the Legend," we used to call him "Young Ben." I remember hearing rumors, later confirmed, that Ben's parents were Christian missionaries and that he had grown up between postings in Mexico and Colombia. Ben was (and still is) a teetotaler, which was rare among climbers in those days. I was the opposite, but we bonded nonetheless when he moved under a rock a little ways up the hill from where I was camped. Later, we even dated the same woman—at different times, of course.

One evening, Ben showed up at Degnan's Deli, aka "the center of the universe"—the gathering spot for Valley dirtbags—and declared that he was going sailing. Considering that he lived under a rock and had never sailed before, this seemed a bold pronouncement. But a few days later, he threw on his pack and hitchhiked down to San Diego, where he managed to talk his way onto a boat heading south to Mexico.

After that first sailing trip, he would occasionally come back to Yosemite to climb between voyages, but that changed when he met a woman in a green sundress with tan legs and long brown hair that she tied in a bun on the back of her head. Early in their relationship, Ben and Danielle moved onto a run-down twenty-seven-foot fixer-upper sailboat on the Caloosahatchee River in Fort Myers, Florida. After a year spent refitting the vessel, they rechristened it *Capella*—the name of a star used in celestial navigation.

While thorough, the refit didn't include any electronics. Ben and Danielle had recently crewed on a boat with a faulty electrical system. They had soldered new contacts and run new wires—but the boat had still nearly burned down. So Ben decided that electricity had no place at sea. Instead, Danielle and he used an all-weather oil lamp in which they inserted red and green theater gels to alert other boats to their position at night. At anchor, they hung a lamp at the top of the mast with a Fresnel lens like those used in lighthouses. A line with a piece of lead on the end served as their depth sounder, and they determined their position through dead reckoning and sighting celestial bodies with a sextant. When *Capella* was ready, Ben and Danielle sailed across the Gulf of Mexico to the Yucatán, down to South America, then up through the Caribbean to Maine.

While climbing sea cliffs in Acadia National Park, they learned that Danielle was pregnant. They put the boat in storage, bought a used car, and drove back to California to move in with her dad. Ben spent two years trying to sell *Capella* as she sat in a boatyard in Fall River, Massachusetts. But she'd been built back in 1967 and there was little interest. Then, in 2005, he heard through the grapevine that his old cave neighbor who had just returned from Pitcairn Island was looking for a sailboat.

"It's the opportunity of a lifetime," I told my first wife after Ben called and named the price—one dollar. The deal was done, although perhaps I should have taken a bit more time to question the price.

A few weeks later, after Ben delivered the boat, I took some experienced sailors out for her maiden cruise on the coast of Maine. As we approached a ledge in the fog near an island called Petit Manan, one of them announced that it was time to tack, a maneuver in which a sailboat turns through the wind.

"Oh, about that," I replied. "This boat doesn't tack."

"What do you mean it doesn't tack?" said my friend, who had once crossed the Pacific Ocean. "Of course it tacks. All boats tack." He grabbed the tiller, yelled, "Ready about," and pushed it hard to leeward. *Capella* rounded up into the wind and the sails went flat. We were dead in the water, or what a sailor calls "in irons," which can be a real problem when it happens to windward of a ledge, which is precisely where we happened to be. I had neglected to tell my friend that the boat's centerboard—a retractable fin that protrudes from beneath the hull to stabilize the boat—had snapped off years ago in the Gulf of Mexico. Ben deemed this to be unnecessary too and never replaced it. In my naïveté, I had figured that if my mentor could sail this boat six thousand miles without a centerboard, then by God, I could putter around the coast of Maine without one too. And so, while *Capella* blew down on those rocks, I fired up the outboard.

As I continued with my sailing apprenticeship on *Capella*, Ben worked on building his dream boat in Danielle's dad's backyard. The thirty-one-foot vessel was a heavy-displacement, full-keeled, gaff-rigged cutter called a Cape George. The project took him five years to complete, and over that time, he and Danielle had three children, all girls. In the evenings, Ben worked as a waiter in Yosemite's Mountain Room restaurant, and he put every spare cent he made into the boat. To save money, he would visit junkyards and shooting ranges to scrounge for lead tire weights and spent bullets, which he melted down on a homemade forge and poured into his keel. When it came time to set up the rig, he didn't have enough money for an actual mast, so he bought an aluminum telephone pole and machined it into what he needed. In 2009, he trucked *Ganymede* (named after the Jovian moon) to the Stockton River, and the Zartman family set off on a multiyear sailing voyage. Shortly after the launch, Ben sent me a picture that he must have taken from the dinghy. Danielle was at the helm, holding the tiller, a baby in a sling strapped to her chest. Two towheaded young girls, ages five and three, sat on the cabin top.

Ben led his family to Panama, where they transited the canal and then caught dengue fever in Colombia. Eventually, they headed north and landed in Newport, Rhode Island. It was supposed to be a short stop to replenish their cruising kitty, but Ben found work as the captain of an eighty-foot schooner, while Danielle homeschooled the three children and did sewing and odd jobs for other boats.

I visited them not long after that, finding *Ganymede* tied up to a snowy dock at a marina in downtown Newport. The Zartman home bobbed in the cold water, wrapped in plastic shrink-wrap. A metal stovepipe poked through the ridge, puffing smoke into the gray February evening. Ben appeared through a small wooden door and led me down into the cabin, where Danielle, wearing a homespun cotton dress, tended a kettle bubbling away atop a cast-iron woodstove. The interior of the boat was about the size of a toolshed, and it was covered in redwood and Doug-fir, every piece of which Ben had lovingly milled. The girls shared a tiny cabin aft with three bunks packed so tightly, they had to take turns getting dressed. How five people could have lived together in such a small space defied my imagination. But when I saw the cozy bunk where Ben and Danielle slept, covered by a homemade quilt, I realized that I had never seen such a perfect picture of domestic harmony.

That spring, the Zartman family finally set off for Newfoundland on the next leg of their voyage. They waited in St. John's for a suitable window to strike out across the Atlantic, but the weather was rough that summer, and after watching a steady procession of low-pressure systems march across the North Atlantic, Ben and Danielle pulled the plug. Changing course, they circumnavigated Newfoundland counterclockwise and then sailed down the St. Lawrence River and through the locks into the Hudson, which carried them back to Rhode Island. Ben found more work as a captain, and in 2015, with his girls growing older and *Ganymede* bursting at the seams, the Zartmans rented a small house and moved ashore.

In recent years, Ben had been making his living as a rigger, fitting out every kind of sailboat from the gilded yachts of Boston Brahmins to the old wooden schooners that still ply the waters of coastal New England. A rig is the name that sailors use for the mast and all the wires and ropes that keep it upright and allow the sails to be hauled up and down and trimmed in and out. A rigger makes and services all of this equipment, from splicing ropes to swaging wires and securing all manner of turnbuckles, shackles, pulleys, and pad eyes.

All of this is to say that Ben is, hands down, one of the most solid and resourceful men I've ever met: whip-smart, calm under pressure, a real-life MacGyver if there ever was one. And for these reasons and more, he would have been my dream partner for the Northwest Passage. But at the same time, I also knew that there was almost no chance that he'd be willing to leave his family and work commitments for a four-month sailing voyage in the Arctic. He had never been away from his family for more than two weeks at a time, and even if Danielle thought that his sailing the Northwest Passage with me was a good idea, he wouldn't do it. I was sure. I admired him for that—and wished I had bit more of whatever that was in myself.

My chest felt strangely tight when I finally phoned him in November of 2020. The words poured out of me like a waterfall as I pitched that the two of us, plus a couple more crew members we'd find later on, would set sail from southern Maine in early June of 2022. Canada had shut down all travel to the Arctic due to the COVID pandemic and no one thought there was any chance it would open that following summer, so I figured 2022 would be the soonest we could go. The good news was that this would give us a year and a half to prep the boat and hopefully find some sponsors. The route I'd plotted would take us to the west of Newfoundland through the Strait of Belle Isle, and then across Davis Strait to Nuuk, the capital of Greenland. From there, we'd intersect with Amundsen's route, work our way north with the help of the Greenland Current, and then shoot west across Baffin Bay into Lancaster Sound.

If we made it through James Ross Strait, which is about halfway through and almost always the ice crux, by mid- to late August, we'd have another month or so to reel off the final twenty-three hundred miles through the heart of the central Arctic, culminating in a long slog west across the North Slope of Alaska. If we pushed hard and the ice gods smiled on us, I reckoned we could make it to Nome, which lies just south of the Bering Strait, before the autumn storm season, when the Sea of Japan starts pumping one low pressure system after another into the Bering and Chukchi Seas. Overall, I figured the roughly sixty-three-hundred-mile voyage would take us about four months, and I broke it into four sections: Maine to Nuuk, Greenland (two thousand miles); Nuuk to Pond Inlet on Baffin Island (eleven hundred miles); Pond to Gjoa Haven on King William Island (nine hundred miles); and then a final slog from Gjoa to Nome across the North Slope of Alaska (twenty-three hundred miles). I'd heard that it was usually necessary to do a fair bit of motoring amongst the ice in the heart of the passage, but I hoped to sail the majority of the time. Besides, the distances between ports and fuel stops were so great that *Polar Sun* couldn't carry enough fuel to motor the whole way, so we'd be forced by necessity to sail as much as possible.

Ben didn't say much as I blabbed away. When I finally wrapped up my pitch and asked him what he thought, there was a long pregnant silence as I waited for him to blow me out of the water.

"Hey, if anyone can pull this off, you're the guy," he finally said.

Whoa! This was not the response I had expected.

As to his own participation, he was noncommittal. He said something about work and family and how hard it would be to get that much time away. I had expected this, and it made perfect sense. I told him as much.

I hung up feeling good that he'd given me his vote of confidence, but I could tell by his response that he was definitely *not* going to be joining me. So I never followed up. Instead, I pitched a few other sailing friends on the idea. Everyone was intrigued, but no one could commit. They all had regular jobs, bosses, mortgages, partners/spouses, children, and pets that they couldn't (or wouldn't) leave behind. A couple of them said that if I could break the voyage into smaller legs involving less time and fly them in and out, maybe they could take part. But it was almost impossible to figure out when I'd be where or if I'd even get there. And the flights into micro airports such as those in villages like Ilulissat, Pond Inlet, and Cambridge Bay were eye-wateringly expensive. I had worked up a rough budget, and just between boat upgrades, food, and fuel, it quickly surpassed $75,000.

I've been on a lot of expeditions over the years and, somehow, have always found someone to foot the bill. In exchange, I've written articles, posed for photos, appeared in films and on television, and tested new products. While climbing Great Trango Tower in Pakistan, I'd even posted live daily updates to the Internet. But as Ben had pointed out during our call, "This isn't climbing, dude. The big sailing companies are only interested in stuff like the Volvo Ocean Race and the Vendée Globe." What they were not interested in, he said, was some random scrappy dreamer trying to sail his old boat through the Northwest Passage.

So there I was: with no funding, no crew, and a boat badly in need of expensive upgrades. By the time spring of 2021 rolled around, I told Hampton that the voyage was off. She'd been remarkably supportive of my crazy plan from the moment I first shared it with her, but I finally decided that I had wasted enough time pushing on this string. She didn't say it outright, but I sensed that she was relieved I was abandoning the project.

But then, almost as if Ben had sensed through the ether that I had given up, I got a text from him. When are you going to call me back about the Northwest Passage it read. We should talk.

I called immediately.

"I talked to Danielle," he said. "She gave me her blessing to go to the Arctic with you."

"What? Are you serious?"

"Yes," he continued. "And I have no contingencies. Even if you don't get any sponsors and I have to pay my own way, I'm in. All I ask is that you buy me a plane ticket home from wherever we end up."

"Deal," I said, deciding not to mention that a few minutes earlier, the whole endeavor had been dead in the water.

We talked for a while about the trip, mostly about what the boat would need to get up to snuff for Arctic duty. I didn't mention it explicitly, but in a roundabout way I let it be known that I would be the captain and he the first mate—despite the fact that he was the more experienced sailor. It was my baby and my boat, and shouldering this ultimate responsibility was a hugely important piece of the puzzle for me. Ben confirmed that he had assumed as much and signing on as the first mate worked well for him. Right before he hung up, he said that he needed something like this in his life to look forward to. Then he added, "I just don't think I'm cut out for chicken farming."

After the call, I practically ran out of my office to find Hampton. She was in our living room sitting by the woodstove. What followed was a typical exchange for us. I'd come charging into the room, all full of piss and vinegar, to share the latest development in my current mad scheme.

"Ben's in," I declared.

Hampton looked up from her computer. "And what are we talking about?"

"The Northwest Passage. Looks like it's on."

She raised her eyebrows. When she's deep in thought, she tends to chew on the inside of her cheek. She started doing that now.

"Have you ever done *anything* with Ben?" she finally asked. "How well do you even know this guy?"

"Oh, I know him super well," I replied. "And he's an incredible mariner."

"I don't doubt he is," said Hampton, "but do you know him well enough to spend four months on a small boat with him?"

I told her most emphatically that I sure did. Yes sirree.

Hampton pensively chewed her cheek a bit more, which got me

thinking too. Back in the day in Yosemite, had I ever actually roped up with Ben? I racked my brain but couldn't think of anything we'd ever done together. Our relationship was that of mentor and mentee, which boiled down to a series of phone calls and emails that were almost entirely sailing related. The last time I'd seen him in person was nine years earlier. But it was Ben Wah, my old cave buddy from Yosemite. The two of us were like peanut butter and jelly, right? What could possibly go wrong?

CHAPTER 2

Pretty Damn Western

Date: June 25, 2022
Position: Davis Strait

The storm spread across Davis Strait like the black wings of a raven. A cold rain, driven by a blistering southeasterly wind, poured into the sea, while a fifteen-foot wave, streaked with tendrils of foam, rose up behind *Polar Sun*. One second, it was looming over my head; the next, it was passing beneath the boat, releasing a roar as it crumbled into a watery avalanche. There was a moment of exhilarating weightlessness; then *Polar Sun* dipped its bow and began flying downward, pulling g's as it picked up speed. I yanked the wheel hard to windward, and the seventeen-ton boat surfed across the face of the wave. Down in the trough, shielded momentarily from the full brunt of the wind, the sail flapped and hung limp for a breath, until along came the next roller, up which *Polar Sun* clawed its way back toward the whitecapped horizon.

I was on watch alone, smack-dab in the middle of Davis Strait, that infamous sea between Greenland and Baffin Island. It was in these cold, deep waters that in 986 AD the Norse first sailed to North America from Greenland, and where in the sixteenth century Englishman John Davis searched for a Northwest Passage between the Atlantic and Pacific Oceans. My three other crew members were below deck huddled in their bunks, lying somewhere between sleep and seasickness. As much as I wanted to join them and crawl back into my sleeping bag, I knew they weren't any more comfortable than I was—and someone had to keep sailing the boat.

Since that first sailing trip to Pitcairn, I had weathered a number of storms, but never as the skipper of my own boat in a sea this deep and committing. Heavy-weather offshore sailing, as I was learning the only way you can, tends to fall into the "type 2" category of fun: god-awful while it's happening; sublime when it's over. But I suppose this was exactly what I had signed up to experience, and as I stared out at that angry ocean, I felt closer to the spirits of the long-lost explorers who had once passed this same way and who had undoubtedly been plagued by the same ferocious elements, the same fears, the same doubts.

It was June 25, 2022, a few days past the summer solstice and three weeks since we'd shoved off from the Saco River in southern Maine. After a quick stop in Lunenburg, Nova Scotia, to clear customs, we cut through Cape Breton Island's Bras d'Or Lakes and then across Cabot Strait to Newfoundland. We reckoned the crossing of Davis Strait to Greenland would take about five days, and we'd gotten about halfway across before running into this storm. Until the gale, the trip had been mostly blue skies, sunshine, and light winds, thanks to an unusually stable high-pressure system that had parked itself over the Canadian Maritimes.

I was thinking about all the twists and turns that had led me to this moment when Ben poked his head up out of *Polar Sun*'s cabin.

"How are we looking?" he asked.

"It's blowing a gale," I reported, proud of the fact that I was giving this report to my longtime sailing mentor. "I've been seeing the occasional gust into the lower forties."

Ben nodded, then turned his gaze to the instrument panel above the companionway, where data for speed over ground, depth, position, wind strength, and direction was displayed. "Looks like it's blowing a bit less than thirty-five," he corrected. "Technically, that's a near gale, not a gale."

He paused for a moment to let the importance of not exaggerating wind strength sink in; then he stepped back to the helm. Despite the stormy weather, he'd come prepared to make good use of his watch on deck. In a small bag that he tossed on the bench, he carried an old-school iPod filled with opera music and a copy of *The Decline and Fall of the Roman Empire* by Edward Gibbon, which apparently he planned to enjoy during pauses in the storm.

With his shaggy, gray-flecked blond hair, kindly blue eyes, and thick beard, he looked a bit like a middle-aged Kris Kringle these days. And he wore a full-length homemade black overcoat he'd sewn together himself. He called it the "Westport Whaler," and its design crystalized Ben's painstaking, pious view of the world. The coat had an insulated liner and an exterior made of waterproof polyester—inexpensive, sturdy, and available at any well-equipped fabric store. At the slightest prompt, he would pontificate on the coat's many merits, like, for example, how it was fastened with buttons—not zippers. Buttons, he would not fail to explain, could be easily repaired at sea. Another favorite touch was how the flap could be buttoned with the overlap to port or starboard, depending on the wind direction. In all, the Westport Whaler perfectly reflected Ben's decided answers to the modern world's waste, expense, and complexity by being inexpensive and homemade yet highly functional, durable, and easily repaired with needle and thread.

What did not matter whatsoever to Ben, though, was the Westport Whaler's beauty, which, in the judgment of the crew of *Polar Sun*, was scant. The coat was bulky and cumbersome, and from behind it looked more like a nineteenth-century funeral dress than foul-weather gear. Its odd appearance seemed to delight Ben, who scorned the aesthetic in all things. Function, thrift, and simplicity were his mantras.

The chief irony of all of this was that Ben had sailed enough ocean miles and climbed enough mountains that he could rightfully wear the contents of a Patagonia catalog. But expensive outerwear was one of his favorite objects of scorn, and he widely viewed the outdoor-gear industry as a racket. When he came on board *Polar Sun*, I had given him a free set of state-of-the-art foul-weather gear. He promptly stashed the suit in his bunk—tags still on—to sell when he got home.

A year and a half had passed since I'd first pitched the trip to Ben, and true to his word, he had dug in hard from that day forth, working with me hand in hand to sort through the hundreds of little details that had to be squared away before we could set off. Among other things, Ben had spearheaded the replacement of all of *Polar Sun*'s running rigging and her lifelines. This entailed several trips from Rhode Island up to Rumery's Boat Yard in Biddeford, Maine, where *Polar Sun* spent the winter, and countless hours of work, which he never billed me for.

Perhaps most impressive of all in terms of his commitment to the project was the fact that he had quit his job as a rigger to come on the voyage with me. He had hoped to take a leave of absence and had given his company months of notice. But when his boss learned of Ben's Arctic aspirations, he made him train his replacement and then let him go. There would be no job waiting for him when he returned.

Ben and I had both been rather naive about the intense stress that being away for four months would put on our household finances. As it sank in for me that a suite of new sails, which *Polar Sun* badly needed, would run me close to $40,000 and as Ben faced down the reality of flushing a well-paying job down the toilet, the need to find sponsorship became an imperative. We were still telling everyone, including our families, that we were doing this "no matter what," but the reality of our voyage was far more uncertain. And from what I'd been reading about the history of Arctic exploration, this humorless scrounge for funds seemed to be the price of entry into the Northwest Passage.

In the midst of all of this, I led an expedition for National Geographic in the winter of 2021 to a cliff-rimmed mesa called a tepui that towered high over the Amazon rainforest in Guyana. The mission was part climbing and part science, with the goal of finding new species of frogs on a remote tepui called Weiassipu. We did manage a significant first ascent, and we found a frog and a snake new to science (although not on the mountain itself), so overall, the expedition was seen by National Geographic as a success. We had filmed it all for a television program called *Explorer*, and when I got home (nursing a nasty infection in my foot that eventually landed me in the hospital), a producer emailed me some lines of voice-over that he needed for the narration.

OUR CLIMBING OBJECTIVE IS DEEP IN THE AMAZON JUNGLE.

TO GET THERE, IT'S A SIX-DAY HIKE TO AN OUTPOST CALLED DOUBLE DROP FALLS—THEN WE HAVE TO HACK OUR WAY THROUGH UNCHARTED JUNGLE TO REACH THE BASE OF THE WALL.

At the end of that email, he had tacked on the following line:

What's the next adventure?!? NG execs are LOVING this footage and your storytelling and are primed to greenlight another adventure.

I knew an opening when I saw one, so I called him right away and pitched my latest grand scheme. But it wasn't just the story of an aging climber who would attempt to sail his old fiberglass boat to Alaska via the Northwest Passage. I knew that wasn't enough, that the story needed more of a hook. And I'd recently found one in the likes of an amateur historian from the Northwest Territories named Tom Gross, who, using extensive analysis of archival Inuit testimony, had narrowed down the possible location of Sir John Franklin's tomb to a thirty-square-mile chunk of territory on the north end of King William Island. In my pitch, I made a point not to exaggerate the likelihood of a discovery, but I'm sure I dangled the possibility that if we did locate Franklin's tomb, it was likely to contain his logbooks—and in the world of exploration mysteries, this was a bit like finding the Holy Grail.

Sure, plenty of people had already tried to find the tomb and those logbooks and come up empty-handed, including Tom, who'd been searching for twenty-eight years. But as far as I knew, none of those searchers had ever gotten to King William Island by sailing there in their own scrappy boat. I hoped that following in Franklin's wake might give me a unique perspective on the mystery—but more than anything, I knew that doing so would make a hell of a story.

I still can't believe that everyone went for it.

The process of getting all the contracts and funding buttoned up involved lawyers and took months, but when it was all said and done, I had made deals for this book, a cover story for *National Geographic*, and an hour-long TV special for *Explorer.* Nat Geo eventually put Ben and me on their payroll, and to make sure that I could sleep at night, I went straight out with the money they paid to charter *Polar Sun* for the summer and bought all the gear we needed. In the months leading up to our departure, I replaced all the standing rigging (wires that hold up the mast), fully overhauled the motor, and upgraded all of the electronics, which included a new chart plotter / radar array, forward-looking sonar (for seeing the bottom ahead of the boat in uncharted waters), and a satellite communication system.

The director and photographer hired by National Geographic, an old friend named Renan Ozturk, took it upon himself to hit up a company called North Sails for sponsorship. Amazingly, they agreed to set *Polar Sun* up with a suite of four new sails—mainsail, genoa, staysail, and gennaker—in exchange for photographs and other PR. And at Renan's insistence, North painted all of the sails, save the gennaker, bright red—so the boat would look like it was on fire when it sailed past all those icebergs.

|||

While Ben got stuck in *The Decline and Fall of the Roman Empire*, I headed down into the cabin, away from the shrieking wind and crumbling waves, where I was assailed by the chaos that reigned below deck. To starboard, hundreds of cans of beans, tuna, Dinty Moore, and Spam rumbled back and forth in two cavernous wooden lockers, while on the other side of the salon, our cookware, stacked to overflowing in the galley sink, clanked and clattered like a toddler kitchen band. On the cabin's floor, a glass jar of green chili hot sauce, sporting bits and pieces of its soggy label, rolled back and forth amidst a scattering of loose macadamia nuts, random footwear, soiled long johns, and a package of pulverized Chips Ahoy! Most unsettling, though, were the noises that emanated from deep within the body of the boat itself. The bulkheads creaked and groaned and a vibration welled up from the keel, making *Polar Sun* feel at once haunted and on the verge of disintegration. As we pitched and staggered through the deep swells, the vibration oscillated to a shudder, then back again over and over.

Amongst this bedlam, I found Rudy Lehfeldt-Ehlinger curled up on the port settee. He was lying in the fetal position with no sleeping bag, wearing boots, foul-weather bibs, and a white wool turtleneck sweater. The settee was so narrow that he could only lie on his side, but it was better than his quarters in the V berth up in the bow, which, given the fifteen-foot swells, treated its occupants to thirty violent feet of vertical movement twice every minute. He appeared to be fast asleep, but when I accidentally brushed against his foot, he opened his glassy eyes.

Rudy's friends call him the "Slotherine," a nod to his preferred slothful state. Stout and burly, he is of medium height with short reddish-brown hair and a lazy eye. The combination of qualities evokes a hibernating bear or maybe a grizzly digesting an enormous meal. It's a deceptive impression because he's a man with freakish strength and endurance and, when they are called upon, uncannily quick reflexes. A year earlier, on that expedition in the Amazon, he had been crossing a small, turbulent river via a muddy log when his feet slipped. For a split second, it looked certain he was going in with a pack full of cameras. Suddenly, he leapt into the air like a spider monkey, grabbed a vine hanging from the canopy, and swung across the water like Tarzan.

Rudy has a technical bent and a tinkerer's hands; he's the kind of guy who builds his own batteries from scratch because nothing commercially available meets his requirements. While Ben loved to preach about his Westport Whaler or why the metric system was so flawed, Rudy's passion was alcohol. Drinking it, yes, but more so the science of making it. He had recently opened his own distillery called Proverbial Spirits back home in Park City, Utah, and he had squirreled away dozens of bottles of his homemade gin and rum in various compartments all over the boat.

National Geographic didn't need Renan to join the expedition until we got to Greenland. Rudy, who would serve as Renan's assistant and "tech cowboy," could have joined us there too, but he's a die-hard sailor, completely immune to seasickness, and I needed another hand on deck for the crossing of Davis Strait. Over the past three weeks, Rudy had made it his mission to acquaint himself with every system aboard. As a result, he'd developed an intimate relationship with *Polar Sun*, and I'd been pleased to discover that we almost always saw eye to eye in terms of the sail plan and how to move a thirty-four-thousand-pound vessel across hundreds of miles of blue water.

As I passed by him now, Rudy rolled over and croaked, "You better check on Mr. Dirt."

Mr. Dirt was the fourth member of my crew, and all five feet, six and three-quarter inches of him were exactly where I'd left him four hours earlier—wedged like a dirty sock into his bunk on the starboard side of the aft cabin, which he shared with me. With the boat heeled hard to port, the only thing keeping him from flying out of his bunk was a piece of canvas known as a lee cloth, which now acted more like a hammock. His barrel chest moved up and down, which meant he wasn't dead. But he was brutally seasick.

Seasickness isn't actually sickness, per se. It happens when our inner ear, where the human balance mechanism resides, sends conflicting signals to the brain. On a moving boat, the inner ear detects all the ups and downs and side to sides, but since our bodies move with the vessel, our eyes register stability. This confuses the brain, and it responds with a cascade of stress-related hormones that lead to nausea, vomiting, and vertigo. Most people's bodies will eventually acclimate to the conflicting signals coming from their senses, but apparently, Mr. Dirt was not most people. I had tried to cure him with a host of motion sickness medicines—including Dramamine and a powerful drug called scopolamine, which had saved me on the Pitcairn Island expedition—but nothing worked.

A blue NAPA Auto Parts five-gallon bucket hung adjacent to his bunk. It sloshed with vomit, and the sour stench of bile permeated the cabin. Mr. Dirt wasn't dead, but when I saw that he'd somehow gotten his hands on one of my prized possessions, I briefly and quite seriously contemplated killing him.

On an expedition, every member usually has a personal item or two that serve a purpose beyond mere function. It might be a photo of loved ones, a lucky rabbit's foot, or a special cozy T-shirt. In my case, it was a neck warmer: sky blue, silky smooth, and given to me by my son Tommy. I wore it all day, every day, and when I was off watch, I pulled it over my eyes like a night shade. I must have left it sitting on my bunk, and in an act of seasick confusion or flaming theft (at the moment, I suspected the latter), Mr. Dirt had taken it. No doubt he had sensed its magical powers. And now I saw, to my horror, that it was stretched over his barf-glazed face, which lay inches from the bucket into which he was making his regular deposits.

Mr. Dirt, aka Erik Howes, was born in 1994, and he found his way onto *Polar Sun* via a crooked path that began with an abandoned trolley car named *Squally*. From the moment he first sat in its driver's seat in a junkyard in Pepperell, Massachusetts, he felt an instant and "electric love" for the machine. Soon thereafter, he purchased it and began converting it into his forever home. Over the next few years, Mr. Dirt kept his many fans up-to-date on his project through his @SmellyBagOfDirt Instagram account. His ethos for the build-out was to use only found or free materials. He harvested most of the trolley's interior woodwork and furniture from a dumpster outside a fancy Boston hotel. Some of *Squally*'s more unique features include a turn-of-the-century tube radio transformed into a utility cabinet, an Egyptian sarcophagus, dynamite boxes, and a sex swing that hangs above the bed.

The realization that Mr. Dirt and I were following parallel tracks dawned on both of us gradually. He was fixing up the trolley in his mother's driveway in Burlington, Massachusetts, while I was refitting a boat on the shore of the Saco River in Biddeford, Maine. Both of us were driven, not necessarily by the machines themselves, but by dreams of where they might one day take us. In the spring of 2022, I posted a story on Instagram asking for help replacing *Polar Sun*'s transmission. It was too heavy to lift out of the engine compartment alone, and the cost to hire someone to do it was exorbitant for me. Besides, if I hoped to be self-sufficient in the Northwest Passage, I needed to know as much as possible about my engine's inner workings. Mr. Dirt saw the post and sent me a note volunteering his services.

A few days later, a brown Jeep Wrangler topped with a cedar-shingled hobbit house pulled into the boatyard. The trolly wasn't yet habitable, so Mr. Dirt had been living in this vehicle house and sleeping on a plank of wood barely wider than a stair tread. By the end of that first twelve-hour workday, we were both covered in grime, and my lower back was so tight, I could barely stand. I tried to thank Mr. Dirt and send him on his way, but he wasn't having it.

"Hey, man, look, this boat is kind of like the trolley, only with sails instead of wheels," he said. "I can see what you're up against here. I'm super into this and what you're trying to do, so I'm not going anywhere."

For the next two days, Mr. Dirt and I worked side by side on *Polar Sun* from sunup to well past sundown. We wired and plumbed a truck heater, replaced impellers and hoses, flushed the coolant, and rebedded deck hardware. At night, we retreated to Hampton's parents' house, where I introduced my new friend as "Mr. Smelly Bag of Dirt." Hampton's mom rechristened him "Mr. Dirt" on the spot, and the nickname has since become so firmly embedded that few people in my orbit know his real name.

When one of my original crew for the leg to Greenland herniated a disc in his back and had to bail from the trip, I immediately called Mr. Dirt and offered him a spot on the boat. As *Polar Sun* set off down the Saco River on a rainy gray morning in early June of 2022, with a dozen people, including Hampton and Tommy, waving from the dock, there he was, sitting in the cockpit next to me and wearing a puffy green jacket that looked like it had been used as bedding by a family of raccoons.

Mr. Dirt had never sailed before, but he had worked for a season as a lobsterman out of Gloucester, Massachusetts. While he didn't exactly have salt water in his veins, he knew the ocean and his way around boats. And he's a good companion—cheerful, low-key, optimistic, and given to long, funny stories about climbing adventures, girlfriends, and wild sex parties. He sports a thick mustache and he's heavily muscled, wicked strong, and hairy all over like a hobbit. And unless there are women in the immediate vicinity, he seems largely uninterested in hygiene. But his best attribute, and the one that drew me to him as a crew member, is a love for hard labor. The dirtier and harder a task was, the more he seemed to delight in it—and there's always a supply of that stuff on a boat.

He did warn me up front that he was prone to seasickness, but I shrugged it off. In the hectic days leading to our departure, a spot of temporary seasickness in one member of the crew hardly registered. I figured, *How bad could it be?*

But now I had the answer: real bad.

It wasn't until he got sick for the first time in the Gulf of Maine that he told me that in his lobstering days he'd barfed so much that his stomach acid had eventually eaten through the lining of his esophagus. Apparently, this is something you can recover from.

Now, as I observed him in his bunk wearing a vomit-stained tie-dye, I wondered if anyone had ever died of mal de mer. Mr. Dirt had been throwing up for days. He lay inert and speechless, took no food, and drank little water. Most of the time, he seemed completely unaware of his surroundings. When I asked how he was doing, he didn't respond.

I could have crawled into my bunk, but every few waves, the sloshing bucket would "burp" some barf into the atmosphere and fill the cabin with the smell of Parmesan cheese. So I headed back to the salon, where I laid my zero-degree synthetic sleeping bag in the narrow passage between the nav station and the engine compartment. And I managed to catch an hour of sleep by lying to myself that the storm was abating.

Perhaps the most rugged aspect of offshore sailing is extreme sleep deprivation. I, for one, hardly ever slept, but when I did nod off, it was the sleep of the dead, and it always seemed to happen right before it was time to go back on watch. As a result, I was often deep in REM sleep when I was awakened, and I almost always remembered my dreams vividly. That night, I was transported to the northwest face of Half Dome in Yosemite, where I was trying to climb a legendary big-wall route called Kali Yuga. While I had never climbed this route in real life, I had always wanted to—mostly because the first ascensionists declined to give it a standard grade, or difficulty rating. Instead, they rated the route as "Pretty Damn Western," or PDW. I loved that term, as everyone knew exactly what it meant, even if it meant nothing: The climbing is committing, hard, and scary. End of story.

When my alarm went off at two a.m. and I swam back to the surface of Davis Strait, the phrase "Pretty Damn Western" was looping in my head like a mantra. And as *Polar Sun* began surfing down the steep slope of another of the thousands of waves that stood between us and Greenland, I realized that Pretty Damn Western was a perfect description for what it's like to sail into the Northwest Passage.

I wasn't the first to follow this path. Many people had passed this way for their own reasons, be it nationalism, personal glory, wealth, or just plain curiosity. And some of them—including the man whose tomb I was hoping to find—had never gone home.

CHAPTER 3
Sir John

Date: 1786–1845
Position: England

It is said that as a youth growing up in Spilsby, England, John Franklin attended every wedding and funeral at the local church, and his father, Willingham, noting his pious nature, pegged him as the one of his twelve children who might take holy orders and become a man of the cloth. And indeed, young John, when asked what he might like to do when he grew up, told his friends, "I'll get a ladder and climb to heaven!" But that was before he made a visit to the seaside hamlet of Saltfleet at age twelve. "That one look was enough," writes Henry Duff Traill in his admiring 1896 Franklin biography. "The boy returned home as irrevocably vowed to a sailor's life as though he had been dedicated by some right of antiquity to the god of the sea." Willingham figured the best cure for sea fever was to ship John off on a merchant vessel bound for Portugal, but the voyage set the hook even deeper. And so, at age fourteen, John Franklin joined the navy and promptly shipped off to the Baltic, where he fought in the Battle of Copenhagen under Horatio Nelson.

Because Franklin's family was prosperous and had connections, he was soon appointed as a midshipman (a rank given to officer candidates) on a voyage of discovery to New Holland under Lieutenant Matthew Flinders, an uncle by marriage. (Until the name was changed in 1824, European explorers referred to the western portion of Australia as New Holland and the eastern as New South Wales.) While surveying and exploring aboard HMS *Investigator*, Franklin found time to study astronomy, celestial navigation, geography, history, Latin, and French. He also read Shakespeare and served as an assistant at an astronomical observatory the crew had erected on a hill in Sydney. Franklin acquitted himself so admirably with his observations that the governor of New South Wales nicknamed him "Mr. Tycho Brahe" after the famed Danish astronomer.

After navigating through the Great Barrier Reef, *Investigator*, which was rotten to its core, began to split apart at the seams. Fearing that the ship might founder, Flinders abandoned the expedition and transferred his crew to a new vessel, which itself was soon wrecked on a reef in the uncharted waters between Australia and Papua New Guinea. For two months, Franklin and the rest of the crew survived on a tiny sand spit, nine hundred feet long by fifty feet wide, subsisting on stores they had scavenged from the ship before it went down. When they were finally rescued, Franklin sailed to Canton (Guangzhou) in China, where he bummed a ride back to England with an East India Company merchant vessel, the *Earl Camden*. While the *Earl Camden* was sailing in convoy with other company vessels in the South China Sea, it was attacked by a squadron of French warships. The lightly armed *Earl Camden* was heavily outgunned, but the crewmen fought back so vigorously that they succeeded in repelling the French. The captain, Nathaniel Dance, later praised Franklin, writing that he "should find it difficult . . . to name anyone who for zeal and alacrity of service and for general good conduct could advance a stronger claim to approbation and reward."

Back in England, Franklin signed on with a new ship, *Bellero-*

phon, which saw heavy action a year later in the Battle of Trafalgar. In the thick of the fighting, *Bellerophon* became entangled with the French ship *L'Aigle*. This allowed another French vessel to maneuver alongside and unleash a withering barrage of cannon and musket fire into *Bellerophon*. Franklin was helping to carry a wounded sailor below deck when a sharpshooter shot the man dead and then barely missed Franklin, who survived by diving behind the mast. By the time the dust settled, *Bellerophon* had lost three hundred sailors, including its captain, but Franklin was lucky and emerged unscathed apart from some hearing damage that would leave him nearly deaf in his later years.

When *Bellerophon*'s captain, John Cooke, was shot down, command fell to Lieutenant William Cumby, whom Franklin greatly admired for his willingness to show friendship and goodwill to even the lowliest ranks of the crew. "When this feeling is evinced on the part of the Commander," wrote Franklin to Cumby, "it seldom fails of producing the best exertions of your companions." As Janice Cavell, one of Franklin's modern biographers, puts it, "Franklin usually felt great loyalty to and friendship for subordinates who were willing to 'strain every nerve,' but he could be sharply critical of those who were not."

After several years of relative stagnation, Franklin next saw action in the War of 1812, during which he was wounded and then later helped storm an American artillery battery on the west bank of the Mississippi River in New Orleans. By the time he returned to England in 1815 at the age of twenty-nine, he had explored a continent, survived a shipwreck, fought in two wars, and survived three of England's greatest naval battles. But none of this could protect him from peace. After Napoleon's defeat at Waterloo, Franklin would fester ashore, on half pay, with nothing to do.

For the next three years, he kicked around Lincolnshire, spending time with his family and acting as a father figure to his many nieces and nephews, especially the children of his sisters Ann and Sarah, who died in 1808 and 1816, respectively. But it was during this interstice that Franklin's and John Barrow's destinies would collide. With Great Britain no longer on a war footing, Franklin decided that his best chance for promotion was to get himself attached to another exploratory expedition. So when he heard that Barrow was planning expeditions to the Northwest Passage and the North Pole, Franklin pulled every string he had, including a connection with Sir Joseph Banks, the president of the Royal Society, who'd been a patron of both Barrow and Franklin's uncle Matthew Flinders.

His machinations were successful, and in January 1818, Barrow placed Franklin as second-in-command, under David Buchan, on an expedition with orders to sail two ships, *Dorothea* and *Trent* (the latter captained by Franklin), between Greenland and Spitsbergen to the North Pole, which, at the time, was perhaps an even greater prize than the Northwest Passage. Spurred on by reports from the Greenland whaling fleet of open water along both the east and west sides of the island extending, in 1817, as far as 80 degrees north, Barrow still believed that the ocean couldn't freeze—or at least not thickly—over the North Pole. And yet *Dorothea* and *Trent* found solid ice to the north of Spitsbergen. Hoping the barrier might be narrow, Buchan ordered the crew to "warp" the ships through the bobbing floes with iron hooks and stout hemp ropes they reeled in with their anchor windlasses. For two days they made what seemed like good progress until Buchan realized that a strong south-flowing current was carrying them backward faster than their forward progress through the ice.

It took nine more days of warping to effect a retreat back into open water, but no sooner had they escaped the ice than they were hit by a ferocious gale. With massive breaking waves threatening to sink the two ships, Buchan ordered them back into the ice. Franklin's lieutenant Frederick William Beechey recorded that

> *the Brig, cutting her way through the light ice, came in violent contact with the main body. In an instant we all lost our footing, the masts bent with the impetus, and the cracking timbers from below bespoke a pressure which was calculated to awaken our serious apprehensions . . . literally tossed from piece to piece . . . the motion was so great that the ship's bell, which in the heaviest gale of wind had never struck by itself, now tolled so continuously that it was ordered to be muffled, for the purpose of escaping the unpleasant association it was calculated to produce.*

Franklin ordered his crew to sail *Trent* deeper into the ice, where the agitation was greatly reduced, and soon the gale passed. *Dorothea* was so heavily damaged by its thrashing in the ice, though, that Buchan called the expedition off and both ships limped back to England.

The Northwest Passage expedition, led by a plucky forty-year-old redheaded Scotsman named John Ross, also returned empty-handed that fall. Ross had sailed west from Greenland across Baffin Bay, which had been discovered in 1616 by William Baffin but then forgotten and erased from the maps of the day. By August, Ross had penetrated deep into Lancaster Sound, and while anchored off the north shore of Baffin Island, he wrote in his log that he "distinctly saw the land, round the bottom of the bay, forming a connected chain of mountains with those which extended along the north and south sides." He drew a detailed sketch and named the range after the first secretary of the Admiralty, John Wilson Croker. If mountains encircled their position, Lancaster Sound was a bay—and *not* the eastern entrance of the Northwest Passage. So Ross turned around and sailed back to England.

Buchan and Franklin, having barely survived their ordeal in the ice, were welcomed as heroes. Not so Ross. Barrow was furious that he hadn't pushed farther into Lancaster Sound. Damn the mountain range; he should have pressed on and spent at least one winter iced into some remote bay. Of course, it didn't help Ross's cause that none of the dozens of men on the two ships could corroborate his sighting of the Croker Mountains. Barrow blacklisted Ross, who would never sail under the banner of the Royal Navy again.

A year later, Barrow sent William Edward Parry (who had served under Ross the year before) back into Lancaster Sound, and in the alleged location of the Croker Mountains, he found nothing but open water. Parry, who would become Barrow's darling, sailed onward for another six hundred miles west, and on September 4, 1819, he crossed the 110 degrees west meridian, earning the £5,000 parliamentary reward. Parry overwintered on Melville Island and returned to England the following year, having linked Baffin Bay and Lancaster Sound with a waterway (later named Barrow Strait and the Parry Channel) that extended to the eastern edge of the Arctic Ocean.

A half century before this, a British explorer named Samuel Hearne had reached a similar longitude more than four hundred miles south of the Parry Channel (but still well north of the Arctic Circle) after following the Coppermine River to its terminus in what is known today as Coronation Gulf. Seven years later, the legendary explorer and navigator James Cook, who was leading his third voyage of discovery in the Pacific, sailed through the Bering Strait to within one hundred forty miles of today's Point Barrow (the northernmost point of the United States), and charted, for the first time, the western entrance to the Northwest Passage. Next came Alexander Mackenzie, who was disappointed when he reached the Arctic Ocean via the river that would henceforth bear his name. He'd been hoping it would spill into the Pacific, which would have allowed him to claim his route as "a Northwest Passage."

Since no one other than Inuit knew whether all these various waterways actually interconnected, it was still an open question in

1819 whether it was possible to sail from the Atlantic to the Pacific over the top of North America. But most explorers of the day, including Barrow, who exercised his geographic curiosity from behind a desk, believed in a Northwest Passage. Whalers, who knew the world's oceans better than anyone, agreed. Off the coast of Alaska, they had found whales with harpoons with marks from the Greenland whaling fleet sticking from their hides. And it was easier to believe they had gotten there from swimming through Arctic waters than from having gone all the way around Cape Horn.

European exploration of the Northwest Passage was initially driven by a desire to reduce shipping costs and time spent getting to and from the lucrative markets in the Far East, and it was dreamed of as a voyage that, in a perfect world, could have been completed in a single season by a ship large enough to be packed with tons of trade goods. But as one expedition after another failed in these ice-filled waters and the true complexity of the Arctic Archipelago slowly revealed itself, Barrow came to understand that the Northwest Passage was a puzzle that needed to be put together one piece at a time. And he worried, not unduly, that the Russians or the Hudson's Bay and North West Companies (which were private and not British governmental entities) might solve the enigma before him.

The irony was that the key to realizing his lifelong dream was much closer than Barrow ever knew, but it was simply beyond his conception that the Northwest Passage had already been discovered by Inuit and their ancestors thousands of years ago. Indigenous knowledge of the local geography could have directed the Admiralty where to go and when and, perhaps more important, how to survive if things didn't go as planned. But the colonial mindset of the time, combined with a healthy dose of racism, made it impossible for men like Barrow to access this knowledge. It was a blind spot that would have fatal consequences for his men in the years to come.

While William Edward Parry probed for the passage along the

seventy-fourth parallel, Barrow launched an overland canoe expedition to the south that would pick up the trail where Samuel Hearne had left off in 1771. To lead it, he tapped John Franklin, who had just returned from Spitsbergen. Franklin would sail to Hudson Bay, travel overland to Great Slave Lake in today's Northwest Territories, then follow the Coppermine River to its terminus in the Arctic Ocean. From there, he was to proceed east, where, if all went according to plan, he'd fill in one of the last blanks on the Arctic map and then rendezvous with Parry. Most of Barrow's budget had been put toward Parry's expedition, so Franklin was sent off on a shoestring with a crew consisting of only four navy men: a Scottish naval surgeon and naturalist named John Richardson; two midshipmen, George Back and Robert Hood; and an ordinary seaman (the lowest rating in the navy), John Hepburn. The rest of the team was to be made up with French Canadian Métis fur traders who were known as voyageurs.

The Admiralty had assured Franklin that the North West Company would supply the expedition with canoes, food, and ammunition, but when Franklin got to Fort Chipewyan on the western shore of Lake Athabasca (in what is today northern Alberta) in the spring of 1820, he found food and men in short supply. The most experienced voyageurs had already been called for—or perhaps they sensed that Franklin's expedition was doomed. Franklin considered most of the sixteen men he was able to recruit to be substandard.

When they reached Fort Providence on the shore of Great Slave Lake later that summer, Franklin met Akaitcho, chief of the Yellowknives First Nation, who had been hired by the North West Company to guide and hunt for Franklin and his men on their canoe journey down the Coppermine. Franklin described Akaitcho as a man of "great penetration and shrewdness," and when he explained that the purpose of their expedition was to find a shortcut for ships traveling from England to the Pacific Ocean, Akaitcho told him that if they insisted on trying to explore the Arctic coast, they would all perish.

But a deal was struck, and after overwintering a second time—on that occasion at a camp called Fort Enterprise—Franklin reached the mouth of the Coppermine on July 18, 1821. Here, he named the ocean bay that stretched before them Coronation Gulf, in honor of King George IV, who was to ascend to the throne in Westminster Abbey the next day. Akaitcho and his men bid farewell to Franklin's party of twenty-one, who set off east in three birchbarks, carrying two weeks' worth of food. Over the next month, as they paddled, sailed, and dragged their canoes across more than six hundred fifty miles of uncharted coastline, hunting as they went, they encountered no Inuit. They did, however, find several campsites and piles of whalebones, which Franklin interpreted as evidence that Coronation Gulf was indeed connected to Lancaster Sound. But Parry never found that link. After overwintering on Melville Island, he had sailed back to England a year before Franklin arrived in Coronation Gulf.

Franklin's plan, if the rendezvous with Parry never materialized, was to retrace his route up the Coppermine. He finally gave the order to turn back on August 21 at a place he named Point Turnagain, which lies on the Kent Peninsula about eighty miles west-southwest of Cambridge Bay. But by that time, winter was coming and their boats, having sustained extensive damage in a series of storms, were no longer seaworthy. Instead, he decided to take a shortcut back to Fort Enterprise across an inhospitable wasteland that Akaitcho had dubbed "the Barren Ground" and warned him against entering.

Franklin had counted on caribou to supply his party with meat on their retreat, but with the early onset of winter, the herd had already migrated south. As they set off into the Barren Ground, shuffling across the razor-sharp limestone shingle that sliced their leather footwear to shreds, game was all but nonexistent. When the pemmican (dried meat mixed with fat) ran out, the only food they could find was a bitter, mushroom-flavored lichen that the voyageurs called "tripe-de-roche."

By October 5, the ground was blanketed in three feet of snow, and Franklin recorded in his journal that "the whole party ate the remains of their old shoes, and whatever scraps of leather they had, to strengthen their stomachs for the fatigue of the day's journey." The leather and tripe-de-roche agreed reasonably well with Franklin's constitution, but not so with several of the voyageurs, who were still carrying ninety-pound loads. It was around this point that several of them, along with the midshipman Robert Hood, began to fail.

Franklin divided the team. Richardson and John Hepburn volunteered to stay behind to care for the weakest men, while George Back and three voyageurs were sent ahead to try to find Akaitcho. Richardson's party eventually left two dying men behind, and as they shuffled across the frozen tundra, trying to catch up to Franklin, a voyageur named Michel Teroahauté split off from the group. When he showed back up, he carried a haunch of fresh meat that he claimed to have scavenged from a wolf kill. Everyone partook, but Richardson soon realized that he had probably just eaten one of the men they'd left behind. Not long after that, while Richardson and Hepburn were out collecting lichen, Teroahauté shot Robert Hood in the back of the head as he warmed himself by a fire; the voyageur later claimed that the firearm had discharged accidentally. Certain now that Teroahauté intended to kill them all and cannibalize their bodies, Richardson, "convinced of the necessity of the dreadful act," shot and killed Teroahauté.

George Back eventually did find Akaitcho, who arrived just in time to nurse Franklin and the surviving members of his party back to health. After another winter at Fort Providence, Franklin returned to England in the fall of 1822, carrying the dubious distinction of having lost eleven of his twenty-one men—a 50 percent casualty rate. In the aftermath of the disaster, local fur traders were appalled by the expedition's poor logistics and planning, and even more so by Franklin's complete inability to live off the land. George Simpson—an explorer who, as the governor of the Hudson's Bay Company (HBC), ruled over a territory comprising nearly half of modern-day Canada—wrote that Franklin lacked "the physical powers required for the labour of moderate Voyaging in this country; he must have three meals p [*sic*] diem, Tea is indispensable, and with the utmost exertion he cannot walk Eight miles on one day."

Simpson was no fan of the British Royal Navy, and a quick review of the mileage covered by Franklin's party shows that Simpson's assessment was overly harsh. Regardless, the British public was hungry for heroes, and Franklin's grisly tales of murder, cannibalism, and, ultimately, his own survival, which he shared in a bestselling book, quickly elevated him into the pantheon of polar explorers. Now a minor celebrity, Franklin found himself in high demand at dinner parties and around London's fashionable drawing room tables. According to his beloved niece Sophia Cracroft, "he disliked notoriety and never sought it," and yet he was visibly amused when the pair was touring the Royal Pavilion in Brighton and someone in the crowd recognized him and called out, "That's the man who ate his shoes!"

In quick succession, Franklin was promoted to captain, elected to the Royal Society, and wed to a young poetess named Eleanor Porden, the daughter of a prominent architect. They soon welcomed a baby girl, whom Franklin doted upon. In a letter to Franklin's sister, Eleanor wrote that Franklin "has had a good deal of practice lately in singing to baby, and were it not for the approaching expedition I should expect to see him come out as a successor to Braham [*sic*]."

Franklin had been tapped once again by Barrow to lead another overland expedition, this one up the Mackenzie River to the Arctic Ocean, where the party would split and head east and west, surveying two of the last tracts of uncharted coastline. But while Franklin planned and prepared, Eleanor slowly wasted away from consumption, or what we know today as tuberculosis. This contagious lung disease, incurable at the time, is said to have killed one in seven of all people who had ever lived, up until the discovery of the antibiotic streptomycin in 1943. As her disease progressed, Eleanor suffered from fevers, fatigue, and a hacking, bloody cough that left her breathless and doubled over in pain. As there was no question as to Franklin abandoning the expedition, the days leading up to his departure in February 1825 became "a race between Death and Duty," as Traill writes, and "the unhappy husband found himself almost longing for the victory of the former, in order that he might be with his wife to the end." But when the day of his departure came, Eleanor rallied and the doctors saw signs that she might yet beat the terrible disease. It might have been, though, that she was just holding on long enough to see Franklin off so she could spare him the pain of watching her die. A few days earlier, she had told him that "it would be better for me that you were gone."

She passed six days after Franklin's departure, while he was somewhere out in the Atlantic writing her one letter after another. In his last, composed at a British naval station on Lake Huron on April 22, just hours before he finally learned that she'd been gone all these months, he wrote, "I daily remember you and our dear little one in our prayers. . . . She must be growing very entertaining, and I sincerely trust she will be a source of great comfort to us, especially to you in my absence. With what heartfelt pleasure shall I embrace both of you upon my return!"

Two years and seven months later, when he did return to London from an expedition that had been remarkable in terms of its success and complete lack of drama, little Eleanor, who was now three, didn't recognize her own father. Franklin wasted no time, though, reestablishing warm relations. "I got her seated on the carpet with her [cousins] forming a circle round me and commenced a game of forfeits," recalled Franklin, "which depended on imitation which delighted the whole party, and Eleanor in particular to such a degree that our intimate acquaintance has been the result and we have been ever since amusing each other."

Eleanor had been living with one of Franklin's sisters, which might have been why he quickly found himself a new bride. Jane Griffin, who ran in the same literary circles as his late wife, had apparently been on Franklin's mind on his Mackenzie River expedition, for he had named a random headland on the north slope of Alaska (Russian territory at the time) Griffin Point. A portrait of Jane, drawn in 1816 when she was twenty-four years old, shows her as a buxom, boyishly handsome woman with a round face and short, curly hair.

Franklin's physical appearance is often described less flatteringly. When he married Jane in 1828 at age forty-two, one of his relatives wrote that "his features and expression were grave and mild, and very benignant; his stature rather below the middle height." Shorty and pudgy with a doughy face, Franklin had gone bald young and his looks often caused people to underestimate him. He tended to make a poor first impression, but according to one of his nieces, she had never known anyone to "improve so much on acquaintance."

Franklin pined to be sent back to the Arctic, but Barrow, frustrated by his repeated and costly failures to find a Northwest Passage, put his discovery service on hold. Franklin was given a choice command of HMS *Rainbow* in the Mediterranean, where he followed the example of his early mentor William Cumby, endearing himself to his crew and leading with such élan that his men nicknamed their vessel "Franklin's Paradise."

But Franklin's leadership style might have been better suited to life aboard a ship than to the governance of terrestrial affairs. In 1844, he found himself back in England after a humiliating recall from Van Diemen's Land (today's Tasmania), where he had served as governor for seven years. He'd been dismissed thanks to the chicanery of a subordinate who had bad-mouthed Franklin to the new colonial secretary. Humiliated and consumed with his desire for vindication, Franklin wrote a pamphlet in his defense and became determined to redeem himself by finding another Royal Navy command.

Barrow, now eighty and at the end of his career at the Admiralty, was looking for a crowning achievement on which to hang his hat for posterity, and he settled on one last, all-out effort to solve the riddle of the Northwest Passage with a state-of-the-art expedition that would set forth from England in the spring of 1845 aboard two former warships, HMS *Erebus* and HMS *Terror*, which had just returned from a four-year expedition in the Antarctic. By now, Barrow had dispatched eight expeditions into the Arctic, three of which Franklin had participated in or led. Between these and a few others sponsored by the HBC, there was now only one piece of territory that still lay in terra incognita—a three-hundred-mile blank on the map deep in the central Arctic in the vicinity of the island named after King William IV.

Franklin, with considerable support from his wife, lobbied hard for the command. In a letter to Lady Jane, he wrote that he was "as active and vigorous as he had ever been, both in body and mind." But the truth was that the man who had eaten his shoes, now fifty-nine, was well past his prime and severely hard of hearing, with a potbelly, a shiny bald pate, and a reputation that had lost a bit of its luster.

Barrow's first choice to lead the expedition—James Clark Ross (not to be confused with his uncle John Ross), who had commanded the recent Antarctic expedition—had turned down the offer. Ross, recently engaged, had promised his fiancée that he would retire from polar exploration. He told the Admiralty that at forty-four years old, he was past his prime for such endeavors, but then turned around and recommended Franklin, who was fifteen years his senior. Barrow's second choice was a young, up-and-coming officer named James Fitzjames, who happened to be a close friend of Barrow's son. But Fitzjames, who was only thirty-one, had never led an expedition nor had he ever been anywhere near the Arctic.

Parry, who had himself retired from the game after a futile attempt to reach the North Pole on foot, might have been the one who clinched it for Franklin. When he was called into the Admiralty for an interview and asked for his opinion on his old friend, he replied: "He is a fitter man than any I know; and if you don't let him go he will die of disappointment."

ǀǀǀ

I was certainly not the first to develop a morbid fascination with the Franklin expedition, and in the months leading up to our departure, as Ben and I planned and prepared for the voyage, I got serious about investigating how we might add our own chapter to this epic saga. I was out to look for the tomb and the logbooks, of course—that was the plan with National Geographic and a goal I very much wanted to accomplish. But there was more too. I couldn't help but compare myself to Franklin and his men: *How would I fare*, I wondered, *if faced with the same dire dilemma of being trapped in the ice with no way out?* And if I followed in Franklin's wake into the Northwest Passage, was I destined to find out?

After I had read about two dozen books on Arctic exploration, I came across one that changed everything. In *Unravelling the Franklin Mystery: Inuit Testimony* by David Woodman, I learned that the author—a former Royal Canadian Navy submariner, ferry captain, and harbormaster turned "accidental historian"—had spent decades trying to locate the lost wrecks of *Erebus* and *Terror*. Before I learned of Woodman's work on the subject, I'd been struck by the fact that most historians seemed to neglect the most obvious source material: the testimony of Inuit who had witnessed firsthand what happened to Franklin and his crew in their final days. Woodman, I learned, was one of the first researchers to use that testimony as his primary source. "Not listening to Inuit," he once said, "would be like showing up to investigate a car crash and failing to interview the people who had witnessed the accident."

In 1975, Woodman was a young undergraduate student at the University of Toronto. One day, as he was passing the return-book stack at the campus library, a title caught his eye: *The Voyage of the 'Fox' in the Arctic Seas*. This book details the legendary expedition, sponsored by Lady Jane and commanded by Francis Leopold McClintock, that set off in 1857, twelve years after Franklin sailed into the Northwest Passage, in a last-ditch effort to figure out what had happened to those 129 men. Two years later, in 1859, the *Fox* expedition found what seems to be the only document left along the way by Franklin, the Victory Point Record. In it, the ship's officers recorded coordinates for the location where they had abandoned *Erebus* and *Terror* on April 22, 1848. For Woodman, seeing that waypoint on the page was like a gift. "I just had to get myself to these coordinates and dig a hole through the ice," he says, "and I'd find the ships. I gave myself two years to get it all figured out."

He dropped out of the university a year later to form a boatbuilding company, and two years after that, he joined the Canadian navy, which stationed him in England. There he pored through the archives at the National Maritime Museum in Greenwich and the British Museum. When he was transferred to Halifax, Nova Scotia, in the early 1980s, Woodman made a road trip to the Smithsonian in Washington, DC, where he dug up the personal papers of Charles Francis Hall, an eccentric newspaperman turned explorer from Rochester, New Hampshire, who became obsessed with the Franklin mystery. Hall is best known for having died under suspicious circumstances during the 1871 American *Polaris* expedition, but before he set his sights on the North Pole, he spent nine years living in the Arctic with Inuit, collecting their firsthand accounts of what had happened to Franklin and his men.

When Woodman showed up at the Smithsonian, he was told that he was the first person in a hundred years to inquire after the Hall collection. He found the papers stuffed haphazardly into a series of dusty wooden drawers, which he took back to the small office the staff had loaned him. Much of the material was indecipherable, but as Woodman painstakingly worked his way through hundreds of handwritten pages, transcribing the material into his own journals, he began to piece together a story much different from the standard Franklin narrative that had stood for more than a century. Before Woodman, most historians believed that Franklin had been buried at sea, that the ships had been abandoned less than a year later, and that by the end of 1848, the crew had perished while trying to reach an HBC outpost.

But Inuit told a strikingly different tale. There was a record of witnessing burials ashore and of remanned ships being visited by Inuit when they were near the shore. One of the ships later sank in deep water while the Inuit watched, and a hospital camp was established in Terror Bay, on the southwest side of King William Island. (McClintock named Terror Bay after the eponymous ship in 1859, but this appears to be a total coincidence since he didn't know at the time that HMS *Terror* had sunk in this very bay. More on this later.) According to the Inuit record, the last hard evidence of Franklin survivors was found in Starvation Cove, on the mainland's Adelaide Peninsula—a walking distance of more than one hundred thirty miles from where the ships had been abandoned. But there was also testimony that spoke of a small group of strange men seen hundreds of miles farther east, as late as the mid-1850s, on the Melville Peninsula, north of Hudson Bay, suggesting that a group of Franklin survivors might have lived longer and made it far closer to getting out than anyone had previously imagined.

"I started to get the sense that the whole story had been glossed over," said Woodman in an interview with the *Vancouver Sun*. "There was a lot more there that nobody wanted to look at because it was not amenable to the myth that they all walked away [from the ships] in 1848; that they stayed in a disciplined party even as the weaker ones dropped off, even if maybe some disreputable guys at the end ate each other. . . . Maybe because I was a naval officer and not an academic—and I'm a contrarian anyway—I began to think [the experts] had got it wrong."

Fast-forward to 1991, when Woodman published *Unravelling the Franklin Mystery*, a nearly four-hundred-page tome written over the course of a decade on a portable Commodore 64 computer and using a fully stocked freezer in his unheated garage as a desk. What he hadn't anticipated was that the publication of the book would lead to his elevation to being the de facto expert on the Franklin expedition, and he soon found himself drawn into a series of expeditions with the goal of unraveling the mystery. Between 1992 and 2004, he led or took part in nine different expeditions to the central Arctic. The first was an airborne magnetometer survey focused on finding one of the wrecks in an area the Inuit call Oot-joo-lik, which lies to the west of the Adelaide Peninsula, more than a hundred miles south of where the ships were reported to have been abandoned. It was here that Inuit spoke of a ship inhabited by a small group of white men who left sweepings from the deck on the ice and went hunting on the land. Later, when the white men disappeared, Inuit cut a hole through the hull. Inside, they found the bloated body of a dead white man with unusually long teeth. Sometime later, they testified, the ship sank in water so shallow that the masts stuck above the surface.

Woodman followed up with two land-based missions focused on finding a tomb reported by an Inuit named Supunger. Then he shifted back to searching for the wrecks. Studying locations indicated by the Inuit, Woodman mapped out zones with the requisite depth and then spent years searching from the air, from ships with side-scan sonar, and on the frozen sea, where he towed a magnetometer on a sledge behind a snowmobile. In 2008, he had two more expeditions on the books, but sponsors never came through. Without funding, Woodman finally threw in the towel. Robert Grenier, the head of Parks Canada's underwater archaeology team, had been pursuing the wrecks on his own, and had taken part in two of Woodman's expeditions, in 1993 and 1997. By 2007, Grenier's funding had also fallen through, but Parks Canada now officially took up the torch, forming a partnership with the Canadian Coast Guard and the Hydrographic Service. At the time, Stephen Harper had just been elected prime minister, and the possibility of finding the wrecks of *Erebus* and *Terror* dovetailed nicely with his political agenda to strengthen Canadian sovereignty in the Arctic.

By 2010, Parks Canada was in frequent contact with Woodman, who was honored to be included in their discussions, plans, and results, even if they frequently ignored his suggestions on where to look. Over the next four years, Parks Canada and their partners surveyed hundreds of square miles and spent millions before they finally found HMS *Erebus* in Wilmot and Crampton Bay, which lies more than one hundred miles south of where the ships had been abandoned. Harper seized on the find as a historic moment for Canada.

I was so impressed by Woodman's work that I decided to track him down. Early in 2021, at a time when I was still very unsure if a Northwest Passage expedition was even possible for me, I called him at his home in Port Coquitlam on the outskirts of Vancouver, British Columbia, where he lived with his wife of thirty-five years. Once he got over the fact that I had asked if we could speak on a Sunday during the NFL playoffs, he agreed to share some of his story with me.

I soon learned that the search for clues in the Franklin mystery was ongoing. He told me that he'd recently been in contact with one of the underwater archaeologists who had been diving on the wrecks of *Erebus* and *Terror.* The latter had been located in 2016 in the aptly named Terror Bay on the southwest shore of King William Island. Like its sister ship, it had been found many miles from where the Victory Point Record said the vessels had been abandoned.

Woodman had been in contact with the Parks Canada archaeologists for years, and as such, he was privy to information that he wasn't at liberty to disclose. But he did tell me that to date no electrifying discoveries had been made. The divers were using remotely operated vehicles to search the interior of the ships, but they'd found the doors leading into Franklin's cabin impeded by debris from the upper deck. Once they were able to work their way into Franklin's cabin, there was a chance they might find logbooks from the expedition in his desk.

In the meantime, there were other mysteries to solve. Two of Woodman's Franklin search expeditions on King William Island had been directed toward finding a burial vault that Supunger had described to Charles Francis Hall in 1866. According to Hall's notes, Supunger had visited the north end of King William Island in or about 1862 with his uncle to "search for things that once belonged to white man [*sic*] who had died" in the area. Supunger claimed that he and his uncle had found a tent, a skeleton of a partially clothed *kobluna* (white man), and a six-foot-high wooden pillar broken off at the top by a *Ni-noo* (polar bear); the pillar had a large decorative ball near its base—a common feature of grave markers at the time. The pillar marked an area where several large stones had been carefully fitted together. After significant effort, Supunger and his uncle managed to pry open the rocks to reveal a stone vault in which they found a knife, a leg bone, and a skull.

"If you're really interested in this mystery, you should talk to my friend Tom Gross," said Woodman. "He's still searching for Franklin's tomb. In fact, I'm sure he's planning to head up to King William Island this summer."

ı|ı

A little over a year after speaking with Woodman and about six weeks before setting off from Maine for the Arctic, I walked into the tiny, metal-sided building that serves as the terminal of the Merlyn Carter Airport in Hay River, Canada. Carter had been a friend of Tom Gross, and I'd later learn that this airport was named in his honor after he was killed by a black bear in 2005 while heading for his outhouse in the middle of the night. Tom, whom I recognized from a photo I'd seen online, stood waiting for me by the baggage claim. His face, ruddy and avuncular, lit up when he saw me, and as we shook hands, I felt like I was reuniting with an old friend, even though it was our first time meeting in person.

Tom and I had been corresponding by phone and email ever since Woodman had put us in touch with each other, and through our mutual interest in the Franklin mystery, we became friends and fellow Franklinites. I had flown to Hay River to meet him on a fact-finding mission and in anticipation of joining him later that summer to search for Franklin's tomb.

On our very first call, he had told me an incredible story. In 2015, when he and three friends had been flying a small plane over a desolate area of gravel eskers somewhere near the northern tip of King William Island, he looked down and saw two black stones standing up vertically on a ridge. "They did not belong there," he told me.

"Somebody had put them there." As he drew closer, he discerned a "perfectly rectangular structure" embedded in the side of the ridge; Tom noted well-built walls, a doorway, and a collapsed roof. The location was several miles in from shore, which explained why no one had previously found it. All other searchers had focused on the area closer to the shoreline. Tom speculated that the sailors had carried Franklin to this spot because the soft sand of the esker and the piles of big flat rocks provided the materials they needed to build a proper tomb for their commander.

Unfortunately, in the excitement of the moment, he and his copilot, Darcy King, failed to record the coordinates on the airplane's GPS unit. Tom assumed the flight path would be easy to retrace, but in subsequent years, he had been unable to relocate the tomb. But he had managed to slowly rule out territory, much like Woodman had done with the magnetometer and sonar, and he now had the location narrowed down to a thirty-square-mile grid. "I've invested about a hundred thousand dollars of my own money in this search," he said. "And I'm so close. If we find it, we'll find the expedition records because they would have buried all that with Franklin. We're talking about the ship's log, photographs, scientific equipment, maybe even letters the crew had written to loved ones back home. It's going to be a treasure trove."

"Well, here's a funny thing," I told Tom. "I kind of have a history of trying to solve historical mysteries myself. If I can get to King William Island, could you possibly use a few more searchers?"

"Could you bring drones?" he asked.

Tom Gross was born in Woodstock, Ontario, in 1959. His family was from the former Yugoslavia, and when the Russians invaded in 1944, both of his parents became prisoners of war. Tom's mother lived in a Russian labor camp for three years, and by the time she was released in 1947, she was severely malnourished and barely able to walk. She never recovered and died when Tom was fifteen.

As a youth, Tom dreamed about exploring the northern wilderness, and the opportunity finally presented itself in 1977 when he got a job building houses in a remote Inuit village on Victoria Island called Holman (now Ulukhaktok). At first, Tom struggled to fit in with the locals, but after a few years, the community accepted him and began to teach him Inuit ways of travel and survival on the land.

Eventually, he moved to Cambridge Bay, a larger community about three hundred fifty miles east of Ulukhaktok. With a population of approximately fifteen hundred people, Cambridge Bay is the largest town in the central Canadian Arctic and a hub for industry and tourism. There, he met an Inuk woman named Susie, who later gave birth to their daughter, Pamela. The relationship didn't last, and Tom raised Pam on his own as a single parent. When Pam grew up, she showed a talent for politics, and after a stint as mayor of Cambridge Bay, she was elected to the Canadian Parliament in 2021 as a representative of Nunavut. At the time of my visit to Hay River, she was serving as Nunavut's deputy premier and minister of education.

In 1990, Tom was working for the Northwest Territories Housing Corporation as a maintenance manager, a job that required him to travel to all the communities in the central Arctic. On one of those trips, he arrived late at the Paleajook Co-Op Hotel in Taloyoak (previously Spence Bay). Everyone had gone home for the night, but they'd left dinner out for him to reheat. Tom was sitting in the dining room alone, having just microwaved his meal, when a program called "Buried in Ice" came on the television. The PBS documentary followed two mid-1980s expeditions that exhumed the bodies of three Franklin sailors who had died during their first winter at Beechey Island and who had been entombed in the ice for 138 years. Their graves had been discovered in 1850 during a search for the missing Franklin expedition. The show includes haunting footage of the men's perfectly preserved bodies after they were melted out of the permafrost. Autopsies performed in field tents on the exhumation site determined that the men had died from a combination of tuberculosis, scurvy, and, possibly, lead poisoning.

Tom, like all Canadians, had learned about the Franklin expedition in school when he was a youth. Now, as an adult, he revisited the story, and perhaps since he was living in the Arctic amongst the Inuit, one of the first books he found was *Unravelling the Franklin Mystery*. On a Friday night not long after finishing Woodman's book, Tom went to bed and dreamed that he found Franklin's grave in Toronto, of all places. The coffin lay inside a sealed glass box under a pile of plastic funeral flowers. When he woke up, the vividness of the dream triggered a memory from his childhood that had been repressed for many years.

As a young boy of maybe five or six years old, Tom had had a recurring dream that he was trapped on a boat surrounded by ice. "I had all my stuffed animals around me, and I could hear the wind howling outside," he explains. "There was a small window in the boat and looking out I could see the land all white, covered in snow, and the boat was frozen into the ice. I knew I was safe there and everything I needed was with me."

Now, suddenly realizing the connection between the two dreams separated by decades, Tom understood that the universe was trying to tell him something. He got dressed and walked down the road to the house of his friend Cathy Rowan, who was Pam's kindergarten teacher. Obsessed as he was, he didn't register how early in the morning it was until Cathy, assuming there was some kind of emergency, opened the door in her pajamas.

"I had this crazy dream," Tom blurted out. "And I'm going to go look for Franklin."

Cathy smiled before replying, "Well, do you have time for a coffee before you go?"

Tom contacted Woodman shortly thereafter. They later met in Kelowna, British Columbia, and became fast friends. It took a few years for the stars to align, but in 1995, they teamed up for their first search expedition together on King William Island. They collaborated off and on over the next several years, and when Woodman retired from the search in 2004, Tom soldiered on without him. When I first contacted Tom, he told me that over the course of the twenty-eight years that he had been looking for Franklin's tomb on King William Island, he had covered more than twelve thousand miles on foot and in four-wheelers, as well as doing hundreds of hours of aerial reconnaissance. "Every time I'm ready to give up on it," he said, "something happens to suck me right back in."

Any misgivings I might have had about Tom being a bit of a wing nut for not locking down the GPS coordinates of the tomb were allayed when he brought me to his home, a split-level ranch on a quiet Hay River side street. His Franklin war room was only about a hundred square feet, but like a detective's evidence board, it was brimming with pin-filled maps, sticky notes, and drawings. One map was overlaid with a matrix of interconnected red lines along which lay hundreds of numbered dots. "Those are the flight paths I've covered since 2015," said Tom. "The numbers refer to GoPro photos taken from each of those locations."

Tom moved over to another wall where a map of the pork-chop-shaped King William Island stretched from floor to ceiling. (The island, which is about the size of Connecticut, is covered in thousands of small lakes and ponds; at its highest, it reaches an altitude of three hundred feet, though most of it is barely above sea level.) The map showed a shoreline incised with bays and inlets along which Tom had placed dozens of different-colored pushpins. Big yellow pins marked the locations in Wilmot and Crampton Bay and Terror Bay where the wrecks of *Erebus* and *Terror* were located in 2014 and 2016, respectively. Red pins marked the places where bodies and graves had been found, including "the Boat Place" in Erebus Bay, where archaeologists unearthed the remains of twenty-three Franklin sailors. From the point where *Erebus* and *Terror* were trapped in the ice of Victoria Strait and then later abandoned, Tom had tacked two lengths of yarn, one green and one red, that vectored out to two different spots on the coast. When I asked Tom what they were, he said, "Everyone takes the Victory Point Record as gospel, but the Inuit tell us that the sailors brought the ships in close to shore, and I have a theory as to where that happened."

Seeing the clues laid out like this allowed me to picture the path that Franklin's doomed men might have followed on their death march across King William Island.

"What's this one?" I asked, pointing to a green pin the size of my thumbnail on the north shore, near a spot called Two Grave Bay. Tom grabbed a leather-bound notebook and began leafing through it. This book, I realized, was his search bible. It was filled with handwritten notes, photos, coordinates, and three decades' worth of doodling and speculation. He stopped at a page in the middle and passed me the book. The right-hand page featured two colored-pencil drawings, made by Tom, that had been glued into the book. At the top, it read, "The Stone House"; below, Tom had written, "Burial Vault Sir John Franklin, scale: ¼"–1'."

Tom explained that a local Inuk named Ben Putuguq from Gjoa (pronounced *Jo-uh)* Haven, King William Island's only settlement, with a population of roughly thirteen hundred, had described a "stone house" to him during an interview in 2004. Ben told Tom that he had come across this structure while hunting on the north side of King William Island. The top drawing was an aerial view. It showed four stone walls, a doorway, and, sitting inside, four rectangular slabs of rock. The bottom drawing was an elevational perspective looking in from the front.

"This is what I saw from the air in 2015," said Tom. "And it matches exactly with the testimony of Ben Putuguq."

The next day, I met up with Tom's friend Darcy King at the Ptarmigan Inn. Elevator music played in the background. A fish tank bubbled next to our table. Darcy owns a company called Landa Aviation and is a lifelong pilot who has spent decades flying around the Northwest Territories and Nunavut. Darcy had been Tom's copilot on the flight when he saw the stone house, and I was eager to ask Darcy what *he* had seen that day.

A thatch of thick gray hair stuck out from under a weathered ball cap, and his bushy eyebrows and blue eyes widened as he started in. "I still remember it clear as day. We had just flown up to check out the cairn at Victory Point and were heading west, about four hundred feet above a long gravel ridge. The light was perfect, coming from behind us a bit, and suddenly, Tom said, 'Well, look at that.' I looked out the window, and right below us I saw these two pillars of dark rock. One was standing straight up, and the other had tipped over like this." Darcy cocked his arm like he was going to salute me, and then continued. "It was a perfect doorway, and you could see right where the lintel had fallen in. Inside of it were these flat rocks that were almost like tables."

"How big was the structure?" I asked.

"Oh, I'd say it was about twelve feet across and maybe twenty feet long, and it was built into the side of a hill. The vault itself was made of lighter-colored rock. The walls were long and smooth."

"Would you bet your life that what you saw was man-made?" I asked.

Darcy bristled. "Oh, guaranteed. There's *no* question. I do surveys all the time looking at animals on the ground, for crying out loud. I've been looking out the window of planes all my life. I've seen a lot of stuff from the air, and I know what I saw."

"So how come you guys didn't get the GPS coordinates?"

Darcy looked down, shook his head, and said, "Three different times, Tom said, 'I know where we are.' And flying as many years as I have, I kind of have a built-in GPS, so it just never came up. I'm kind of embarrassed that we forgot to hit the button. But if you put me in a plane right now, I bet I could fly us pretty close to the spot."

Tom and Darcy went back the next year with a GoPro strapped to the wing of the plane but were shut down by fog and couldn't relocate the tomb. It was the same story in 2017. In 2018, they landed a float plane on a lake near Cape Felix, the northernmost tip of King William Island. They set up a base camp and searched on foot for days, but found nothing. The next year they went in on ATVs, but again came up short. Then COVID hit and shut down all travel in the Arctic.

"It's weird, really, the way the vault just disappeared on us," said Darcy. "I can't explain it. I'm not a superstitious person at all, but there's some weird shit going on up there. It's spooky almost. Tom has always said that the Inuit knew where the tomb is, that there were cairns leading right to it. But they didn't want anyone to find it because the shamans put a curse on Franklin's men, so they knocked down the cairns and erased the trail."

CHAPTER 4

Povl

Date: June 27, 2022
Position: Nuuk, Greenland

There was no fanfare or greeting committee when *Polar Sun* slid into the inner harbor in Nuuk, Greenland, after a storm-tossed five-day crossing of Davis Strait. It was twelve forty-five a.m., and the port, snug and packed tight with an assortment of banged-up fishing boats, glowed under the twilight of the midnight sun. To the east, a three-hundred-foot-long red ship, with the name *Royal Danish Company* painted on its side, was parked alongside a metal-sided industrial building. In the other direction, a long wooden staircase led up a mossy precipice, beyond which craggy peaks, with glaciers spilling down between them, formed a jagged skyline. As we slipped along-side an industrial wharf, I took a deep breath and inhaled the distinct mélange of rotting fish, salt, and diesel.

Ben headed straight for his bunk, but Rudy and I poured cock-tails to toast our Greenlandic landfall, and Mr. Dirt, who doesn't drink, raised a water bottle adorned with a homemade sticker of a skiing peacock. A few hours earlier, after we had sailed into a chan-nel between the mainland and a wind-blasted slab of rock that formed a small island called Qaarusulik, the seas flattened out, and for the first time in days, it was possible to move around the boat without holding on with two hands. Mr. Dirt, sensing that the worst of his ordeal had passed, appeared in the companionway, his bearded face shiny with dried vomit.

"Where are we?" he asked. "And what day is it?"

His eyes were still glassy and lifeless now, and by the way he sat with his shoulders slumped forward, staring mindlessly at his water bottle, it was clear that a part of him was still out in those waves in Davis Strait.

"How *are* you, Baby Dirt?" I asked.

"I can't believe we're actually here," he said. "You know, when we were in the worst part of the storm, I made the mistake of looking down into the trough of one of those waves, and it scared me so bad, I got the shaky-leg thing. At one point, I actually said to myself, 'Well, it's been a good life.'"

"To a good life," I said, raising another toast.

And it was. It had taken us twenty-three days to sail roughly two thousand miles from southern Maine to Greenland. During the crossing of Davis Strait, which was by far the crux of the voyage thus far, the four of us had been cooped up in a fiberglass shell the size of a storage container, focused single-mindedly as a team on one thing only: to sail *Polar Sun* as quickly and safely as possible from Labrador to Greenland. We might not have always seen perfectly eye to eye on how best to accomplish our goal, but rarely had I ever enjoyed such intense communion with both the elements and my fellow man. And if Davis Strait had been a test of our worthiness to enter the Northwest Passage, I suppose one could say that we had passed.

In the morning, a friendly customs officer showed up. He barely noticed our 12-gauge shotgun and the three dozen bottles of liquor, but he did take issue with the twenty-five pounds of raw sugar we'd cached in various compartments throughout the boat. The sugar prohibition was an odd one and might have been due to an antique right of import held by the Danish Crown, but after we convinced our new friend that Ben had an ungodly sweet tooth—and that we had no intention of selling all this sucrose on the black market—he cleared us in. His only request was that we check in with the Greenland Coast Guard and keep them apprised of our whereabouts.

I had just seen the agent off the boat and up a steep ladder to a paved parking lot when a dented Mitsubishi station wagon hooked a U-turn and parked right in front of me. A thin, bowlegged sprite of a man wearing a red wool beret jumped out and strode in my direction.

"Jens," he said with a lilting Danish accent as he extended a bony hand. "Jens Kjeldsen. Actually . . . it's Jens *Erik* Kjeldsen. We all use our middle names here. Where you from and where you to?"

Jens had come down to the harbor to work on his own boat, and when he spotted *Polar Sun* tied up to the town wharf, he'd immediately swung by to say hello. When I told him that we had sailed in from Maine and were headed to Alaska, he smiled. "I thought that might be the case. You see that double-ender on the other side of the harbor?" he said, pointing toward a forty-foot sailboat with a white hull and a canoe stern. "That's *Kigdlua.* My wife and I sailed her around the world in 2018 and '19."

"Which way did you go?" I asked.

Jens smiled even wider. "Why, via the Northwest Passage, of course. And we're setting off on another big voyage here in another month or so, but this time we're going south."

Later, I'd learn that Jens salvaged *Kigdlua*, which means "burning fire" in Greenlandic, from where it had sunk in a nearby fjord. He bought the wreck from an insurance company for ten kroner ($1.40 USD). He paid forty cents more than I had for *Capella*, but in the process, he ended up with a vessel worthy of sailing around the world.

Jens and I climbed down the ladder into *Polar Sun*'s cabin, and I offered him the best seat by the woodstove, where he started right in with his story. He'd first come to Greenland from Copenhagen as a young man to complete his obligatory Danish military service. After an epic snowbound trek across Disko Island, just off Greenland's west coast, he fell in love with the land and an Inuk woman named Dorthe, to whom he had now been married for more than forty years. For most of their lives, they had lived in Aasiaat, a small fishing community three hundred fifty miles north of Nuuk. They survived doing "odd jobs," including running a charter boat that offered iceberg tours in Disko Bay. "Aasiaat is where my heart is," said Jens, "but we moved to the big city nine years ago to be closer to our children and grandchildren." Currently, he was working as a carpenter, which explained the homemade tool pouch he wore on his belt. What he didn't mention, and I wouldn't learn until much later, was that he had retired as a judge in Greenland's criminal civilian court just a month earlier, at age sixty-seven. For some reason, he didn't think this fact worth mentioning.

What Ben and I were most interested in was the ice, of course, and how the breakup was progressing in Baffin Bay and the Northwest Passage. Jens hadn't paid much attention to all that, he told us, but it had been an exceptionally cold winter over a lot of the Arctic. The previous summer, though, had marked the lowest ice extent ever, which meant that we'd mostly be dealing with what Jens called "first-year ice." He figured that if the summer followed its usual pattern, the passage would be navigable starting around mid-August, which gave us another six weeks to get from Greenland into the heart of the Arctic.

Jens said that back in 2018, he and Dorthe had followed the Amundsen route and had not encountered any serious ice until they'd gotten to Peel Sound, on the west side of Somerset Island. They had found a lead of open water near shore and had been working their way south when, in an instant, the wind shifted, and the pack closed around *Kigdlua*'s fiberglass hull like a vice. "There was nothing we could do, so we just floated with it," he said. "Luckily, it was going in the right direction." Eighteen miles and thirty-six hours later, the wind shifted and spat them back out into open water. Jens told the story as if getting stuck in pack ice were no more serious than jamming his car into a snowbank, and his account gave me the distinct impression that there were sailors who knew how to deal with ice and others who had no clue. We were not included in the former category, of course—not yet at least.

It wasn't long before the subject turned to the Franklin expedition. Over the next hour, I told him everything I knew about Dave Woodman and Tom Gross and about our plan to search for Franklin's tomb on King William Island. Jens alternately wrung his hands and stroked his beard, and I often caught him staring into space with a faraway look in his eye. At one point, he asked for a scrap of paper and took a few notes with a carpenter's pencil he fished from his tool pouch. By the time I'd finished my story, evening was setting in and Jens looked tired. I followed him up the companionway and out onto *Polar Sun*'s deck, where we stood in silence, looking out over the harbor.

"I've got someone I want you to meet," he said after a pause. "I'll swing by tomorrow to pick you up."

Late the next morning, Jens and I rattled up the hill above the harbor in his Mitsubishi Galant, and as we crested the height of land, the metropolis of Nuuk splayed out across the narrow plain below us. With its brightly colored Scandinavian-style apartment blocks dotting the hillside and modern ten-story glass-walled buildings silhouetted against a backdrop of soaring, glacier-draped peaks, the city looked like something out of a fantasy film. Cars, trucks, and buses clogged the streets, which surprised me, given that Greenland has no road network connecting its towns and villages. All this hustle and bustle extended no farther than the city limits. If someone wanted to explore the country outside the comforts of Nuuk, their choices were limited to foot, four-wheeler, helicopter, or boat.

Nuuk, which is Greenlandic for "cape," sits on a spit of land nestled between the second-largest fjord system in the world and a toothy granite peak called Sermitsiaq. The city and its environs are home to around twenty thousand people—most of whom are Inuit. This accounts for roughly a third of Greenland's total population, which has grown every year since Jens arrived in the late 1970s. This growth, he told me, is due in part to generous government incentives that encourage people from Denmark and other countries to immigrate to Greenland.

Nuuk's recorded history goes back more than a millennium to 986 AD, when a group of Icelanders led by the Erik the Red, the father of Leif Erikson, established a settlement known as Vestribygð (the Western Settlement) a bit inland from Nuuk's present-day location. Erik the Red had set sail in search of a new home after being banished from Iceland for killing one of his neighbors. He eventually landed on the southwest coast of an unknown island that he named "Greenland." The name was a ploy, of course, given that 80 percent of his new homeland was covered by one of the world's largest ice caps.

This ice cap acts like a giant refrigerator, and its cooling effect is so strong that it actually impedes prevailing westerly winds, deflecting them north. As a result, Greenland's west coast enjoys a relatively mild climate and mostly ice-free seas that allowed Viking ships to sail without restraint between Greenland and Europe, where they exchanged valuable goods like walrus ivory for salt and sugar. Along the fringes of the ice cap, the Norse found bountiful pastureland where they built sod houses, established farms, and raised cattle, goats, and sheep. Around the same time, a second, larger Norse settlement known as Eystribygð (the Eastern Settlement) was established about three hundred miles south of Vestribygð near present-day Narsarsuaq. At the height of its prosperity, Eystribygð had some four thousand inhabitants, including Erik the Red himself. One of its farms, called Gardar, had stalls for a hundred sixty cows.

Erik the Red and his countrymen did not "discover" Greenland, though. Archaeological evidence suggests that the first people arrived in the northwestern part of the island approximately five thousand years ago. While today we like to lump all Arctic people into a single group we call "Inuit" (which in the Inuktitut tongue means "the people"), the native inhabitants of the region were part of at least four distinct cultures known as Independence I and II, Saqqaq, Dorset, and Thule. (Researchers are not in complete agreement on these subdivisions, and some of the earlier groups of people are often referred to simply as pre-Dorset.) The Saqqaq lived in western Greenland from approximately 2000 BCE to 800 BCE, while the Independence I and II culture settled in the northern part of Greenland, now known as Peary Land, from 2400 BCE to 1300 BCE. The Dorset arrived around 800 BCE and lived in small bands along the island's west coast for the next thousand years. They disappeared from the area around 300 CE, only to return four hundred years later, this time settling in the far north.

The same favorable climate that gave the Norse a toehold in western Greenland drew the Thule (ancestors of modern-day Inuit), who followed bowhead whales and other marine mammals across the Arctic, arriving in northwestern Greenland around 1200 CE. There is no conclusive evidence that the Thule had any direct contact with the Dorset, but it's likely that the Thule adopted new technologies by salvaging harpoon heads and carvings from abandoned Dorset camps. The Thule period is defined as the time up until the first contact with Europeans in the eighteenth century, at which point

Inuit (called Eskimos at the time) enter the historical record. But according to the Avataq Cultural Institute, which has created a detailed chronology of Arctic cultures, Thule and Inuit can be considered interchangeable.

In 1986, a tuft of frozen hair was excavated from an archaeological site in the village of Qeqertarsuaq (formerly Godhåvn). For twenty-five years, the sample moldered in a plastic bag in storage at the National Museum of Denmark, until it was discovered by a team of scientists from the University of Copenhagen who sequenced the surviving DNA. The results suggest that the hair had come from a Saqqaq man who had lived in Greenland four thousand years ago. He likely had brown eyes, black hair, and "shovel-shaped teeth." The most interesting finding, though, was that his closest relatives were the Chukchi people of eastern Siberia, which proved a stunning theory anthropologists had long entertained: that Saqqaq culture migrated from west to east across the high latitudes of the Arctic as the people tracked marine mammals and herds of caribou and musk oxen. By logic, it is safe to assume that they followed a route roughly tracing what we today define as the Northwest Passage, by boat in the summer and by foot and dogsled across the ice the rest of the year.

When the Norse arrived at Vestribygð at the end of the tenth century, they found extensive signs of settlement, including stone tent rings, tools, and midden piles—but no people. The extent of interaction and contact between the Norse and the Dorset/Thule during the centuries when they overlapped in Greenland is unknown, but recent genetic studies suggest that they did not interbreed. By the time the Thule worked their way south along the west coast to Vestribygð, the Norse settlement was facing collapse due to a disastrous combination of deforestation, soil erosion, and climate change. The Icelandic sagas contain an account from 1379 that paints a grim picture of relations between the Norse and the Thule. "The *skraelings* assaulted the Greenlanders, killing 18 men, and captured two boys and one bondswoman and made them slaves." In Norse, the term "*skraelings*" roughly translates to "wretches."

While the exact date when the Norse settlement collapsed is unknown, there is little doubt that the final blow was dealt by the effects of the Little Ice Age, an era of global cooling that began in the fourteenth century and persisted for more than five hundred years; during that time, glaciers advanced across Europe and North America. By the beginning of the fifteenth century, shipping in the North Atlantic had been severely curtailed and archaeological excavations in Vestribygð paint a horrific picture of the settlement's final days. The topmost layer of debris was filled with the bones of animals that would normally not be eaten, including small birds, newborn calves, lambs, and even dogs. This, along with other evidence, points to a famine of fatal proportions.

The Thule survived, of course, and eventually took over the site of Vestribygð. They didn't interact again with European visitors until the late-sixteenth century, when John Davis, the namesake of Davis Strait, landed in southwestern Greenland and kidnapped Inuit and brought them back to England as human pets; there they were paraded as long-lost descendants of Greenland's original European settlers.

The myth of a lost colony of Christians in Vestribygð compelled Greenland's first missionary and entrepreneur, a Danish-Norwegian named Hans Egede, to sail to Vestribygð in 1721. But instead of a surviving tribe of Christians, he found a thriving Inuit community that showed him the ruins of the Norse settlement and shared oral testimony about the Vikings' final days. In 1728, Egede built a church, founded a Lutheran mission, and established the trading port of Godthåb on the site of present-day Nuuk. Today, two-thirds of all Greenlanders are Evangelical Lutherans, with many of the rest following other forms of Christianity. Only a small minority still practice traditional animistic beliefs like shamanism.

Since the founding of Godthåb in the early 1700s, Greenland has been under continuous Danish rule. In 1953, Greenland was reclassified (along with the Faroe Islands) as an autonomous country within the Kingdom of Denmark. The first Greenland Parliament was formed twenty-five years later in 1978. Self-governance was expanded in 2009, and the Greenland constitution now contains a clause that ensures Greenlanders the right to declare independence when they are ready. According to Jens, there is still a heavy dependency on Denmark for subsidies, national defense, and monetary policy, and the dream of full sovereignty will take some time yet before it is fully realized.

After driving past the Hotel Hans Egede and a lively shopping mall, Jens turned and navigated through Nuuk's historic waterfront. The wind had been screaming steadily since we arrived, whipping waves that broke against a long stone pier that sits completely exposed to the north and west. "I know it doesn't look like much of an anchorage," said Jens, "but you have to understand that back when the first sailors landed here, their boats didn't have engines, so they chose harbors that you could sail in and out of."

We pulled up outside a small red house on a side street near the harbor. The yard was cluttered with whale vertebrae, empty buckets, fishing nets, and other detritus. A man of middle height with a thick beard, wild gray hair, and crooked yellowing teeth opened the front door and invited us inside. This was Jens's old friend Povl Linnet, who introduced himself in the slow and strained English dialect of someone who speaks three or four other languages.

While Povl and Jens chatted in Danish, I took the liberty of poking around. The kitchen, warm with the smell of baked bread, opened into a living room that felt a bit like a Greenlandic Aladdin's cave. Built-in wooden cabinets stretched from floor to ceiling on three sides, overflowing with books and a colorful menagerie of tchotchkes. Other surfaces, perhaps chairs and side tables, were covered with an eclectic mix of soapstone carvings, figurines, stuffed animals, scrimshaw, a boomerang, and dozens of antique glass bottles. The bleached skull of what might have been a musk ox sat on a high shelf, and dangling in front of a picture window that looked out over Nuup Kangerlua (formerly Godthåbsfjord) were two translucent balls made from some sort of animal skin. I gazed open-mouthed at this trove of treasure, the collection of Povl's lifetime in the Arctic.

I had edged my way in to examine an intricate wooden model boat when I felt Povl standing behind me. "Is this your boat?" I turned and asked. That morning, Jens had pointed out Povl's vessel to me in the harbor. It was unmistakable, with a plumb bow and stern, tiny pilothouse, and communist red hull that reminded me of Popeye's cartoon ship, the *Olive*. Povl nodded and explained that *Auvek*, which means "walrus" in Greenlandic, is a wooden whaling cutter built in the 1940s. It's primarily a motor vessel but has a mast and can fly three sails as a backup if the engine ever dies. Its most distinctive feature, though, is the large harpoon cannon on its bow. From amidst the clutter, Povl found an empty brass shell casing and handed it to me. It was two inches in diameter and still smelled of gunpowder. I wondered if the cannon had once fired a harpoon into a whale. Povl told me that he was retiring to Denmark in a matter of months, and he would travel to the Baltic aboard *Auvek*. "I would never leave her behind," he said. I looked around and decided not to ask what he was going to do with everything else.

Povl drew my attention to a small wooden table in the corner handsomely decorated with a mosaic of variously shaped pieces of dark wood. "I made this table from wood that I salvaged from the *Fox*," he said.

I froze. Even to an amateur Arctic historian, the *Fox* was an almost mythical vessel. Lady Jane Franklin had purchased the *Fox* in

1857 and dispatched it to the Northwest Passage in a last-ditch effort to find her husband and his men, who by that point had been missing for a decade. During that time, more than a dozen expeditions had set off in search of Franklin, but no one had yet looked in the vicinity of King William Island, and Lady Jane, knowing that the Admiralty had instructed her husband to look for channels leading south toward this area, had a hunch that this was where they would be found.

"The *Fox*?" I replied. "The boat Francis Leopold McClintock sailed when he found Franklin's Victory Point Record in 1859?"

I looked at Povl and Jens, who were smiling and eyeing each other knowingly. "It's a good story," said Jens. "Let's sit down and you can hear it over a coffee."

Jens and I found seats at the kitchen table while Povl disappeared. A few minutes later he came into the room dragging a five-foot length of rusted steel cable as thick as my wrist, with a thimble spliced into its end. It was the bobstay—a rope or chain used to support a boat's bowsprit—from the *Fox*, the first of several artifacts from this famous ship that he would bring out over the next hour.

Lady Jane's instincts had been prescient, and two years after sailing from England, McClintock's lieutenant, William Hobson, found human remains as well as the Victory Point Record stashed inside a metal cylinder in a cairn near the northern tip of the island. The record contained two entries, one from May 1847 and a second added a year later. Together, they sketched the outlines of a tragedy.

The first installment stated that the expedition had overwintered at Beechey Island (the note said that this had been in 1846–1847, which we know to be an error, since it had to have been the year before this) and then, in September 1846, the ships were iced in fifteen nautical miles north-northwest of Victory Point at 70 degrees 5 minutes north, 98 degrees 23 minutes west. The first note was signed: "Sir John Franklin commanding the Expedition. All well." The second note, written into the margins of the first on April 25, 1848, stated that *Erebus* and *Terror* had been deserted three days earlier at the same location where they'd been icebound for a year and a half. By this point, they had lost nine officers and fifteen men, including Franklin, who died on June 11, 1847—just two weeks after the first note had been written. The final line read: "And start on tomorrow 26th for Backs Fish River."

When McClintock returned to England in the fall of 1859 with this intelligence, Lady Jane gifted the *Fox* to him out of gratitude, but for reasons that remain obscure, McClintock declined, and the famous ship went back to work. In the early 1860s, the *Fox* did surveying between the Faroe Islands and Greenland to help in the laying of the first North Atlantic telegraph cable. Later, it was sold to the Danish Royal Greenland company to run supplies from Denmark to Nuuk and beyond. In 1912, after decades of loyal service, the *Fox* was scuttled in the mouth of a small cove near Godhåvn (now Qeqertarsuaq).

Povl found the wreck in the mid-1980s. At the time, he was running a small ship-repair business out of Aasiaat, and he had been called in to fix a pontoon bridge in Qeqertarsuaq. He saw the wreck poking out of the water not far from the town dock, and some local Inuit, who had known about it for decades, told him that it was the *Fox*. Ice had destroyed most of the ship, but the boiler was still intact, as was the whole bottom part of the hull, which had been built with English oak and Scottish elm, sheathed with copper, and held together with bronze rivets. Knowing that the ice would eventually finish the job and reduce what remained to splinters, Povl began diving on the wreck after work and salvaging everything he could. He told me in great detail about the days he had labored underwater to detach the entire drive train, including the shaft and its massive bronze propellor, which he winched to the surface with the help of a local fishing boat. When he finally got it all ashore, he wrote a letter to the Aberdeen Maritime Museum in Scotland, in the city where the ship had been built. It took a few years, but they eventually raised enough money to ship the propeller, still attached to the shaft, to Aberdeen, where it now sits in a place of honor inside the museum's front entrance.

Inspired by his own recounting of the tale, Povl dug deeper into his hoard, bringing books, charts, and maps to the kitchen. When his labored English failed him, he reverted to a direct dialogue with Jens in Danish. The words "Franklin" and "Whale Fish Islands" punctuated the story as Povl unspooled a faded, dog-eared nautical chart across the table. The paper was weather-beaten and covered in pencil markings from decades of Povl's navigating the west coast of Greenland the old-fashioned way—without GPS, chart plotter, or radar.

The three-hundred-fifty-mile section of coast between Nuuk and Ilulissat, where we planned to cross Baffin Bay into the Northwest Passage, is a dangerous maze of poorly charted shoals, inside passages, fjords, and islands. For centuries, successful navigation through this labyrinth depended entirely upon the advice of local mariners, and as I had yet to plot my own course to Disko Bay, I was planning to rely upon the same. As Jens and Povl walked me up the coast, I followed along, dropping waypoints into an app called Navionics on my phone.

When they got to Disko Bay, Povl pointed to a tiny speck south of Qeqertarsuaq called Blubber Island. "This buoy," he said, tapping the chart with a crooked finger, "is made out of the funnel from the *Fox*. It's painted red; you can't miss it." A few miles south of the strange marker was a small cluster of islands marked as Kronprinsens Ejland, aka the Whale Fish Islands. "Franklin anchored here before he set off for the Northwest Passage," said Povl, pointing to a small cove nestled inside this little archipelago. "There's an abandoned whaling station with a good anchorage. If you look around onshore, you might find the iron bollards that Franklin used to fix their stern lines."

I could have sat in that cozy kitchen and listened to Jens and Povl for days, but my mind kept getting tugged away by the long list of repairs that *Polar Sun* needed before we could set off from Nuuk.

When I stood up and told Povl that I had to get going, he handed me a slip of paper with his telephone number and said to call if I needed anything. As I walked back to the boat, an eclectic mix of loose ends swirled inside my head, including our broken generator, a contaminated fuel tank, the crew, my family, the wreck of the *Fox*—and Franklin.

As I came down the hill to the harbor, my back buffeted by a cold, driving rain, Mr. Dirt zoomed by on a two-wheeled kick scooter, followed closely by a young woman whose face I couldn't see under her hood. When he saw it was me, Mr. Dirt skidded to a stop, and he and his new friend padded over to say hello; apparently, the dating app Tinder works in Greenland. The woman, I now saw, was a brunette in her early to mid-twenties, with striking large brown eyes that sparkled mischievously. She was soaked to the bone, with water dripping off the tip of her nose. When I nodded toward Mr. Dirt's new whip, wondering where it had materialized from—not to mention how he had gotten so good at riding one of these things—he explained that he'd found both of the scooters "in an irrigation ditch." You had to give the guy credit: He knew how to keep things interesting.

Later, when he got back to the boat, he showed me what he'd been working on that morning while I had visited with Povl. Mr. Dirt had completely dismantled the generator, and in the process, he'd discovered why it had filled the cabin with choking black smoke when we'd fired it up in Davis Strait to charge the batteries. The previous owner had connected the injector to a fuel line that was slightly too large. During the storm, the seal had broken and diesel had sprayed onto the exhaust manifold, soaking a piece of insulation that Mr. Dirt now held in his hand. Thankfully, it never got hot enough to catch fire, but it had driven all of us out of the cabin.

Mr. Dirt and I spent the next two days combing the city for an eighth-inch to a three-sixteenth-inch coupling, but as everything in Greenland is metric, we were out of luck. Finally, I called Povl. He arrived at the boat within an hour, where he studied the sections of hose gravely, muttering in Danish. Then, without a word, he climbed up the companionway and drove away in his beat-up gray van. I stared after him blankly. Mr. Dirt shrugged. The generator was kaput, Greenland was metric, and a man I barely knew had just driven off with the key parts in his pocket.

The next morning, I was sitting in the cabin drinking coffee, working on the day's to-do list, when Povl reappeared in the companionway. He was holding a tiny fitting the size of a pencil stub in his calloused hands. I put on my glasses and held it up to the light. It was made from copper and looked like it had been polished on a grinding wheel.

"Where the heck did you find this?" I asked him.

"I made it," he said, grinning.

Povl had bought two pieces of copper piping, soldered them together, then ground down the ends until they were exactly the right diameter. Mr. Dirt installed the new coupling under Povl's supervision, but the generator still wouldn't start. Povl took a seat on the upturned blue NAPA bucket—the same bucket that Mr. Dirt had so rudely violated in Davis Strait—and began tracing fuel flow through the motor. As he removed different hoses and fittings, he hollered for me to turn the mechanical fuel pump on and off with its switch on the electrical panel. At some point I got sidetracked into a conversation with Ben and apparently missed my cue when Povl yelled, "Off!" Mr. Dirt would later tell me that he saw the injector shooting a stream of diesel into Povl's mouth as if it were a dentist's Waterpik. A volley of Danish curses erupted from the workroom. When Povl staggered into the salon, he was covered in an oily sheen like a honey-glazed ham. Scowling and muttering, he climbed the companionway and left. The generator lay in pieces and my eyes watered from the powerful diesel fumes that now permeated the air inside the boat.

I assumed we'd never see Povl again, but the next morning, he returned to the boat as if nothing had happened. We apologized profusely for the incident the day before, but Povl just smiled and went back to work on the generator. Overnight, he had downloaded the owner's manual and figured out that the problem had to be that the glow plug wasn't getting enough voltage. He ran a new wire and switch, and voilà, the generator fired up on the first try.

|||

The storm we'd ridden in on, which continued to rage for the next week, finally cleared on July 1, and the sun broke through the gray overcast for the first time since we'd left Labrador. Sadly, though, it was the day that Mr. Dirt was scheduled to leave the expedition and fly back to New England. His departure filled me with a mixture of gratitude and dread. Despite his seasickness, he'd served admirably as *Polar Sun*'s mechanic, deckhand, woodchopper, dishwasher, cabin boy, gofer, laborer, and raconteur. And he'd shared my cabin for the past month. Most nights, as we had lain across from each other, we shared not just stories, but our dreams and anxieties, and I'd come to rate him as a true friend. I would miss that hairy little hobbit in the months ahead.

At the airport, we swapped him for four fresh faces—Renan Ozturk and his wife, Taylor Rees, plus my wife, Hampton, and Tommy, who was now six. With my family aboard, I felt like I'd just had an appendage reattached. From the moment I'd first envisioned this expedition, my biggest worry wasn't the ice, the cost, or the fact that *Polar Sun* is made of plastic. What concerned me the most was the time away from home. Ben and I had estimated that the voyage could take up to four months, and in a lifetime of going on expeditions, I'd never been away for more than half that. Hampton and I had seriously considered the possibility of having her and Tommy join the crew for the whole trip, but after a lot of discussion, we decided it was too dangerous a mission for a child. And there was no way she was coming without him. Our compromise was to have the two of them join the expedition in Nuuk and spend a few weeks with us sailing up the west coast of Greenland. In Ilulissat, they would fly to Denmark, while the rest of us headed deeper into the Arctic. Hampton, who loves adventure as much as I do, agreed that this Greenlandic family reunion would break up my absence enough that the whole enterprise was doable for us as a family. Tommy snagged Mr. Dirt's bunk, where he lined up his Legos, plastic dinosaurs, and stuffed animals, while Hampton squeezed into mine.

Renan was now joining for the duration of the expedition, which he and Rudy would document with photos and video for National Geographic. Taylor, like Hampton and Tommy, was joining only for the next leg to Ilulissat, and then she'd head off to the Atacama Desert in Chile for her next assignment. She and Renan had been married in 2016, and they run a high-end adventure-film company called Expedition Studios. A year earlier, they had codirected that tepui film, which was only the latest in a string of expeditions that Renan and I had done together over the years—from Borneo to Chad to Mount Everest—usually with me in front of the camera and him behind it.

Renan has had a lot of close scrapes but perhaps none where he came closer to the edge than an incident that occurred on the expedition to Weiassipu tepui. After weeks of living in what we called "mud world," we were camped in portaledges hanging about seven hundred feet up the wall in a driving rainstorm. Everyone was soaked and demoralized, and I thought we had given up on trying to find that damn frog, when I saw Renan and a Venezuelan climber named Federico "Fuco" Pisani gearing up for one last search. It was well after dark when Alex Honnold and I watched them set off across a narrow, vegetated ledge, and I noticed that they were not carrying a rope.

Renan told me later that he was following Fuco, filming with a camera in one hand, clutching clumps of vegetation with the other, as the frogs, invisible to the eye, chirped all around them. At a steep bulge, Renan hung his full weight off a tuft of needle grass. It held for a second, then ripped out of the wall, sending him ass over teakettle. I had just lain back onto my sleeping pad, about two hundred feet away, when I heard someone yell, "FUUUUUCK!" I recognized Renan's voice, and by the modulation of the scream, I could tell he was falling. He still doesn't know how he did it—it might have been nothing more than pure luck—but after tumbling about fifteen feet, he somehow stuck a perfect gymnast's landing in a trough of mud mere feet from the lip of the overhanging cliff. And in classic Renan fashion, he'd kept the camera running the whole time.

The morning of July 3 was a whirlwind as the new crew got settled in and we prepared *Polar Sun* to head north. Jens and I went to the gas station to fill our jerry cans with diesel while Ben worked on replacing the block and tackle on our mainsheet—one of its pulleys had shattered in a fifty-five-knot katabatic gust on our approach to Nuuk. Below deck, the rest of the crew feverishly dug for places to stash the camera equipment and groceries that covered every horizontal surface, while Tommy, oblivious to it all, staged a mock battle on the salon table between T. rex and Indominus rex. We were almost ready to depart when I heard the now familiar jangle of Povl's gray van pulling up to the dock. I figured he'd come to bid us farewell, and I was wondering how the crusty mariner would respond to a hug when he abruptly handed me a thick roll of charts—the same well-loved stack he'd unrolled on the kitchen table when we'd plotted our route north a few days before.

"I want you to have these," he said. "I don't need them anymore."

I stood there holding the heavy bundle in my arms, dumbstruck by his generosity, trying to hide the tears that were pooling in my eyes. I pulled myself together and said: "Thank you so much, Povl. But I can't possibly take these. I mean, they have all your notes, and we already have most of the paper charts we need."

"No," he said firmly. "You need them more than I do. And they'll be useful. Take them. Please."

I was moving in for that hug when he reached back into the van and handed me a book—*Eskimo Diary* by Thomas Frederiksen. I opened it to a page marked with a green sticky note and there was a colored-pencil drawing of *Auvek*; it had its mainsail up and was towing a smaller boat at the end of a long line. Here was another story, still to be told, from this extraordinary man. But before I could inquire, Povl said, "I stopped by earlier when you weren't here and gave something to Ben for you." Without saying anything more, he turned around, jumped into his van, and drove away.

Back inside *Polar Sun*, I found a green-tinged bolt, covered in rust, sitting on the nav table. It was about eight inches long and as thick as my thumb. It appeared to be bronze and looked very old; its button head still bore impact marks from a heavy hammer.

"What the heck is this?" I asked Ben.

"Povl said it was a rivet from the *Fox*," he replied. "He thought you'd appreciate it."

CHAPTER 5

The Whale Fish Islands

Date: July 5, 2022
Position: west coast of Greenland, between Maniitsoq and Disko Bay

Polar Sun, dwarfed by soaring granite walls, slid silently up the canyon-like fjord. I had engaged the autopilot and stepped out onto the deck, where I stood in mute awe of the scene unfolding before us. On the starboard rail, Hampton, Tommy, and Taylor sat with their legs dangling just above the water, where they pointed and cried out as we passed beneath massive waterfalls that cascaded from the heights. Some were thousands of feet tall, and where they poured into the sea, clouds of fulmars, cormorants, and kittiwakes circled in the salty mist. Every so often, one of them would dive-bomb into the froth, emerging, to Tommy's delight, with a silvery fish clutched in its talons.

Following a few faint pencil markings on Povl's charts, we'd taken our time working our way north along Greenland's west coast, sticking mostly to fjords and inside passages and anchoring in secluded coves where we crowded together in the cockpit to watch sunsets that sometimes lasted all night. With a crew of seven it took some coordination to get around below and above deck, and it didn't take long before all the carefully stowed provisions and camera gear were strewn everywhere as if a bomb had gone off. But everyone seemed to have settled in and made themselves comfortable, and since there were so many bodies aboard and we were sailing for only a day at a time and usually not in the open ocean, we didn't even bother with a watch schedule. Rudy and Renan were hard at work filming everything, but for Ben and me, it felt a bit like a vacation.

It was on Povl's recommendation that we had sailed sixteen miles out of our way down that waterfall-lined, dead-end fjord. At its head, we anchored in milky water at the snout of a glacier that tumbled into the sea from between a row of spiky peaks that rimmed the skyline to the east. Ben brought us ashore in the dinghy, and the Synnott family, tied together with an old halyard, spent the afternoon jumping across crevasses—and keeping a wary eye out for polar bears.

That night, our fifth since leaving Nuuk, we found our way into a small cove called Agpamiut, which Povl had marked on the chart with a tiny asterisk. There were no soundings but the water was gin clear, and as we carefully worked our way behind a small, rocky island in the center of the bay, I could see kelp-covered boulders sliding by just beneath our keel. When the depth sounder read three feet, I put us in reverse and stopped the boat. Ben released the anchor, and as the chain rattled through the hawsepipe, the *clackety-clack* echoed off the surrounding hills.

It was nearly midnight, but we were now only sixty-six miles from the Arctic Circle, and the flaming orange sun still hung a finger's width above the horizon. I'd hoped the crew would sit with me for a bit, maybe share a drink and commune in the enchantment of Povl's secret garden, but it had been a long day, and within minutes everyone slipped away to their bunks. So I sat alone in the cockpit, and as I reveled in the complete silence and the reflections in the water of the surroundings peaks, a little voice inside my head whispered, *Fish.* Hopping up, I slipped my rod out of its holder on the stern and arced a silver spinner through the twilight, landing it right where a tiny stream gurgled into the cove. The lure had barely broken the surface when the rod bent over like a spring sapling. This was something big. I fought the fish in a silent frenzy of give-and-take until Rudy showed up, drawn by all the stomping on deck.

As I pulled the fish alongside, he hoisted it onto the deck with a single deft motion. It was nearly two feet long, muddy green, and speckled like a trout with a white underbelly and catfishlike lips. As we turned the fish over in our hands, I saw the distinct beard, or "barbel," of an Atlantic cod protruding from the proud lower lip. In our hands we held a survivor of the once abundant schools of "the fish that changed the world." I thought about throwing it back, of course, but it had swallowed the lure, and I remembered a fisherman in Newfoundland telling me that wild, fresh-caught cod was "the best-tasting fish in the world." If we did catch one, he said, we should cook it up "whatever way is the least healthy."

And so that's what we did.

We crossed the Arctic Circle the next day at 1:37 p.m. It was July 7—thirty-six days since our departure from Maine. Everything to the north of us, an area of about 5.5 million square miles encircling the North Pole, was "the Arctic." The majority of this territory is covered by the Arctic Ocean, but it also includes pieces of the United States, Canada, Russia, Norway, Sweden, Finland, and, of course, Greenland. Mariners have a long tradition of celebrating first crossings of the equator and the Arctic Circle. Those who cross the former are known as "shellbacks" and the latter as "blue noses." When our GPS indicated that we had crossed the line, I cut the engine and let the boat slowly drift. As a weak sun struggled to burn through the haze overhead, we all stripped down and jumped off the stern, hooting and hollering as we briefly plunged into the 42-degree water. Back on deck, Rudy handed me a bottle of whiskey. As the liquor burned inside my shivering body, I smiled, knowing that the real business of this adventure had just begun.

I awoke early the next morning to the sound of footsteps on deck and the unmistakable sensation of the boat being underway. It took me a minute, but then I remembered that we were tied to the dock in Sisimiut, the second-largest town in Greenland, with no plans to go anywhere for a couple days. "What the heck is going on?" I grumbled as I hauled myself up the companionway. On deck, I found Ben and a couple of young Greenlandic men moving our dock lines so a dented gray metal boat with a cannon mounted on its bow could squeeze between us and the wharf. We had parked in its spot.

After some deft maneuvering, the whaling ship *Qarsoq Aidan* tucked itself neatly between *Polar Sun* and the dock. Three Inuit deckhands and the skipper quickly went to work on the blood-soaked deck, moving and sorting dozens of cinder-block-sized chunks of red meat. It was a gory scene, and the air felt thick and smelled of blood and rust. One of the men, who was tall and thin with short black hair and looked to be in his late twenties, wore a black sweatshirt and a backward-facing baseball hat. When I asked him about the meat, he told me that it was from a minke whale they'd harvested forty miles offshore in Davis Strait. He reckoned it weighed about twenty thousand pounds.

I thought that whaling had been banned worldwide since the 1980s, but this man informed me that it was currently allowed in Greenland under a quota system. Their boat had a permit to harvest ten whales in 2022, and this was their first of the season. He hoped there would be more. The main market for the meat was the local grocery store, where the minke would fetch about $7,000. The smile on his face indicated that this wasn't a bad haul. As we chatted, his mates covered the whale with a green tarp, gave us a quick wave, and disappeared up the ladder onto the dock. Later, after they had returned and loaded the meat into crates of ice for transport up the hill to the supermarket, I found a pot-roast-sized chunk sitting on our rail.

What to do with the meat? I wondered as I took the hunk below and put it in the icebox. On one hand, I didn't want it to go to waste, and I'd never eaten whale, so I was curious how it would taste. On the other, I was ethically torn about eating such a beautiful and threatened creature. But hunger and curiosity bested whatever qualms I felt, and that evening I fried up a small steak in a skillet down in the galley. I offered it to everyone aboard, including Hampton and Tommy, but there were no takers. The meat was tender, with a delicate flavor I would describe as much closer to beef or venison than fish. I savored every morsel.

In 2010, the International Whaling Commission began allowing limited aboriginal whaling in Greenland as long as it's done humanely and sustainably and only for local subsistence and cultural needs. Many conservationists, however, believe that Greenlandic whalers violate the spirit of the quota when they sell whale meat to local stores and restaurants, where tourists and non-Inuit consumers can purchase it. When we were shopping for supplies in Sisimiut, I found a fifteen-foot-long cooler filled with *hval* selling for about $8.50 per kilogram—far less than the frozen beef, pork, and chicken imported from Denmark. *Hval* is a staple of the Greenland diet and an important element of Inuit culture.

Whaling, when pursued on an industrial scale, is yet another example of how advances in maritime technology enabled colonial powers to nearly destroy a resource that indigenous people had managed sustainably for thousands of years. By the time the Thule arrived in Greenland around 1200 CE, they had developed several new technologies that allowed them to be more successful in the Arctic than their predecessors. Unlike the Dorset, the Inuit used bows and arrows and dogsleds, and they were master mariners who invented the kayak and also a larger, skifflike craft known as an umiak. These boats expanded their hunting range and enabled them to harvest a rich assortment of marine mammals, including belugas, walrus, and seals. But it was the massive bowhead whales and the plentiful and reliable food source they offered that allowed the Thule and Inuit to build communities hundreds of people strong.

Umiaks could be thirty feet long and five feet wide, and they could comfortably carry up to twenty people. The hulls were made from sewn skins of bearded seals or walrus stretched over whalebone and driftwood frames and fastened with sinew and hide lashings. Umiaks empowered Inuit to move from coastal whaling out into the open ocean. The whales were captured by a harpooner who stood over the bow, ready to strike with a long wooden (or ivory) spear tipped with detachable bone, stone, or metal points designed to twist into the animal's flesh. The tips were attached to lines made from seal hide that were in turn tied to animal bladders or entire sealskins filled with air like balloons. These floats would drag along behind the whale, marking its location and exhausting the animal until the umiak's crew could catch up and fatally wound it with lances.

Whale meat, skin, and blubber (known as muktuk) provided much-needed protein and omega fatty acids, which helped compensate for the lack of plant-based foods in the Arctic diet. The oil was used for heat, cooking, and light. Baleen, a sort of sieve composed of keratin found in the mouths of whales, provided material for fishing line and baskets. Whalebones, particularly the ribs, were even used to frame houses. A single bowhead whale could be sixty feet long, weigh 120,000 pounds, and provide enough food to feed a village for a year, but because of the relatively low population density and the difficulty of killing a whale, each community harvested only a few animals each season. As a result, the whale and Inuit (as well as their predecessors) coexisted symbiotically for thousands of years.

But these magnificent creatures would not remain insulated from the wider world for long. Basque sailors were the first Europeans to pursue whaling commercially, beginning in 1059 when they passed a regulatory measure to concentrate the sale of whales to the city of Bayonne in southwestern France. The Basques dominated whaling in the North Atlantic for the next five hundred years. By the 1500s, they had spread across the Atlantic to establish their main base in Labrador's Red Bay. European whalers harvested baleen and used whale teeth for scrimshaw, but their primary interest was the blubber, which could be rendered into a versatile oil used in lamps, lighthouses, soap, textiles, paints, and much more. Before the discovery of petroleum, whale oil lubricated the machinery of the industrial revolution and represented the original oil boom.

By the 1730s, bowhead and right whales had been fished into near extinction in North American coastal waters. In search of new hunting grounds, whalers pushed north into Davis Strait and the Arctic. For the British, the economic incentives of whaling fueled a growing interest in the Northwest Passage, and a number of early voyages used whaling revenue as a way to fund Arctic exploration. After the War of 1812, whaling entered a period of unparalleled growth, driven largely by the American fleet. From the 1820s to the 1840s, American whalers spread across the world, discovering new fisheries in Africa and the Pacific and on both sides of the Arctic. In the process, they charted thousands of miles of hitherto unknown territory and fueled the economic lifeblood of towns like Nantucket, New Bedford, New London, Sag Harbor, Charleston, and San Francisco. At its peak in 1846, the year that Franklin was beset in the ice off King William Island, the United States whaling fleet numbered 735 ships, out of a total of nine hundred worldwide.

American whalers not only dominated the industry; they were some of the best mariners in the world. They were the first to map the Gulf Stream, the mighty current that flows northward up the Atlantic Ocean from Florida to Virginia, then turns east toward Europe. British customs officials were baffled as to why their ships often took two weeks longer than American vessels to cross the Atlantic. The reason was that the American captains knew about the Gulf Stream, while the British did not. In fact, the Americans told the British about the current, but according to the eighteenth-century Nantucket whaler Timothy Folger, they were "too wise to be counseled by simple American fisherman."

There is little doubt that every species of whale would have been hunted to extinction had it not been for the invention of kerosene in 1846, followed by the Pennsylvania oil boom about fifteen years later. Whale oil as a light source was quickly rendered obsolete, but there was still enough of a market for the meat, baleen, and bone that a small whaling industry persisted on the West Coast of the United States. That market ended in 1973 when Congress passed the federal Marine Mammal Protection Act, which made it illegal for anyone to kill, hunt, injure, or harass marine mammals in U.S. waters. Meanwhile, many depleted species of whales came under the additional protection of the federal Endangered Species Act.

In 1986, the eighty-eight member countries of the IWC enacted a full worldwide moratorium on commercial whaling. The only exceptions are for aboriginal subsistence whaling, which takes place in Greenland, Russia, the island of Bequia (in the Caribbean), and coastal Alaska. Norway, Iceland, and Japan, all of which still have commercial whaling industries, are the only nations that oppose the IWC ban.

While exact data is scarce, studies show that the whale population in Greenland has rebounded significantly due to the global moratorium. The country is now balancing the right for indigenous groups to harvest whales on a subsistence basis against the impact this has on the popular whale-watching tours based in Nuuk and elsewhere. At the very least, the whale population in Davis Strait and Baffin Bay is protected by a well-enforced quota system, and whalers like those I met can reap the centuries-old economic, cultural, and health benefits of eating local fare.

ǀǀǀ

It was midnight on July 11, and the Arctic summer evening held only a wisp of darkness as I navigated *Polar Sun* through a complicated and poorly charted archipelago of small islets on the south shore of Disko Bay. Over the past two weeks, on our three-hundred-mile trip up the coast from Nuuk, we had crossed paths with a few icebergs, but thanks to the north-flowing West Greenland Current, ice had been scarce—so far. But we were now only sixty miles from Jakobshavn, the fastest-moving glacier in Greenland, which surges forward up to a hundred thirty feet a day and produces approximately 10 percent of all icebergs spawned from the Greenland Ice Cap.

As we pushed north into Disko Bay's phantasmagorical wilderness of ice, Tommy sat beside me in the cockpit, and we played a game conjuring the icebergs into images of dinosaurs, crab eyes, and medieval castles. Sometimes a chunk of ice would calve, rumbling into the water as the mother berg, now with a new center of gravity, rotated and rose into the air, revealing a new waterline and more of its mysterious, gleaming underbody. It was the colors, though, that made it impossible to look away. When the sun shone directly into the bay, the light reflected off the facets of the ice in infinite shades of blue and green, like a polar disco ball. When the sky was overcast, as it often was, the bergs still glowed teal and emerald as if lit from within.

Every year, Jakobshavn—known in Greenlandic as Sermeq Kujalleq—sheds the equivalent of twenty thousand Great Pyramids of Giza into the thirty-seven-mile-long Ilulissat Icefjord. The weight of all this ice, an estimated 300 billion metric tons, is so massive that it's causing the landmass beneath the ice cap to rise, a phenomenon that glaciologists call "isostatic rebound." Between 1850 and 2010, the Jakobshavn retreated twenty-five miles, but then in 2016, its withdrawal slowed dramatically, and it actually began to gain ice and snow mass. Theories abound as to why—localized cold seawater currents, increased snowfall, underlying topography, and more. The bad news is that most scientists think the slowdown is temporary. The good news is that it will likely be thousands of years before the Greenland Ice Cap melts altogether; but once it does—and many scientists believe that we have already passed the tipping point—the world's oceans will rise by approximately twenty feet, causing entire countries to disappear like the lost city of Atlantis.

A study in the journal *Nature* from August 2022 predicts that, even if global temperatures can be stabilized at current levels (which is extremely unlikely), the amount of ice pouring into the ocean off Greenland will cause sea levels to rise somewhere between ten and thirty inches during Tommy's lifetime. This is enough to displace billions of people and flood millions of acres of farmland. Jason Box, one of the authors of the paper, says that the Greenland Ice Cap already has "one foot in the grave."

The privilege for us was the opportunity to visit a place, while it still existed, where you could actually see and feel the seismic forces that have shaped our world. And it felt important that I should pass on what I saw to my readers. From the deck of *Polar Sun*, Tommy and I watched an iceberg the size of a Walmart supercenter being born in the Ilulissat Icefjord. And when he looked up at me, his blue eyes wide with wonder, it hit me that he might be a member of the last generation to witness what Pierre Berton described as "great frozen mountains . . . sculpted by a celestial architect . . . coruscating in the sun's rays, each one slightly out of focus as in a dream."

ı|ı

After a brief stopover in the town of Aasiaat, we followed another Povl chart north toward the Franklin anchorage in the Whale Fish Islands. Weaving our way between house-sized icebergs, we turned into an enclosed basin where the abandoned village of Imerissoq came into view. The settlement had been deserted many years before; its history was obscure. Now it was no more than a collection of graying, ramshackle wooden buildings that lay scattered across a scrabbly hillside.

And yet this lonely spot charged my imagination because it was the first place on our route where I knew with certainty that we were crossing paths with Franklin and his men; they had anchored here in July 1845 on their way into the Northwest Passage. After years of research, plotting, and planning, I felt for the first time that I had finally found my man. Our arrival was almost 177 years to the day after that of the Franklin expedition, and to mark the occasion, I wanted to drop the anchor in the exact spot where Franklin had. Knowing that *Erebus* and *Terror* had both drawn about sixteen feet, I kept a close eye on the depth sounder, and when there were two or three feet more than that below our keel, I called out, *"Let go!"*—like Franklin might have done—and our seventy-seven-pound lead-tipped Spade anchor sank to the bottom and bit hard into the mud.

As the boat settled, I took a look around. A few small ice chunks were orphaned along the shoreline, and a lone seal popped its head out of the clear water to check us out. Polished granite boulders rimmed the shoreline, and a small stream, choked with lush hummocks of grass, trickled down through the remains of the village.

Ben and I rowed ashore in the dinghy early the next morning.

Conditions were calm, the anchor was well set, and *Polar Sun* was secure for the time being. The rest of the crew—including Hampton, Tommy, and Taylor, who would be flying home soon—was still drinking coffee and eating breakfast. They would come ashore in the next wave. I figured Ben would join me to explore the island and the old ghost town, but he said he was going back to "watch the ship." Since the arrival of our new crew, I had appreciated having him aboard *Polar Sun* even more. He alone helped maintain an unspoken sense of order (not to mention safety) on board, especially now that Rudy was focused on helping Renan document the voyage. Even rowing across the harbor, I could tell that half his mind was back on *Polar Sun*—no doubt making mental calculations about our fuel levels, water consumption, and running repair list. And when he got back to the boat, he'd probably strike up a conversation with Tommy about his favorite dinosaurs or maybe challenge him to a game of Uno. His mother hen instinct was highly evolved, and he played with Tommy more than I did.

I, on the other hand, didn't want to miss the chance to stretch my legs and enjoy the quiet solitude of the island. *Polar Sun* and her list of anxious needs could wait.

"Join me," I said. "Let's go look around."

"Nah," he said. "You go. I'll keep an eye out. Give a wave from the beach when you're ready for a pickup."

As he shoved off, I wondered why he wasn't more interested in exploring the island. Was it simply that he was looking for a break from me? Maybe I was projecting, but I sensed a subtle but uneasy tension that hung in the air between us. And I knew exactly when it had started.

Three weeks earlier, we'd been motoring north in light air and drizzle up the Strait of Belle Isle between Newfoundland and Labrador, when the engine sputtered and then died. A couple days before this, we'd sailed the snot out of *Polar Sun* in a ripping twenty-five-knot southwesterly, and I figured that all the sloshing around had broken loose some of the sludge that forms inside diesel fuel tanks. I removed the filter, and sure enough, it was black and clogged with debris. *Easy fix,* I thought as I plugged in a fresh one and bled the fuel line. The engine fired up on the first try, and we got back underway. But the tide had turned against us, so we decided to duck into Red Bay, the historic port that once served as the base of the Basque whaling fleet.

As we approached the L-shaped public wharf with Ben at the helm and me on deck holding the dock lines, I remember thinking that he was coming in exceptionally hot. When I glanced back at him, he gave a winning nod as if to say, *All good here, skipper.* I thought. *Well, okay. He's done this a thousand times, so I guess he knows what he's doing.*

When we were maybe about twenty feet from the wharf, Ben threw the transmission into reverse and goosed the throttle to slow us down. I heard the engine rev loudly, then silence. Our motor had died. Again. We hit the wharf hard a few seconds later and ground down its face. Knowing that we were about to T-bone a 90-degree wall a short way ahead, I jumped off onto a greasy, algae-covered ladder with a dock line in my hand. The rope burned through my fingers as the ship kept rolling, but just as I was about to lose it, I managed to loop it over a cleat. It came taut and *Polar Sun* crunched to a stop. There was a moment of stunned silence. Then Ben yelled out, "Whadja do that for? I could have easily made the turn and gone back out."

I looked down at him in disbelief. Then I gazed forward at the bulkhead that lay a short distance ahead. *Is he fucking with me?* I wondered. It sure didn't look to me like he could have made that turn. And besides, if that had been his plan, why hadn't he turned right when the engine died?

I stood at the top of the ladder, chest heaving, hands on my knees. As the adrenaline faded away, equal measures of embarrassment and anger rose up inside of me: embarrassment at having been berated in front of Rudy and Mr. Dirt; anger at Ben's high-speed approach, which I believe had been ill-advised given the dodgy engine. At this point, we were only two weeks in, and in those early days of our expedition, I was still in open awe of Ben. I consulted him on most every decision, large and small, and he was always more than happy to advise and opine on the best course. Invariably, we agreed. But it was slowly dawning on me how much he valued his own experience. To my mind, he tended to his office as the most knowledgeable man on the boat a bit too carefully. And owning mistakes did not seem to be part of his DNA.

It's a not-so-hidden secret to anyone who knows me that I am, shall we say, "sensitive" to criticism? It's a shortcoming, I know. My only excuse is that I grew up with a father whose main interaction with me was to criticize almost everything I did. Being thrown under the bus, when it seemed to me like I had just saved our asses, triggered a feeling that reminded me of being a wounded kid. But I bit my tongue and ducked below without saying anything. We still had a *long* way to go, and I knew things would escalate quickly if I started arguing with Ben.

I'm sure it was more in my head than in his, but ever since Red Bay, things had felt different between us, and I was pretty sure that the hierarchy aboard *Polar Sun* was beginning to rankle him. I'd known from the beginning that it might be awkward to be the captain of this expedition, considering Ben had so much more experience as a mariner than I did. Maybe I had been naive, but I figured that we'd operate like my teams always had on Himalayan trips, where someone took on the role of "leader" even if it was just a name on the permit. When it came to making big decisions on those expeditions, we had always worked things out democratically, even voting on occasion, if that was what it took to reach a consensus. Sailboat voyaging should operate similarly, I reasoned, and thus far, I had endeavored to ensure that we worked as a team on which each crew member had an equal voice. That said, I know I'm not always as easygoing as I like to think I am, and I've had friends and family tell me that I can sometimes be a bit overbearing.

Alone in Imerissoq, walking among the boulders, I was brooding on this subject when I came upon the remains of an old storehouse. The walls had long since disintegrated, but the posts and beams that once supported the roof still stood straight and true. From where I imagined the building's front door once stood, the faint outline of a grassy path led down a steep slope to the water's edge. I stepped down to the shore, where I found what I'd been looking for: an iron rod, about an inch and a half thick, drilled into the Greenlandic bedrock. An oxidized brass fitting capped the end. Leaning in, I could make out a faint arc of now illegible letters and numbers. I sat down and rested my hand on the bollard. From where I sat, I could see *Polar Sun* lounging idly in the bay a few hundred feet away. And I tried to imagine this place when *Erebus* and *Terror*, perhaps lashed by a thick hemp rope to this very rod, had sat exactly where my ship did now.

It was a scene I didn't have to conjure entirely from my imagination because James Fitzjames, skipper of Franklin's flagship, *Erebus*, and third in command overall of the expedition, had sketched it on July 7, 1845. I had a copy of this pencil drawing in my notes and had consulted it on our way into the Whale Fish Islands to ascertain where Franklin had anchored his ships. In the illustration, you can see the Danish storehouse, very much intact in those days, down by the shore. *Erebus* and *Terror* lie at anchor just beyond.

Franklin's ships were bomb vessels, which, as the name implies, means that they were designed as floating platforms to support cannons for shore and shipside bombardments in battle, and *Terror* had indeed seen its share of action. Both ships were heavily built to absorb the recoils of the cannons and bow-mounted, muzzle-loaded mortars. Rigging-wise, they were schooners, each with three masts and a long bowsprit.

Terror, the smaller of the two ships, weighed in at 325 tons and was 102 feet in length overall. She had been built in 1813 and immediately seen service in the War of 1812, where she fought in several battles, including the bombardment of Stonington, Connecticut. When she got back to England, she was laid up for a number of years before sailing to the Mediterranean and then to Hudson Bay in 1836, where she became trapped in the ice near Southampton Island. On the morning of February 18, when the thermometer read −33 degrees Fahrenheit, the crew awoke to a nightmare scenario, which George Back, who'd been part of Franklin's disastrous Coppermine River expedition in 1821, later described in a letter to the Royal Geographical Society. "The waves of ice, 30 feet high, were rolled towards the ship which complained much. The decks were separated, the beams raised off their shelf pieces, lashings and shores, used for supporters, gave way; iron bolts partially drawn; the whole frame of the ship trembled so violently as to throw men down." At one point, the ship was lifted by the ice nearly forty feet above the sea. The stern post shattered, which certainly would have been the end of a less solidly built ship.

When she was finally released from the ice after a ten-month imprisonment, Back sailed her home across the Atlantic. According to his report, the ship was taking on five feet of water an hour, and the crew survived only by constantly manning the bilge pumps. After barely making it to the north coast of Ireland, they ran her aground on a beach at the head of a fjord called Lough Swilly, and when the tide went out, they were able to see the full extent of the damage for the first time. "It was found that upwards of 20 feet of the keel, together with 10 feet of the stern post, were driven over more than three and a half feet on one side, leaving a frightful opening astern for the free ingress of the water," wrote Back. "The forefront was entirely gone, besides numerous bolts either loosened or broken; and when, besides this, the strained and twisted state of the ship was considered, there was not one on board who did not express astonishment that we had ever floated across the Atlantic."

But *Terror* had survived her thrashing in the ice, and only two years later, she would set off on an even more daring expedition to Antarctica.

Erebus, built in 1826, had a few more feet than *Terror* overall, a thirty-foot beam, and a displacement of 372 tons. Before she was enlisted to carry Franklin and his men into the Northwest Passage, *Erebus* spent three seasons (accompanied by *Terror*), from 1839 to 1843, exploring the Antarctic under the command of James Clark Ross. On that expedition, neither ship had an engine, and Ross felt strongly they should have stayed that way. But more and more, naval sailing vessels were being fitted with auxiliary steam engines, and John Barrow and William Edward Parry, who'd been tasked with overseeing the ships' refit for the Franklin expedition, felt that such mechanical propulsion was all but essential for navigating through the narrow leads and channels that Franklin would invariably encounter in the pack ice. Historians debate exactly which engines were used, but Peter Carney, a British mechanical engineer who has done extensive research on the subject, believes that they came from two locomotives from the London and Croydon Railway. At the Woolwich Dockyard on the River Thames, the fifteen-ton, twenty-eight-horsepower motors were "marinized," then lowered into the holds of the two ships and connected to thirty-two-foot-long iron shafts. The props were seven feet in diameter and designed to reside in wells in the hull that allowed them to be removed and brought inside the ships to prevent damage from collisions with ice and to reduce drag while sailing. It was estimated that in calm waters, the engines could push the ships along at three or four knots.

Parry ordered the vessels strengthened and reinforced with extra diagonal cross planking on the decks, a second layer of oak and copper sheathing around the hulls, and a plate of thick cast iron bolted to the bow for bashing through ice. Before being slipped back into the water, both *Erebus* and *Terror* were painted black and given a distinctive yellow boot stripe at the waterline, which showed not only how heavily they were laden but how they sat on their trim.

As with icebergs, perhaps the most significant measure of a sailing vessel is how far it projects beneath the water. *Erebus* drew seventeen feet, *Terror* a foot less—numbers that some, including John Ross, believed were far too much draft to carry into the uncharted shoals of the Northwest Passage. McClintock's ship, the *Fox*, was half the size of *Erebus* and drew six feet less. Amundsen's ship, *Gjøa*, drew just six feet—the same as *Polar Sun*.

When *Erebus* and *Terror* arrived in the Whale Fish Islands on July 4, 1845, they were accompanied by the supply ship HMS *Barretto Junior*, which lay alongside *Erebus* when Fitzjames made his sketch. At the time, the crew was hard at work transferring three years of supplies aboard *Erebus* and *Terror*.

These stores, all listed on the ships' manifests, included enough coal to power the steam engines for twelve days; thirty-six hundred pounds of soap; twenty-seven hundred pounds of candles, brass oil lamps, wicks and their glass globes, spare clothing, rigging and spars, wolfskin blankets, sailcloth, and munitions; and three live oxen (*Barretto Junior* had started with eighteen, but the rest had died on the way). The food supply (these amounts were roughly split between the two vessels) included 36,487 pounds of biscuits (packed in tins to protect against weevils), 136,656 pounds of flour, 64,224 pounds of salted beef and pork, 23,576 pounds of sugar, 9,450 pounds of chocolate, plus eight thousand variously sized tins of meat, soup, and vegetables. To ward against scurvy, a horrible disease caused by a deficiency of vitamin C (more on this later), they carried 9,300 pounds of lemon juice, and every day, the sailors had to down their dram in the presence of an officer. Additional antiscorbutics included pickles (which, unbeknownst to them, didn't contain any vitamin C), cranberries, walnuts, raisins, and mustard seed. They got their drinking water, much like we did aboard *Polar Sun*, with a desalinator that scrubbed the salt from seawater through an ingenious steam-distillation process.

Many of the sailors smoked, so the Admiralty had victualed the expedition with more than 7,000 pounds of pipe tobacco. (Franklin had always satisfied his nicotine habit with snuff but quit shortly before the expedition departed.) And a sailor in the British Royal Navy didn't have to quit drinking just because he was heading off to sea. To maintain morale—without undue impairment—each man was carefully allotted four ounces per day of 135-ish-proof West Indian rum. Three years' worth for 129 men worked out to 3,600 gallons, with another 5,000 gallons of beer thrown in for good measure. All in, each sailor's daily food allowance added up to about three pounds of grub, which, by any measure, would have been more than enough to meet their caloric needs.

Mealtime was always signaled by the ship's bell: three p.m. for the crew in the main forward mess hall and an hour later for Franklin in his private mess in the stern, where he would usually dine with Fitzjames and two or three other officers. The officers all supplied their own monogrammed sterling silver cutlery: two forks and three spoons, one each for soup, dessert, and tea; and they supplemented the Admiralty's stores with their personal larders, which included specialty items from companies like Fortnum & Mason such as cocoa, high-quality teas, and wines. Franklin's meals might have begun with a selection of cheeses, crackers, smoked herring, oysters, salamis, and fresh-baked bread. An entrée could have been tinned beef, ham, or fowl, followed by turtle or oxtail soup. Franklin was a lifelong teetotaler, but his officers likely enjoyed their meals accompanied by fine claret. Dessert, perhaps a pudding or cake, might have been paired with sweet wine as a digestif.

Franklin's cabin, which stretched across the entire stern of the ship, served as a war room of sorts, and it was here that the officers often gathered. Shortly before the expedition set off from England, the *Illustrated London News* published a woodcut of this stateroom based on a drawing by Fitzjames. The perspective looks aft, where light streams in through three large windows framing a wooden locker. To starboard, two cabinets marked "Pacific Charts" and "Home and Arctic" are built into the wall. The interior of the cabin is covered in varnished golden brown paneling, and on the wall hangs a portrait of Franklin's wife, Lady Jane. Franklin had made room in his private quarters for the expedition naturalist and assistant surgeon, Harry Goodsir, twenty-five, to set up his own desk. And it was here that Goodsir worked on a scientific paper—"On the Anatomy of Forbesia"—a study of a colorful broad-leafed plant he had found in Imerissoq. Before Goodsir left on the expedition, his father had written to him: "I would earnestly recommend to you to keep a correct account of all your private thoughts and observations; and above all don't sleep too much."

We know from a long letter that Franklin wrote to Lady Jane while they were anchored in the Whale Fish Islands that he vented to his officers about his ill treatment in Van Diemen's Land. But apart from these gripes, the tone of the letters posted from the Whale Fish Islands paints a picture of a group of men who were enjoying themselves and feeling confident in the ultimate success of their endeavor. Twenty-two-year-old third mate Edward Couch—whom Fitzjames described as a "little black haired, smooth faced fellow, good humored in his own way . . . never in the way of anybody, and always ready when wanted"—wrote to his parents that Franklin was "an exceedingly good old chap" whose Sunday sermons were better than those of "half the parsons in England." Charles Osmer, the ship's purser, wrote to his wife, "The more I see of our worthy chief, the more I like and admire him, in that he is deservedly beloved by us all."

But of them all, perhaps no one was more bullish than Fitzjames. In a letter to John Barrow Jr., son of the Second Secretary to the Admiralty, Fitzjames wrote: "We hear that this is supposed to be a remarkably clear season . . . we intend to drink Sir John's health on the day we go through Behrings Straits—If we get through this year, we shall have to land somewhere or other to discharge some cargo." Some of that cargo included Fitzjames himself, for he had hatched a scheme to be dropped off in Siberia, whence he would travel overland back to London. Two months before the expedition set off from England, Fitzjames wrote to a friend that "in whatever year we do get through, the month will be August or September, so that these will be the times to go at once to Oshotk [*sic*], and start off for St. Petersburg. . . . Sir John tells me that he has thought of such a journey for one officer." Fitzjames figured that his trek across Asia would take less time than that required for the ships to round Cape Horn, and he almost certainly let himself get carried away a bit with anticipation of the welcome he'd receive when he arrived back in England. In his book *James Fitzjames: The Mystery Man of the Franklin Expedition*, William Battersby points out that this ambitious circumnavigation would have instantly elevated Fitzjames into the elite club of exploration's all-time greats, eclipsing even Franklin, who would have only circumnavigated the Americas, not the world.

It was a wildly ambitious scheme, but Fitzjames had the mettle to pull it off. In the spring of 1845, the thirty-one-year-old was a decorated war veteran, hero, swashbuckling explorer, and practical joker with a penchant for colorful storytelling. He was born in Brazil in 1813, the illegitimate son of the British ambassador, Sir James Gambier. His mother remains unknown, but Battersby speculates that she was probably an unmarried daughter in a Portuguese noble family or possibly even a member of the royal family itself. Fitzjames was later adopted by a prominent family in England who provided him with an impeccable education. He went to sea at age twelve, under the recommendation of his cousin, Admiral Robert Gambier. When he enlisted in the navy, Fitzjames falsely claimed to have been born in London, thus concealing the truth about his origins.

In February 1835, Fitzjames joined an expedition that would explore the Euphrates River as a possible trade route between England and the Middle East. While loading supplies for the voyage onto a ship in Liverpool, a customs official fell overboard and was swept into the strong current of the River Mersey. Fitzjames ripped off his overcoat, dove into the icy water, and pulled the drowning man to shore by his hair. Newspapers all over the UK reported on the incident, hailing Fitzjames as a national hero.

While stationed in Baghdad eighteen months later, he left the Euphrates expedition to deliver a bag of mail to London, a thousand-mile journey by small boat, camel, and foot during which he was robbed and held hostage by a sheikh. This epic trek would later prompt Fitzjames to describe himself as "the best walker" in the Royal Navy.

A few years later, he fought in the Egyptian–Ottoman War, commanding a bold amphibious assault on Beirut that inspired an Egyptian general to put a bounty on his head. And in the First Opium War when Fitzjames led a rocket brigade against several Chinese forts, a sniper shot passed through his arm, lodging close to his spine. While recovering from surgery, he submitted a proposal to the British plenipotentiary asking permission to walk home from China across the Tibetan Plateau, Central Asia, and the Middle East, via the ancient

Silk Road. Unluckily for Fitzjames, the Foreign Office denied his request. Had it been granted, he probably would have arrived in England too late to join the Franklin expedition.

There was one officer, though, who was conspicuously absent from Franklin's table, and that was Francis Rawdon Crozier, captain of *Terror* and second-in-command overall. Crozier, a forty-eight-year-old Irishman, had enlisted in the navy at age thirteen. In 1812, he joined the crew of HMS *Briton*, which sailed into the South Pacific by way of Cape Horn with orders to hunt down a U.S. warship called *Essex*, which had been wreaking havoc on the British whaling fleet. When *Briton* arrived in Chile, *Essex* had already been captured, so the crew pressed on to the Galápagos and the Marquesas. On their way back to Chile, they made landfall at Pitcairn Island, where Crozier met Thursday October Christian, son of Fletcher Christian, leader of the *Bounty* mutineers. It had been twenty-five years since the infamous mutiny, and there was now only one surviving mutineer, John Adams, who presided over a small community of the descendants of the original crew and their Polynesian wives.

When he got back to England, Crozier was put on half pay, and like Franklin, he eventually found his next assignment with Barrow's discovery service. In 1821, he sailed with Parry in search of the Northwest Passage, and it was on this expedition that he formed a lifelong friendship with James Clark Ross. While overwintering outside the village of Igloolik, to the north of Hudson Bay, Crozier worked on his Inuktitut and in the process became friendly with a ten-year-old boy named Aglooka, which in English means "he who takes long strides." The connection must have been strong because the boy's relatives decided that he should swap names with Crozier, which is an old Inuit tradition. For the rest of his life, the boy told any Europeans he met that his name was "Cro-zhar," and Inuit began calling Crozier "Aglooka."

In 1839, James Clark Ross chose Crozier to be his second and captain of *Terror* on the Antarctic expedition. They never found the south magnetic pole, but they conducted magnetic observations and eventually managed to penetrate the ice barrier surrounding the continent, where they discovered two volcanoes they named after their ships, as well as the Ross Ice Shelf, the world's largest body of floating ice, which later served as the starting point for the Scott and Amundsen South Pole expeditions in 1911.

In between their voyages into the Antarctic, Ross and Crozier overwintered in Hobart, the capital of Van Diemen's Land, where Franklin was serving as governor. To honor the visit of the esteemed polar explorers, Sir John and Lady Jane organized a formal ball to be held aboard *Erebus* and *Terror*. The ships were decorated with awnings, flags, and floral wreaths, and two hundred mirrors fastened to their topsides reflected the glittering lights of the town. The three captains, dressed nattily in their full navy regalia, stood in the receiving line alongside Lady Jane and Franklin's niece Sophia Cracroft, who wore elegant, wide-shouldered evening gowns. After dining aboard *Terror*, everyone shifted over to *Erebus*, where they danced the galop, the polka, and the Viennese waltz. Of course, no one would have believed it if someone had prophesized that in a few years' time, Van Diemen's Land's governor would lead 128 men to their doom aboard this very same ship.

And yes, there was love in the air that night, for Crozier had fallen hard for Sophia, who must have looked lovely in that flickering lantern light after he had spent months battling ice in the Antarctic. But unfortunately for Crozier, the feeling wasn't mutual. Sophia had eyes only for James Clark Ross, whom Lady Jane once described as "the handsomest man in the navy." There are rumors that Crozier proposed and was rejected by Sophia before he sailed off once again into the ice. After that, they didn't see each other for a few years, but it seems her hold on Crozier's heart only grew stronger in the absence. When Sophia returned to England with the Franklins in June 1844, she received a letter from Crozier that is believed to have contained a marriage proposal. No record of her response exists, but we know that she once again rebuffed his "application," as she called it.

Crozier was crushed. He took a year's leave of absence from the navy, and while in Dublin, he wrote to Ross, who'd been feeding him information about Sophia from Lady Jane. Crozier said that he was "quite reconciled" with the humiliation he'd suffered and sorry for the "bother" he'd caused his old friend. And he hoped, with God's blessing, to keep "clear of all such scrapes in the future."

It was during this gloomy time, while Franklin was lobbying hard for command of Barrow's Northwest Passage expedition, that Crozier was offered the job by the first lord of the Admiralty, Thomas Hamilton. By this point, Crozier had been on five expeditions to the high latitudes, and he was one of the most experienced polar explorers in the world. But according to his biographer Michael Smith, he was "a modest, unassuming man who never sought the limelight," and because of his state of mind, it seems he just didn't have the heart to take on the role of being the public face of such an important expedition. So he turned it down, later writing to James Clark Ross, "I am, in truth, still of opinion of my own unfitness to lead."

Having been turned down by their first and second choices, the Admiralty offered Franklin the job. Maybe Crozier thought that working under Franklin might put him in better stead with Sophia; when asked to captain *Terror*, he said yes. In *Captain Francis Crozier: Last Man Standing?* Michael Smith writes that Crozier was daunted by the responsibility of having ultimate command of the expedition and felt more comfortable and better suited to serving in a supporting role. And yet he harbored serious misgivings about Franklin's fitness to lead such an ambitious enterprise. According to Pierre Berton, while the ships were fitting out, Crozier complained to a friend that "he [Franklin] is very decided in his own views but has not good judgement." And he later confided to Ross that he wished his old friend had been in command, in which case he'd have "no doubt as to our pursuing the proper course." Franklin, he worried, would "blunder into the ice."

Perhaps as evidence of this poor judgment, Franklin allowed Fitzjames to select the rest of the officers, instead of giving this important job to Crozier, who had far more experience. Fitzjames, of course, filled the ranks with men he knew from his past military service. His choices led to an officer corps with almost no ice experience. But he also, inadvertently, created a cliquey environment that made Crozier feel like an outsider, and this was probably the reason he hardly ever took Franklin up on his offers to dine aboard *Erebus*. "All goes on smoothly but James dear I am sadly alone, not a soul have I in either ship that I can go and talk to . . . no congenial spirit as it were," Crozier wrote to Ross from the Whale Fish Islands. "I am generally busy but it is after all a very hermitlike life—Except to kick up a row with the helmsman or abuse Jopson at times I would scarcely ever hear the sound of my own voice."

Lady Jane, either intentionally or not, reinforced this divide when she commissioned a photographer named William Beard to make daguerreotype portraits of the officers—but only those aboard *Erebus* plus Crozier. It's an interesting quirk of the Franklin expedition because these fourteen photographs (which were sold at auction in 2023 for more than a half million dollars by Franklin's descendants) have in some ways become the face of the expedition. And there is something unmistakably powerful that I feel when looking into the eyes of these men, knowing about the hard deaths they would later suffer.

The photographs were taken on the deck of *Erebus* over the course of three days, shortly before the expedition sailed. Beard kept one set, Lady Jane another. They weren't published at the time, and it wasn't until 1851, when Lady Jane was trying to rally the British public to the cause of searching for her husband, that they first appeared in the *Illustrated London News*. All these years, the photos, which are each two and three-quarters inches by three and a quarter inches, have been stored in a book-form Moroccan case. They're remarkably detailed, and the hatbands, buttons, and epaulettes have been hand-painted with shell gold.

Franklin, on the top left, wears a cocked hat and a jacket stretched tight by his bulk; it reminds me a bit of the Westport Whaler. He looks ill, which he was, recovering at the time from a bout of the flu. His eyes are dark and sunken, like those of a Dickensian character, which, in a way, he eventually became. Fitzjames comes next. Of all the portraits, his is the sharpest. He holds a telescope in one hand, his hat in the other. He is clean-shaven, with tousled curly locks and a hint of a receding hairline; his mouth wears the faintest hint of a smile. Crozier's portrait has a foggy, faraway feel to it. He wears a flat hat with the brim pulled low, and his shoulders are slumped, as if he's either leaning against something—or carrying a burden of melancholy from which he can't escape. Graham Gore, in his mid-thirties, is in the third row down; rakishly handsome, he sits with his arms and legs crossed. He served aboard *Terror* in the Antarctic. Two years after his portrait was taken, Gore signed and deposited the Victory Point Record, alongside Charles Des Voeux (who is pictured in the second row), in that cairn on King William Island. Charles Osmer, the ship's purser, in the bottom row, was forty-six at the time of this portrait. He too had sailed to Pitcairn Island, with Frederick William Beechey, in 1825. Initially, he made a poor impression on Fitzjames, who noted that he was a "stupid old man." But like Franklin, Osmer improved upon acquaintance, and Fitzjames later amended his opinion, calling him "delightful" and "as merry-hearted as any young man, full of quaint, dry sayings."

What the portraits don't show, though, are any of Crozier's

officers, nor any of the other members of the crew, who have always remained faceless. But it's possible to draw at least an outline of these men, including their physical appearances and how they ended up on Franklin's crew, thanks to a set of Royal Navy description books that were sent back to England aboard *Barretto Junior* when the Franklin expedition departed the Whale Fish Islands. The navy kept these books for the simple reason that they needed physical descriptions of their sailors so they could hunt them down if they ever went AWOL. These books are housed in the UK's National Archives in Kew. Thankfully, the work of analyzing them has already been done by a naval historian named Ralph Lloyd-Jones, who cross-referenced the muster books with census records and parish registers to create a sketch of the men who sailed with Franklin.

We know that a total of 129 men set off into the Northwest Passage with Franklin, but the expedition originally left England with twenty-six more than that. Of those, twenty-one got off in Dundee, Scotland, to join HMS *Perseus*, and another five were deemed unfit for service in the Whale Fish Islands. The *Erebus*'s armorer and the *Terror*'s sailmaker were described by Crozier as being "perfectly useless either at their trade or anything else," and two other able-bodied seamen were dismissed with no official reason cited. Of those who remained, sixty-seven would sail aboard *Erebus* and sixty-two on *Terror.* The former had thirteen officers, the latter eleven. The warrant and petty (noncommissioned) officers had job descriptions like engineer, boatswain (in charge of equipment and crew), carpenter, sailmaker, blacksmith, armorer, caulker, and cook. The captains of the foretop and the maintop were in charge of rigging and sails aloft, while the captain of the forecastle ran the crew's quarters. In addition, there were stewards who served the captains and officers, stokers who fed the steam engines, and marines who were in charge of security. The rest were listed on the ship's muster as able-bodied seamen—a catchall for those who did the work of sailing *Erebus* and *Terror*, and keeping them shipshape.

The oldest noncommissioned member of the crew was Richard Wall, forty-five, from Yorkshire; he served as the cook aboard *Erebus*. Wall had extensive high-latitude experience, having previously served in the Arctic on John Ross's 1829–1832 *Victory* expedition and then later with James Clark Ross in the Antarctic. Both the Rosses rated him highly, and he would have been an invaluable resource to Franklin. According to the muster books, Wall stood about five feet, five and a half inches and had "small features, blue eyes, and a sallow complexion with dark hair." The youngest member was Thomas G. Evans, age seventeen, although E. J. Helpman, the clerk in charge of *Terror*, listed Evans's age as eighteen, since he wouldn't otherwise have qualified for polar service. A year earlier, while serving aboard a naval steam vessel called *Styx*, Evans was described as being "only 5 feet 3½ inches tall, of florid complexion with grey eyes and dark brown hair."

Crozier's steward—Thomas Jopson, twenty-seven, the only member of the entire crew Crozier had recruited himself—had also served on the Antarctic expedition. He was the son of a London tradesman and, like many of the crewmen, was of middle height, at five feet, five and a half inches tall. He had brown hair, hazel eyes, and fair skin. He lacked tattoos but did have a distinguishing feature in the form of a scar on his right leg.

Perhaps the tallest member of the crew was one John Hartnell, twenty-five. Charles Osmer recorded his height as five feet, eleven and a half inches and noted that he had black hair, hazel eyes, and a sallow complexion. In hindsight, his pale, unhealthy skin tone was likely a sign of underlying disease. Another unique thing about Hartnell was that he sailed alongside his younger brother Thomas.

Harry Peglar—described as standing five feet, seven and a half inches tall, with brown hair, hazel eyes, and, again, that sallow complexion—was thirty-seven years old when the expedition set off, and he'd been at sea for twenty-five of them. During that time, he had transited the Atlantic three times, spent a year chasing slavers on the west coast of Africa, and sailed to India and then China, where he saw action in the First Opium War. Now he served aboard *Terror* in the important role of captain of the foretop.

In *The Arctic Grail*, Berton dismissed Franklin and his men as bumbling incompetents who sailed off blindly to their doom. But the description books paint a different picture: Sixty-two of Franklin's crew had sailed in the Arctic before, nine were veterans of the recent Ross-Crozier Antarctic expedition, and each ship had an "ice master"—James Reid with *Erebus* and Thomas Blanky with *Terror*. Blanky was forty-six at the time and had first gone to sea at age eleven. He'd worked as a Greenland whaler and then as a mate on a timber hauler before joining the Royal Navy in 1824. Prior to setting off with Franklin, he had served on three previous Arctic expeditions, making him one of the most experienced Arctic hands on the expedition. "It is both foolish and insulting," writes Lloyd-Jones,

> *when Franklin, his officers, and men are dismissed as innocent, inexperienced victims (or even villains) of a hubristic state that sought to somehow conquer the natural world itself. They were very well aware of the dangers that they faced, very well qualified to undertake them, and tragically unfortunate in their lack of success. Failure in no way diminishes their heroism.*

While the crew labored at stuffing *Erebus* "as full as an Egg," Franklin scratched away on what would turn into a tightly spaced fourteen-page letter to Lady Jane. He shared most everything he could think of, including that their monthlong passage from Scotland to Disko Bay had been favorable and "attended with strong breezes, and these generally from the west & SW so that in making our way across we were led to the North and even carried to within 60 miles of Iceland before we could get past Cape Farewell—but we did not see Iceland. It would have been contrary to the long experience of the Greenland Seamen if we had gone round Cape Farewell unattended by a gale." Franklin noted that Inuit in Imerissoq were comfortably dressed and "well taken care of by the Danes," and he was "delighted that many of them read their Bibles, and that the Children are taught at a school to read & perhaps to write." He was put off, though, by "the odours that surround their residences."

As Franklin sat in this little harbor preparing for the Northwest Passage, his primary concern—which I also now shared—was the next leg of the voyage across Baffin Bay into Lancaster Sound, a perilous course that, then as now, is entirely dependent on the state of the ice. The Danish inspector of the whaling station was away from his post, but Franklin learned from a carpenter that the past winter had been severe. Nonetheless, the spring breakup in Disko Bay had occurred at the normal time in early May. This boded well for their prospects of crossing Baffin Bay into Lancaster Sound. Even more favorable was the report that British whalers had already made their way up the Greenland coast to the Woman's Islands (in the Upernavik Archipelago), a further 5 degrees north.

As the expedition prepared to depart, *Barretto Junior* took on the crew's letters to their friends and families back in England, Wales, Scotland, and Ireland. This correspondence included Harry Goodsir's paper on forbesia, Fitzjames's sketches, and dozens of other heartfelt notes that have all been collected in a moving book entitled *May We Be Spared to Meet on Earth.*

"Of this we are certain," wrote Franklin in that final letter to Lady Jane, *that the two ships will have on board three years supply of provision fuel & clothing. I mention this the more particularly that you may not have the slightest apprehension respecting our welfare though we should have to winter twice, and with respect to this point, let me entreat you & Eleanor [Franklin's daughter, now twenty-one years old] not to be too anxious, for it is very possible that our prospects of success and the health of the officers & crew might justify our passing a second winter in these regions. If we do not succeed in one attempt to try in other places, and through Gods blessing we hope to set the question at rest. . . . To the Almightys care I commit you & dear Eleanor. I trust He will shield you under his wings and grant the continual aid of His Holy Spirit—again that God bless and support you both is and will be the constant prayer of your most affectionate Husband.*

Early on the morning of July 13, 1845, which was, by pure coincidence, the same day that *Polar Sun* set off from the Whale Fish Islands 177 years later, *Erebus* and *Terror*, with a complement of 129 ill-fated men, weighed anchor and sailed off into the unknown.

PART 2

IN FRANKLIN'S WAKE

CHAPTER 6

Qikiqtaaluk (Baffin Island)

Date: August 4, 2022
Position: somewhere in Baffin Bay

Polar Sun rose and fell in a long, easy swell as we motor-sailed northwest from Greenland out into the middle of Baffin Bay. The wind blew lightly from the north, where a thin, ropy band of stratus hovered a few fingers above the horizon. In the gap between the sea and these dark clouds, the sun sparkled, lighting the quilted sky above in purple, pink, and blue pastels.

Ever since we'd left Maine, there hadn't been a day when *Polar Sun* wasn't under an avian escort. The pair of birds that accompanied us now had white bellies, gray backs, and stubby yellow bills. With quick, shallow wingbeats, followed by long glides, they soared and swooped in perfect synchronization, flying high above the mast, then disappearing as they zipped past the bow—almost as if daring each other to see who could cut it the closest. Occasionally they tilted their bodies like fighter jets and nipped the water with their wing tips. It made me happy to imagine that they were trying to put on a good show—not just for each other but for me too.

That quiet morning felt like one of those rare moments when life delivers on the promise that at least once in a while, reality should match the ardor of one's dreams. For years, I had dreamed about sailing deep into the Arctic, and now that it was actually happening, I felt a certain sense of self-actualization in the knowledge that we all have the power to shape the paths of our destinies if only we can settle on what we really want to do with the limited time that we have. Lost in this reverie, I looked up to see a wall of fog stretching across the water like a curtain; it hadn't been there just moments before. Minutes later, when *Polar Sun* pierced this woolly veil, a thick and clammy mist drew tight around the boat and the temperature dropped 20 degrees. I pulled up my hood and turned to the radar, which showed several dark red blobs—icebergs, I assumed—floating ahead of us in the ether. Ducking my head out of the enclosure, I looked around for my two feathered friends, but they were gone.

For three days, Ben, Renan, Rudy, and I had been sailing northeast from Greenland toward a waypoint that was supposed to mark the edge of a chunk of sea ice the size of New Hampshire. To assist ships navigating through Arctic waters, the Canadian Ice Service (CIS) publishes daily ice charts throughout the summer. The one I'd downloaded that morning showed a giant lobster claw of ice extending two hundred miles into the bay from the southeast shore of Baffin Island. The chart segmented the ice into sections marked in red, orange, yellow, and green, and each color was labeled with a letter that corresponded to "eggs" that filled the chart's margins. The eggs, in turn, were divided into horizontal segments with numbers denoting ice concentration, thickness, and size. We were currently sailing into a zone labeled "7," which meant that ice covered seven-tenths of the surface. We'd known for weeks this ice was out there, and to get to Pond Inlet, our next port of call, we had to sail around it. And as it slowly drew closer and closer, I became intensely curious what seven-tenths ice might look like in the flesh. I'd been told by some old Arctic hands never to venture into ice greater than three-tenths or four-tenths, and even this density would be undertaken at our peril. All I knew for now was that what we'd already experienced in Greenland was supposed to be one-tenth—and it had seemed hairy enough.

As *Polar Sun* ghosted through the fog, I knew that this monstrous sheet of frozen ocean lurked somewhere nearby. Soon "bergy bits" and "growlers" began to appear alongside. The former are just what they sound like, while the latter are ice chunks that can be as big as a house but lie so low in the water that they're often invisible. "It's not the icebergs you need to worry about," an old fisherman in Newfoundland told us. "They show up well enough on radar. It's the growlers that give us old mariners gray hair." Growlers get their name from the sound they make when they roll around in the surf, which is sometimes the only way you can tell that they're out there. "If it's dark or foggy or, God forbid, both, don't let your boat go faster than five knots," said the fisherman. "At that speed, you'd at least have a chance if you hit something."

As our speed hung steadily above five knots and the frequency of these chunks increased, so too did the prevalence of "brash" ice, which looked like someone had dumped the contents of a slushy machine into the water. The Newfoundlander had warned me that brash ice wouldn't hurt the hull, but it could pulverize the rubber impeller if it got sucked into the engine's raw-water intake. As *Polar Sun* sliced through an Arctic Slush Puppie, it sounded like sandpaper was rubbing against the outside of the hull. And this noise, unlike anything I'd ever heard before at sea, must have set people's teeth on edge, because the crew soon emerged from below to see what the fuss was about.

When the fog lifted as suddenly as it had appeared, we gazed out at a six-foot wall of jagged ice that heaved up and down in the swell like a marine lung. We had found a frozen ocean, hundreds of miles from any land. Ice stretched as far as I could see to the north and south, and it felt like we had sailed to the edge of the world. And as we inched into this alien landscape of ice, fog, and roiling water, I couldn't help but wonder what sailing in this part of the world must have been like centuries ago for the early explorers who had lacked all the techno-wizardry we had aboard *Polar Sun*. As the rivet from the *Fox* rolled slowly back and forth inside the nav table, I thought about Francis Leopold McClintock and how close he'd come to losing his ship in these very same waters.

|||

By the late 1850s, McClintock was considered the most accomplished Arctic traveler in the British Royal Navy. Born in County Louth, Ireland, in 1819, the son of a customs officer who had a dozen children, McClintock had gone to sea at age twelve and seen duty in the Irish Sea, the English Channel, Newfoundland, Bermuda, the Caribbean, Brazil, the Río de la Plata (on the border of Argentina and Uruguay), Cape Horn, and the Sea of Cortez. McClintock's family was not well-connected, so he had to earn promotion through hard work and merit—a slow and uncertain path to advancement in the nineteenth-century Royal Navy. But throughout his life, McClintock benefitted from astonishing physical strength and a substantial reserve of intellectual horsepower and curiosity. He made lieutenant in 1845, and four years later, on the first expedition launched by the British Admiralty in search of the lost Franklin expedition, he and James Clark Ross (who had come out of retirement to search for his good friends Franklin and Crozier) man-hauled two small sledges five hundred miles around the north and west coasts of Somerset Island.

The 1849 Somerset Island expedition marked the arrival of "sledging" as a primary means of exploring the Arctic; think dog-sledding, only with men instead of huskies. Sledges varied widely in design, but the ones favored by McClintock were six to eight feet long, half that in width, and they were constructed of wood with metal runners. Food, fuel, stoves, tents, and clothing were stacked atop the frame, covered with canvas and furs, and lashed down with rope. The beauty of the system was that the design could be easily adapted depending on the size of the load and the number of haulers. Sledging proved a simple and effective way to explore vast tracts of the Arctic, but it was also brutal and exhausting. While the land might look relatively flat and smooth on a topo map, in reality, travel over this horizontal realm often meant fighting through deep snow in biting winds or whiteout-inducing glare when the sun did shine. Invariably, it also involved climbing up and over chaotic pressure ridges in the ice, which could stand as tall as thirty feet. In the spring, the ice was often covered in water and slush and incised with watery crevasses called leads. As a means of exploration, sledging brings to mind Winston Churchill's quote about democracy as "the worst form of Government except for all those other forms."

As a sledger, McClintock was unmatched. He delighted in the struggle and was infamous amongst his peers for his ability to maintain a stiff upper lip, no matter how grim things got. Charles Parry (no relation to William Parry), who sailed with McClintock in later years, wrote that he

> *could not have conceived so much calmness to have been the property of any one man. In the greatest difficulties, and under the most aggravating circumstances, his face would not alter a muscle, and except occasionally a little quiet chuckle and a rub of the hands, he would show no symptom of noticing changes in weather, position of the ice, or other intensely interesting matters. . . . No outward show of anxiety, no nervous irritability, no unnecessary noise, ever betokened an anxious mind.*

I like to imagine that after years of running into the invisible barriers of class that blocked his naval career, it was in harness on the ice that McClintock found his true path forward.

In 1851, after visiting the site on Beechey Island where the Franklin expedition spent its first winter, McClintock set off on foot down Barrow Strait toward Melville Island, where Parry had overwintered in 1819. This time, he covered seven hundred seventy miles in eighty days. In 1853, on an expedition led by Vice Admiral Henry Kellett, McClintock traversed more than thirteen hundred miles in 105 days, during which time he charted lands unknown even to Inuit.

In the spring of 1857, McClintock and the crew of the *Fox* set off in search of Franklin and his companions. Even today, the first crux of any voyage into the Northwest Passage is finding a way through the pack ice that covers Baffin Bay well into the summer sailing season. In the early-nineteenth century, whalers discovered that the Greenland Current hooked west at around 74 degrees latitude, and its warm waters sometimes opened a "northern passage" across the top of Baffin Bay. Other years, the ice held fast in the north, but if the southern portion of the pack broke away, a narrow "middle passage" sometimes opened. Of course, in those days, there were no satellites or Canadian Ice Service, and captains like McClintock had no choice but to take their chances, pushing into the pack and hoping for the best.

After hiring an Inuk driver and picking up thirty sled dogs in Disko Bay, McClintock probed for a middle and then a northern passage, neither of which opened that summer. By the first week in September, the *Fox* was three-quarters of the way across Melville Bay on Greenland's northwest coast when the ice closed around the ship and froze solid like quick-set cement. One day the ship had been pushing north through thick pack ice, following the occasional lead; the next, the ocean that surrounded them was as immovable as the Greenland Ice Cap.

Over the next eight months, the pack ice, and the *Fox*, drifted 1,385 miles south down Davis Strait. When the ice finally began to break up in April 1858, McClinotck and his crew found themselves in the Labrador Sea—closer to Newfoundland than to Greenland. As a storm blew in and colossal floes of ice pummeled the *Fox*, McClintock pointed the ship's iron-reinforced bow directly into the swell. He later wrote that "the shocks of the ice against the ship were alarmingly heavy," causing the vessel to shake violently and its bells to ring. Fortuitously, they found a massive iceberg and followed in its wake as it tore a path through the tumbling floes. When they finally emerged from the "villainous pack" into open water, McClintock wrote in his ship's log: "After yesterday's experience I can understand how men's hair had turned grey in a few hours."

| | |

With the *Fox*'s Baffin Bay experience in the front of my mind, I turned *Polar Sun* north, and for the next hour, we followed the floe edge, probing for its tip. Pack ice, I soon realized, was a different beast entirely from anything we'd experienced yet. As we crept through the cold, we passed dirty mounds of ice that rose several feet above the ocean's surface, where they melted into the gunmetal gray skies above. Occasionally, amongst the windrowed pressure ridges that ran to the horizon, an iceberg jutted up from the pack like a frozen tombstone.

After an hour, the ice loosened and cracks of open water appeared to port. I turned into a narrow lead, which led us through a mass of ice boulders and, eventually, to clear seas. But the ice wasn't quite ready to let us go. As soon as I pointed the bow toward Pond Inlet, on the northwestern tip of Baffin Island, the wind started blowing twenty knots from exactly the direction we needed to go. The port tack (with the wind coming over the left side of the boat) would have pointed us back toward Greenland, so I chose the starboard tack, which led us down the other side of the lobster claw we'd spent days trying to put behind us. And as the waves crashed against the ice, with *Polar Sun* skimming along its edge, I finally handed the helm over to Ben and went below to see how much of my own hair had turned gray.

For the next twenty-four hours, we sailed *Polar Sun* hard in the heavy seas, usually within sight of the ice, which never failed to incite a foreboding sense of impending doom. Wave after wave slammed into the bow, causing *Polar Sun* to occasionally bury her nose in the troughs, sometimes all but stopping us dead in our tracks. In these conditions, maneuvering around the boat was like riding on the back of a bucking bronco while climbing across a jungle gym.

Heavy weather always changed the mood aboard. Small talk stopped. The crew shared details about the conditions and the sail plan at watch changes but little else. We each retreated into our own private worlds, listening to the boat, gauging the wind, always tensed for the next pitch and roll, wondering and worrying. By some miracle, Ben managed to cook at least one hot meal for us every day, which tended to be the only time when the crew would gather. Renan, who spent most of his time looking at things through the lens of a camera, occasionally donated his meals to the NAPA bucket.

On our fifth day out from Greenland, the skies cleared, and as we bathed in sunshine in the cockpit, a U-shaped valley, framed by towering granite monoliths, materialized off to the west. On either side of this amphitheater, snowfields and glaciers poured into the turbulent sea. Deep in Baffin's mysterious interior, rocky peaks and the edge of what appeared to be an ice cap shimmered under a pale blue sky. Orienting myself with the chart, I realized that we were gazing upon Buchan Gulf, an area I'd flown over in the late 1990s while hunting for unclimbed cliffs.

Paging through a Canadian Sailing Directions booklet, Ben

found a reference to a possible anchorage near a small island called Nova Zembla, which lay about fifteen miles away. I had slept so little since leaving Greenland that I could barely form a coherent thought, and I salivated at the idea of dropping the hook and getting a decent night's sleep, so I set a waypoint. But we made it only a few miles before running up on a vast field of loose pack ice. Renan launched the drone, which served the purpose of capturing aerial footage while at the same time giving us a bird's-eye perspective on the ice that lay before us. The tiny display on the controller showed that the pack extended for about ten miles, with small channels of open water between the floes. In a few spots, the ice crowded in more densely and looked impassable. *What are we looking at?* I wondered.

Ben and I had both been advised never to enter ice of greater than three-tenths' or four-tenths' concentration in a fiberglass-hulled boat. As we hovered over Renan's shoulder, Ben declared that this ice was only one-tenth or two-tenths, and it wasn't a big deal. "We're going to be maneuvering through tons of ice later on," he said. "We might as well get some experience now."

"How can you be so sure it's only one- or two-tenths?" I asked him.

Ben replied that he knew the difference between ice we could make it through and ice we couldn't. He just did. And while Ben leaned in with knowledge and confidence, I pushed back because my internal radar flashed red—even if I didn't know why. The truth is, I'm extremely cautious by nature. Well, at least compared to other professional climbers and adventurers. My goal is never to eliminate risk, but I'm obsessed with understanding it. If you know what you're up against, the right decision generally clicks into place. Back in college, when my buddies and I got seriously into climbing frozen waterfalls, we would sit around arguing about the tiniest details: the ideal diameter of rope, plastic versus leather boots, which gloves gave the best grip on a frozen ax handle, how exactly to sharpen a pick to get the deepest penetration in the ice. Tweaking the details just right not only put you in the position to succeed; it made you feel like you were actually better. It created confidence, and there is nothing more valuable than that when you're hanging hundreds of feet in the air off a half inch of steel.

A lot of it seems silly in retrospect, but it was the birth of a mindset that I have followed ever since—a dogged focus on collecting all the necessary information, sifting through it, and hoping that the right answer will rise to the surface. And in climbing, it almost always did. But here, in the ice, things were different. My radar was going off—not because I didn't like the risk or our odds, but because I didn't understand what I was up against. Ben had tons of knowledge but not about ice. And the fact that he was so confident about something I knew he didn't understand made me wary. My mind turned to a long string of what-ifs that had no immediate answers. What if we committed to pushing through the ice and the wind shifted? What if the ice closed around us like it had to the *Fox*, *Erebus*, *Terror*, and so many other ships? As I looked around at the jagged, floating chunks surrounding us, it wasn't hard to picture *Polar Sun* jammed between them like a peanut in a nutcracker. The thought almost made me sick.

In the end, Ben's confidence and stubbornness prevailed and we pointed the boat into the labyrinth of ice. Maybe it was inevitable. Initially, I had enjoyed the game of dodging around the ice floes. Gradually, though, the leads of open water began to narrow, and our route became so circuitous that in places we were forced to follow long, looping paths, at times scraping between the floes.

After an hour or so, we entered a small basin of open water. Seeing that the way was clear ahead, I increased the throttle. Almost instantly, I heard an alarming *CRACK* and the boat shuddered and seemed to twist along its axis. I looked astern and saw two chunks of thick pale gray ice pop to the surface and spin off behind us. A hidden floe had been lurking just under the waterline. In the short battle between fiberglass and the frozen ocean, *Polar Sun* had won, cracking the floe cleanly in half. A few inches thicker and it might have gone the other way.

Around ten p.m., we broke into dark open water. As I handed the helm to Ben, I couldn't shake the feeling that we had taken an unnecessary risk. But what worried me more was Ben's confidence—overconfidence, in my opinion. He had been right this time. But, what if he'd been wrong?

ı|ı

After a fine night's rest at Nova Zembla and another full day of sailing along Baffin Island's cliff-lined coast, we turned west into Tasiujaq Sound (formerly Eclipse Sound), a narrow inlet that separates Baffin from Bylot Island to the north. The wind funneled down the sound from behind us, so we put up the "Whomper," a bright yellow spinnaker-type sail made of lightweight material and designed for downwind sailing in light air. In the narrow confines of the sound, the waves mellowed and *Polar Sun* found her happy place. As we glided along toward Pond Inlet, which now lay only forty miles away, everyone sat out on deck admiring an extended sunset that painted the surrounding hills and mountains in alpenglow.

After the sun slipped behind Bylot Island, I noticed to our north an enormous ship with a low-slung red hull and a giant grinning shark's mouth painted across its bow. Twin anchors dangling from the hawseholes formed the eyes; a five-story superstructure dotted with the windows of crew cabins and topped with radar arrays perched on the stern. Our AIS (automatic identification system) named the vessel as the *Nordic Qinngua*, a bulk carrier en route to Milne Inlet to take on a load of iron ore from the Mary River Mine. As all seven hundred fifty feet of her blew past and *Polar Sun* rocked like a metronome in its wake, I could feel the low rumble of the freighter's massive diesel engines, which lingered long after the ship had passed.

As we neared Pond Inlet, we glided past a smattering of small cabins and hunting camps on the outskirts of town. Before long, a row of buildings came into view. Dozens of houses stretched up to the top of a small hill, where several large fuel tanks lorded over the sound. Pond Inlet didn't look like much, but the lights were on and, after seven days out in the wilds of Baffin Bay, it felt like we'd arrived in New York City.

At one a.m. on August 10, we pulled into a brand-new public dock and tied up next to a small red sailboat called *Regina*, which I recognized from our time in Disko Bay, where I had chatted with its skipper, Gregor. The sun had set hours earlier, but the perpetual Arctic twilight made it possible to see without a headlamp. The dock was so new that it didn't have a ladder, so Ben lassoed a cleat, and Renan monkeyed his way up with a rock-climbing move called a heel hook. As soon as the boat was secure, a bicycle gang of curious local kids gathered on the wharf overhead.

A teenage girl, maybe sixteen or so, knelt in the gravel watching us silently. She wore a shiny blue down jacket and eyeglasses, with a pair of sunglasses perched on her forehead. Her friend, about the same age, wore her hair in two jet-black ponytails that trailed down to her waist. With them was an older boy in a black hoodie striking a cool pose astride his mountain bike. Two younger boys, about ten and twelve, bristled with excitement.

The girl with two pairs of glasses asked, "Where did you come from?"

"Greenland," I replied.

Everyone's eyes grew wide.

The older boy, who carried himself as the leader of the posse, told us his name was Justin. "How was it out there?" he asked. "Was it scary?"

I looked up at him and raised my eyebrows. He smiled and nodded to show that he understood.

"Don't your parents care that you're out so late?" asked Ben. "Isn't there some sort of bedtime?" There were no adults in sight.

"No bedtime in the summer!" they all chimed in unison.

Justin wanted to know what kind of smartphones we had, and all of them wanted to come aboard. But it was late, and we were fried.

"Come back tomorrow, and we'll give you a tour," I promised.

By this point, we'd created a fair bit of commotion, and I hoped that Gregor might appear to say hello. I could hear *Regina*'s diesel heater cranking away, and I could picture him down in his cabin, cozy and warm. Gregor, a Pole, was attempting a solo crossing of the Northwest Passage in memory of his late grandmother, Regina, whose face was printed on an enormous decal pasted across the bow of the boat just above the waterline. When I'd first seen Regina's smile in Disko Bay, I couldn't help but chuckle at the fact that her face was going to take a serious beating if Gregor ran into any ice.

When I came up on deck the next morning around eight a.m., *Regina* was gone. The Baffin Island bicycle gang, though, was right where I'd left them, sitting on their bikes, eyeing *Polar Sun* hungrily. "He went that way," said Justin, pointing toward Navy Board Inlet—the direction of the Northwest Passage.

I climbed up onto the dock, gave Justin a fist bump, and set off to run some errands in town. Shuffling down a potholed dirt road along the water's edge, I passed drab one- and two-story houses in various states of disrepair. Gone were the bright colors and orderly streets of Nuuk and Sisimiut, which had offered a soothing contrast to the harsh environs. Pond Inlet had an unmistakable weight to it, a fatigue, as if the small settlement were slowly losing its battle against the steady polar winds. From what I could tell, there wasn't much time or appetite for color here.

Halfway up the hill, I spied two Inuit women and a young child walking down the street toward me. The younger woman, wrapped in a heavy parka with a fur-lined hood, grabbed her child's hand and scurried across the road. I waved, but she turned away. The older one kept her course. She looked to be in her late fifties and was bareheaded, with salt-and-pepper hair framing deep wrinkles on her brow. She glared at me as I approached. When I was a few feet away, she stepped into my path and hardened her gaze even more. I raised my hand in a friendly greeting, but before I could say anything, she leaned forward and spat three crystal clear words: "Go. Fucking. Home."

Shocked, I spun past her as she continued to scream obscenities at me. Looking back from a safe distance, I noticed a city-sized cruise ship in the harbor that I hadn't seen before. Black Zodiac inflatable boats ferried passengers to the shore. *Maybe she thinks I'm one of the cruise ship people,* I thought, trying to console myself about what had happened. Deep down, though, I knew the woman didn't give a shit what kind of boat I'd come in on. After generations of nonnatives imposing their will and way of life on the people of Pond Inlet, she had reason to be wary of outsiders.

Inuit have lived in Tasiujaq Sound for more than four thousand years, moving nomadically with the seasons and always maintaining a balance with the natural world that allowed the wild animal populations to thrive. But the whalers who arrived in the 1820s didn't care about conservation, and over the next half century, they nearly wiped out Tasiujaq Sound's ancient fishery. Next came the HBC, which arrived in Tasiujaq in the 1920s to trade for furs and narwhal tusks. They were followed by Christian missionaries, with their mandate to turn Inuit away from their age-old animistic beliefs. In 1964, the Canadian government began funding missionary groups who removed children from Pond Inlet to attend the Churchill Vocational School in Manitoba. The purpose of this "residential school" was to assimilate Inuit youth into Western culture; it was part of a nationwide network that indoctrinated as many as 150,000 First Nations, Inuit, and Métis Nation children—many of whom suffered physical, emotional, and sexual abuse at schools like Churchill. A shocking number of them simply disappeared and never returned to their families.

In 2021, a mass grave containing the remains of 215 children was purportedly discovered in British Columbia on the grounds of a former residential school. Some believe that there are hundreds, if not thousands, of such graves yet to be identified. That same year, the Canadian government established the National Day for Truth and Reconciliation to honor the thousands of indigenous youths who died in the residential-school system due to abuse, neglect, and disease. In January 2023, the government agreed to pay $2.8 billion Canadian to settle a series of lawsuits after a national commission determined that the residential-school system had been equivalent to "cultural genocide."

On my way back to the boat, I passed a well-kept official-looking building with green metal siding, red window trim, and a sign that read, "Sermilik National Park Administration Office"—in English, French, and Inuktitut. Curious, I stepped inside. A middle-aged man with a beard and shoulder-length blond hair dusted with gray hopped up from behind a desk and introduced himself as Darrell Makin, park ranger. Pond Inlet is a small town and Darrell had seen *Polar Sun* on the wharf through his office window. "I was hoping you guys would stop by," he said. "Welcome to Pond Inlet."

Darrell led me to a conference room in back and spread a map on a long table. Sermilik National Park (*sermilik* means "place with glaciers" in Inuktitut) was established in 1999. Darrell showed me how the park stretched across Bylot Island and most of the Borden

Peninsula, which lies west of Navy Board Inlet. The park is a popular destination for skiing and mountain climbing. Most years, visitors traverse Bylot Island on skis—an expedition that typically takes about two weeks.

The map also included scattered parcels marked in purple that were labeled as "Inuit Owned Land" (IOL). Darrell said that Pond Inlet was quiet right now because many families were out in these areas at their summer camps, hunting, fishing, and collecting bird eggs. The IOL parcels had been carved out in 1993 as part of an agreement between the Inuit community and the Canadian government. The agreement granted Inuit title to 19 percent of a new 700,000-square-mile Inuit-governed Canadian province called Nunavut ("our land" in Inuktitut), which officially came into being in 1999. Nunavut is the largest territory or province in Canada (roughly the size of Mexico) and one of the most sparsely inhabited regions on Earth. The IOL gives Inuit unrestricted hunting access to these lands, many of which have been used by the same families for generations. It also offers them mineral rights and guarantees them a role in the management of these parcels and a share of the royalties if any of them are developed for commercial purposes.

As we talked, Darrell pointed out the window toward a two-hundred-foot-long vessel with a sleek dark gray hull that was anchored just west of the cruise ship. He explained that this ship was a former research vessel converted to luxury yacht. It had recently anchored in Tay Sound near some IOL and angered locals who believed that the ship's engine noise drove away the wildlife they'd gone to hunt. "Narwhal and beluga are very sensitive to sound," said Darrell, "and there is concern that all the new shipping is screwing up their migration patterns."

Between the bulk carriers like the *Nordic Qinngua*, the cruise ships, the luxury yachts, and sailboats like *Polar Sun*, a record amount of shipping was taking place in Tasiujaq Sound and all across the Northwest Passage. And it's certain that as the climate in this part of the world continues to warm, the problem will only get worse. That evening, I had a conversation with a local man named CJ, who claimed that no narwhals had been caught in the sound since 2020, and I'd later learn that twenty-two cruise ships called in Pond Inlet that summer. But according to Darrell, it was the Mary River Mine that was having the biggest impact on the local environment and wildlife.

In 1962, two Canadians—mining engineer and prospector Murray Watts and pilot Ron Sheardown—spotted several large black circles on the ground while flying over the interior of Baffin Island, about a hundred miles southwest of Pond Inlet. Surveys later determined that those black circles contained some 780 million metric tons of iron ore. In 1986, Baffinland, a subsidiary of the company that would later become the Luxembourg-based mining giant ArcelorMittal, secured a claim to the area and began exploration and development of the mine. At the time, most mining in the Canadian Arctic was for gold and diamonds, which came out of the ground in small enough quantities that they could be shipped out as air cargo. But exporting millions of tons of ore required the construction of a port facility and a sixty-mile tote road over fragile permafrost. The location and high cost would typically have marginalized such a remote mine, but Mary River Mountain was no ordinary prospect: the ore inside it was 64.7 percent iron, making it some of the purest on Earth.

In 2014, after a positive recommendation from the Nunavut Impact Review Board (NIRB), Canada's federal minister of aboriginal affairs approved the megaproject, and operations began the following year. The original lease allowed for the removal of 3.5 million tons of ore per year, an amount later increased to six million tons. By the time we arrived in Tasiujaq Sound, the mine's contribution amounted to nearly a quarter of Nunavut's GDP—even though the company ran at an annual loss due to the extreme cost of operations in such a remote location. According to Baffinland, they needed to increase output to make the mine commercially viable. To do this, they proposed building a railroad across the tundra and doubling the number of bulk carriers. Several years of contentious public hearings followed.

The mine is located within IOL, and with billions of dollars in potential royalties and tax revenues expected for the Nunavut government and Inuit organizations that own the land, most people assumed that the powers that be would rubber-stamp Baffinland's request. But a group of seven Inuit hunters from Pond Inlet had a different idea. In February 2021, they blockaded the mine's airstrip and tote road, effectively trapping seven hundred miners in the work camp. The protest attracted international attention and support from the World Wildlife Fund and Greenpeace, who staged their own demonstration at the headquarters of ArcelorMittal in a show of solidarity with the hunters.

A story about the protest in Baffin's only newspaper, the *Nunatsiaq News*, had more than eighty comments on the paper's forum. "My husband is being held hostage up there," wrote J. Gray. "With no fresh supplies coming in or new staff, things are gonna get ugly quick."

A user named Rule of Law responded: "These people do not represent anybody except themselves. Who elected them? This also shows blatant contempt for the rule of law."

"I'm a lot more worried about people's wellbeing than the 'rule of law,'" countered an anonymous poster. "Residential schools were operated according to the 'rule of law.' Mothers who refused to send their children were breaking the law. Instead of complaining about people not 'respecting' the law, maybe we should be asking what laws we need to change to create a more just and fair society."

A spokesperson for Baffinland claimed that the hunters were demanding to be recognized as their own Inuit association and to be paid royalties accordingly. One of the protesters, Naymen Inuarak, told the *Nunatsiaq News*, "We would like to see actual negotiations with the most impacted communities and have us involved right away. We've been ignored way too long."

In May 2022, three months before we arrived in Pond Inlet, the NIRB recommended denying the mine's request to increase output, citing "potentially significant adverse impacts on vegetation and freshwater, leading to adverse socio-economic effects on Inuit harvesting, culture, land use, and food security in Nunavut," as well as "adverse eco-systemic effects on marine mammals and fish, caribou, and other terrestrial wildlife." Baffinland immediately filed a grievance and petitioned Northern Affairs minister Dan Vandal, hoping he would overrule the decision. In November, Vandal rejected Baffinland's appeal. By then, the mine, anticipating the worst, had already laid off eleven hundred workers.

The controversy over the Mary River Mine is an example of the dilemma faced by communities across Nunavut. Many towns, including Pond Inlet, are sitting on literal gold mines of untapped natural resources with massive potential to create jobs, stimulate the economy, and raise families out of poverty. But industrial development comes at a heavy cost to the environment, the wildlife, and the ancient Inuit hunting-and-fishing culture. As one person commented in the *Nunatsiaq News* forum, "It is almost impossible to have the benefits of a Western lifestyle including petroleum vehicles and modern hunting rifles and central heating in houses, but still claim allegiance to the old ways. They are mutually incompatible."

ııı

Back at *Polar Sun*, I met up with our newest crew member: Jacob Keanik, a sixty-two-year-old Inuit guide and former wildlife-conservation officer. Jacob had flown in from his home in Gjoa

Haven, an Inuit community on King William Island several hundred miles to the west of Baffin, deep in the heart of Nunavut. He was a close friend of Tom Gross, the Franklin expert from Hay River whom I'd visited the previous spring. Over the years, Jacob had accompanied Tom on six Franklin search expeditions. Of middle height and build, Jacob wore jeans, a black North Face puffy coat, and a rabbit-fur-lined leather bomber's cap. Gray bangs hung down over his forehead, framing bushy eyebrows, high cheekbones, and a thin black mustache. I had already met Jacob on another expedition just a few weeks before—more on that soon—and he had already been given a tour of *Polar Sun* and moved into the bottom bunk in Ben's cabin. But as I watched him struggle to find a home for his giant hard-sided suitcase, I was reminded that this was his first time on a sailboat. When Renan and I had first invited him to fly to Pond Inlet and join our crew, he jumped at the opportunity, telling us that he didn't know a single Inuk who had ever sailed through the Northwest Passage. I was grateful that he'd agreed to join our crew. With him, we had something the Franklin expedition had never had: an Inuit crew member with a lifetime's experience adapting, surviving, and thriving in one of the harshest environments on Earth.

Jacob, the youngest of nine children, was born in 1960 on the shore of McNaughton Lake on the Adelaide Peninsula, about a hundred thirty miles south of King William Island. His parents adhered to a seasonal calendar, hunting caribou, musk oxen, and polar bear in the summer; spearing Arctic char in the rivers in the late summer and fall; and sealing on the coast in the winter and spring. During the long winters, they lived in igloos that were lit and heated with seal-oil lamps.

When Jacob was five, Canadian authorities had forced the family to move to Gjoa Haven so the children could receive formal education. They were given a small plywood-sided house and a modest allowance, but the money wasn't enough for them to afford the imported food sold at the HBC store, and the hunting around Gjoa Haven was poor. At school, Jacob struggled to fit in. "I had caribou clothes—caribou pants, caribou mitts, caribou everything," he says. "The kids teased me about it because they had new clothing that came from the south." The Christian missionaries that ran the school christened him Jacob, but the name given to him by his parents, which he still uses amongst his family, is Nevekitok—an Inuktitut word used for shamans who have the power to change the weather.

His parents eventually returned to McNaughton Lake, but Jacob stayed in Gjoa Haven, where he trained as a conservation officer. His tasks included tranquilizing polar bears, measuring them, and taking blood and fur samples. He then turned to working as a private hunting guide and as president of the Nattilik Heritage Society, a local museum that focuses on Inuit culture.

As Jacob struggled with his luggage, Ben told me that the bicycle gang—Cain, Kian, Annabelle, Rickie, and Justin—had come aboard earlier for a visit. It was low tide at the time, and the drop from the dock down to the deck was about ten feet, which they had navigated with a rope ladder I'd made out of a spare dock line. According to Ben, the kids only wanted to be down in the cabin, where they spent an hour poking around before making themselves comfortable on various seats around the salon. They each got a pack of Welch's Fruit Snacks and a Polaroid photo from Renan; then everyone joined in on a rousing chorus of the *SpongeBob SquarePants* theme song.

Now, as dusk fell, the bicycle gang had scattered, except for Justin, who was sitting on the edge of the dock. He had his hood up and was staring at Jacob, who stood on the aft deck puffing on a cigarette and watching a heavily laden bulk carrier chug east toward Baffin Bay. The mountainscape across the sound that stretched off into the interior of Bylot Island looked like a painting, with ashen layers brushing against a lumpy sky. Patches of light illuminated the clouds, and dark streaks of rain slanted down into the valleys, where fog poured into the sea like vapor from a witch's cauldron.

Jacob drew on his cigarette and then said something in Inuktitut to Justin, who also turned his gaze north. As we all stared into the Northwest Passage, I contemplated how we each had a unique relationship to this magical place: different yet connected, not only through a coincidence of time and place but through the brutal and complicated history of colonialism. My time in Pond Inlet had shown me that as much as I might see myself as different from those who had always exploited Inuit in the name of progress and civilization, the truth is that exploration and exploitation are still as closely tied as ever. The tragedy is that the climate change wrought by an increasingly industrialized and imbalanced world will have outsized consequences for Inuit—before it comes for us all.

For several minutes, we enjoyed a comfortable silence as Jacob smoked and the boat bumped rhythmically against the dock.

After a while, Jacob turned to me. "The land here, it's something to see. The mountains are so high, and the ice caps . . ." He trailed off. After some moments, he added, "I'm very happy to be on this boat, to have this opportunity. Thanks again for inviting me."

CHAPTER 7

Consider Your Ways

Date: August 14, 2022
Position: Beechey Island, Parry Channel

The rocky ground was bare and battered by wind and sea spray, and the only traces of vegetation were a few patches of moss that clung to the gravel burial mound. Years ago, someone had removed the original tombstone and driven into the earth a board with a bronze plaque that read:

SACRED TO THE MEMORY OF JOHN HARTNELL, A.B.
OF H.M.S. EREBUS
DIED JANUARY 4TH 1846
AGED 25 YEARS
"THUS SAITH THE LORD OF HOSTS,
CONSIDER YOUR WAYS"
HAGGAI, I.,7.

I stared down at the ground, hands in my pockets. It was cold. It was always cold in the Arctic. Seems obvious, but you can never escape it, except in the ship's cabin with the stove blazing. I shuffled my feet and took in my surroundings. To the north rose three more graves like this one. Beyond, a skinny, curving isthmus of gravel connected this grave site to Devon Island, where dark clouds obscured the tops of rolling hills. In the bay, *Polar Sun* lay at anchor, smoke puffing from its chimney.

Polar Sun beckoned, but I lowered my head and tried for a moment of silent communion with Hartnell. I'm not much for solemn occasions or trite emotion, but this morbid hunt for Franklin and his dead men was beginning to affect me. Maybe it was the hard overcast, the desolate brown hills—or the searching cold. Hartnell was twenty-five years old when he died here during the Franklin expedition's first winter in the Arctic. I felt a pang of sadness as I thought of my own son who was the same age. *How would it feel,* I wondered, *to suffer and die in this lifeless desert that even Inuit have forsaken?*

Thanks to heavy wind and clear water in Lancaster Sound, we had made a fast passage from Pond Inlet to Beechey Island, covering the three hundred miles in about forty-eight hours. For half of that time, we tracked west along the south shore of Devon Island beneath horizontally banded cliffs rising hundreds of feet out of the ocean like the battlements of a medieval fortress.

We were now nearly 75 degrees above the equator, so far north we'd left behind almost all vegetation. Waterfalls poured from high cliffs into the frothing sea, and every so often we passed mysterious inlets that snaked into the island's interior. Here and there, tongues of tired-looking glaciers spilled through gaps in these ramparts, and where they met the shoreline, milky plumes of fresh water drew ripples in the sea. Renan put his drone up and took aerial photos, which showed our little ship with its bright red sails running along the edge of a landscape so desolate that NASA had chosen it as a site to test Mars rovers.

When we had motored into the bay that lies on the east side of Beechey Island around midnight on August 13, the setting sun radiated a soft light along the skyline of a distant plateau. The Franklin expedition had anchored in this same spot in September 1845. Their last contact with the outside world had been back on July 26, exactly two weeks after departing the Whale Fish Islands, when two British whaling ships, *Prince of Wales* and *Enterprise*, found *Erebus* and *Terror* moored to a large iceberg in Baffin Bay. Captain Dannett of *Prince of Wales* later reported to the Admiralty that the Franklin expedition had erected a collapsible wooden hut on the summit of the iceberg, from which they were taking astronomical observations. A boat with seven of Franklin's officers, including one who is believed to have been James Fitzjames, visited Dannett and his crew aboard *Prince of Wales*. Dannett later recorded that the men were "in remarkable spirits expecting to finish the operation in good time."

According to the whalers, 1845 was an unusually mild summer, and Dannett concluded that *Erebus* and *Terror* likely made good time on their passage across Baffin Bay. But when Franklin got to Beechey Island, we can assume that while following the directions he had been given by the Admiralty, he found the seas to the south and west blocked by ice. The Victory Point Record tells us that he next headed north, in search of the Open Polar Sea. They made it as far as 77 degrees before turning back and circumnavigating Cornwallis Island, which lies to the west of Beechey and is now the location of an Inuit community called Resolute. As the ice closed in, Franklin brought *Erebus* and *Terror* into the bay where *Polar Sun* now lay at anchor.

By the time the bay froze solid in late October or early November, Franklin and Crozier would have already ordered *Erebus* and *Terror* to be put into "winter quarters." After retracting the propellors, removing the rudders, and stripping the sails, the spars and the running rigging were struck down on deck. The sailors chinked the gaps in the ships' planking with oakum and tar to seal out icy winter drafts, and they banked snow along the hulls and across the decks for insulation. Spare sails draped over the booms formed enclosures over the deck, and wooden poles driven in the ice marked the route between the ships to help seamen find their way back and forth in the endless blizzards and darkness. They finished preparations by digging large holes through the ice, each topped with a wooden tripod and a pulley system so the crew could quickly form a bucket brigade if a fire broke out aboard.

Franklin's crew watched the sun set for the final time that fall on November 6, 1845; it would be three months before it reemerged. Two months later, twenty-year-old John Torrington died, followed three days later by Hartnell. William Braine, a thirty-two-year-old marine, the final casualty of that long winter, passed away on April 3. Franklin's men must have labored for days with pickaxes, pikes, and shovels to dig graves into Beechey's frozen gravel. After they lowered the coffins into the ground, Franklin, who was deeply religious and held a service in his cabin for the men every Sunday, would have read a passage from his personal Bible. I presume that he must have personally chosen the verse that adorned Hartnell's headstone, and he might have also considered the one that precedes it: "Ye have sown much, and bring in little; ye eat, but ye have not enough; ye drink, but ye are not filled with drink; ye clothe you, but there is none warm; and he that earneth wages earneth wages to put it into a bag with holes."

ı|ı

As I stood over Hartnell's grave, my mind flashed to the ghoulish image of his body, which I had seen in the book *Frozen in Time: The Fate of the Franklin Expedition* by Owen Beattie and John Geiger. The book tells the story of two scientific expeditions that exhumed the remains of Torrington, Hartnell, and Braine in the mid-1980s. Beattie, an anthropologist from the University of Alberta, led the expeditions. The first task was to disinter the coffins, which were buried four to six feet deep. After prying off Torrington's coffin lid, they found his body sealed in a block of ice, better preserved than an Egyptian mummy. Beattie set up a mobile medical lab by the gravesides to perform autopsies, take X-rays, and collect tissue and clothing samples for analysis back in the laboratory.

On the second expedition, in 1986, the researchers followed the same process they had with Torrington, melting away the ice encasing Hartnell with buckets of hot water. First, his nose appeared, then a single hazel-colored eye. (Beattie presumed that the missing eyeball had been damaged when an earlier Franklin search expedition dug up the graves in 1852.) Next came the mouth, with the lips pulled back in a sneer, and it was at this point that the expedition's photographer, Brian Spenceley (who happened to be Hartnell's great-great-nephew), captured the haunting image that I had seen in the photo insert of *Frozen in Time*. It's a singular photograph that captures, better than any words ever could, the hardships that Franklin and his men must have faced that first winter, with conditions so brutal that their bodies even now haven't yet returned to the earth.

After hours of pouring hot water into the coffin, the researchers eventually exposed the entire body. Hartnell had been wrapped in a blanket. His head, swathed in a wool stocking cap, rested on a small pillow with a frilly edge and filled with wood shavings. When they pulled off the hat, a shock of dark auburn-tinged hair spilled out. Hartnell lay with his hands crossed, left over right, his skin almost perfectly preserved. He wore three shirts but no pants. His outermost shirt was made from cotton and decorated with blue pinstripes and shell buttons. The initials "T.H." and the year "1844" were monogrammed in red thread on the hem. Beattie speculated that the shirt belonged to Hartnell's younger brother Thomas, who was also a member of the crew.

Thomas, of course, would have been at his older brother's deathbed and at the burial. And as we know from that muster book, he was towheaded, with freckles and fair eyes, and he stood several inches shorter than his brother at five feet, eight and a half inches tall. We can surmise from letters written to John and Thomas from their family members that they were a close-knit family. Their father, a shipbuilder, died in 1832, when John, the eldest of five children, was twelve years old. John began his career as a shoemaker and might have gone to sea because of an injury to his right wrist. Two years after his death, his mother, Sarah, having no idea that he was already dead and buried on Beechey Island, penned a Christmas letter to her two sons. "My Dear Children It is a great pleasure to me to have a chanse to write to you," she wrote. "I hope you are booth well I assure you I have many anx[i]ous hours about you . . . if it is the Lords will may we be spared to meet on earth."

It was unusual for two healthy crew members to die so early in the expedition. Torrington and Hartnell had been healthy enough not to be dismissed from the expedition in the Whale Fish Islands, and yet here they were, only six months later, dead and buried. When Beattie and the other researchers removed Hartnell's shirts, they were surprised to discover a Y-shaped incision in his abdomen. Beattie attributed this previous autopsy to a surgeon on *Erebus*, but it had actually been conducted in 1852 by Peter Sutherland, the physician aboard *Isabel*, another ship that Lady Jane had privately financed to search for her husband. Regardless, the original autopsy appeared to have been done in haste. The organs were piled haphazardly into the chest cavity, and the breast plate stitched back in place upside down.

Beattie determined that Hartnell, like Torrington and Braine, had died from pneumonia, possibly a complication of tuberculosis. And of course, it's hard to imagine a more fertile incubator for TB than the damp, overstuffed, coal-dusted confines of a nineteenth-century wooden sailing ship filled with dozens of men, a dog, and a pet monkey named Jacko. The real bombshell, though, wouldn't come until later, when lab analysis suggested that all three sailors had elevated levels of lead in their tissues. This led to a hypothesis that the toxic metal might have contaminated their canned goods, leeching into the meat, soup, and vegetables they consumed.

The theory that lead poisoning might have been a significant factor in the tragic fate of the Franklin expedition, proposed for the first time by Beattie in *Frozen in Time*, made headlines around the world. Later, though, when the lead levels found in tissue samples taken from Torrington, Hartnell, and Braine were compared to those of nineteenth-century British sailors exhumed from a cemetery in the West Indies, the concentrations were similar. This suggests that Beattie's findings were likely due to lifetime exposure and not to acute poisoning during the expedition. In 2016, a study of John Hartnell's thumbnail showed that his lead levels were similar to those of Harry Goodsir, the expedition's naturalist who died more than two years after Hartnell on King William Island. It stands to reason that if the lead poisoning had resulted from the expedition cans, as proposed by Beattie, those who died later should have had more elevated levels of the metal in their bodies. But that doesn't appear to have been the case.

|||

Earlier that day, before I'd found the Franklin expedition graves, we took the dinghy to the eastern end of the island and landed near the remains of an old derelict wooden building. A weathered plastic plaque bolted to a metal pole told the story of Northumberland House, built in 1853 by the crew of HMS *North Star* with wood scavenged from a shipwrecked whaling vessel. The building was constructed as a refuge for survivors of the Franklin expedition, in the unlikely event they found their way back to Beechey Island, and it served as a haven for wayward sailors until it collapsed in the early 1900s. A photo of the building taken in 1875 shows a rectangular cabin, twenty feet wide and sixteen feet deep, with a hipped roof, a shed awning, and a flagpole in the front yard. As I walked around the ruins, I stumbled onto a mound of coal where a smithy had once stood. Figuring that we could always use more fuel for *Polar Sun*'s woodstove, I grabbed a handful and put it in my bag.

Up the hill stood a worn wooden pillar about six feet tall, adorned with bronze plaques commemorating not Franklin or his men, but some of those who had died in search of them. This place, where the Franklin expedition had spent their first winter, was discovered by American and British searchers in the summer of 1850. After spotting a monument from the water, the men landed in a small boat below Beechey's "dark and frowning cliffs" and ran up the beach, where they found a cairn constructed out of seven hundred gravel-filled tin cans. In the days before the telegraph, cairns served as the mailboxes of the Arctic, and the crew excitedly tore it apart, emptying every tin and even excavating the ground around the area in search of a written record. A few hundred feet above, on the craggy summit of the island, they found two more cairns. One held an upright wooden pole, visible from miles out in Barrow Strait, that had lured them in. Frustratingly, these too were empty of any documents.

While the cairns and the graves proved that *Erebus* and *Terror* hadn't sunk in the deep, stormbound waters of Baffin Bay, as some in England had speculated, they left more questions about the fate of Franklin and his men than they answered. Why hadn't Franklin left a note in one of the two hundred airtight metal cylinders supplied by the Admiralty? He was, after all, under orders to leave records of the expedition's deeds along the way, and we know that he was icebound in the bay on Beechey Island for at least six months—more than enough time to pen a simple note describing his intentions going forward.

A few feet away from the Franklin monument, someone had buried dozens of the old tin cans in the shape of a cross, and countless more were strewn across the landscape. I bent down and picked one up. It was about six inches in diameter, covered in rust, with cracks down the sides, and someone had jabbed a knife repeatedly through the bottom lid. The creases held faint traces of protective red paint, and thick strips of lead solder held the seams together. One bead of lead on the bottom of the can was embossed with a tiny number. The can I held, which had once provided a meal for one or more Franklin crew members, was number one hundred.

When this site had first been discovered in 1850, it included the foundation of a storehouse on which timbers supported a roof that had likely been fashioned from a spare sail. Inside, the searchers found sacks of coal, wood shavings, bits of clothing, and rope. Outside, they found a shooting range, wooden tubs that might have served as a wash station, and a pair of knitted cashmere gloves held down by two small stones. Both were for the left hand, with a faint heart embroidered into each palm.

Renan and Rudy busied themselves filming me as I poked around, while Jacob wandered off up the hill with our shotgun. He seemed more interested in finding wildlife than he was in shrines to European explorers. I wondered how it all must have appeared to him—these people from down south in their boat tromping all over the island, staring at the things left behind by other interlopers, taking pictures of everything, seeing nothing, picking things up and walking off, blundering around like children. Admittedly, I understand little about Inuit or their culture, but Jacob's apparent uninterest in our attempt to solve the Franklin mystery made me question if everything we did was more proof that we weren't really any different from all the white people who had come before us.

I hadn't yet figured out that the Franklin graves were located a mile farther down the beach, and as I wandered around looking for them, I found a half dozen monuments to random people who had had little to do with Beechey Island or even the Arctic aside from a desire to be buried near the first to die on Franklin's doomed expedition.

One held an epitaph on a worn metal plate sporting a faded photo of a man with a receding hairline; he wore a coat and tie and peered at me over a pair of dark-rimmed glasses. The inscription read: "Dr. Maurice Francis Coffey, PhD, 1919–1975. This cairn contains the mortal remains of . . . a great man who rests with the brave men who helped bring this Northland into being. . . ."

I continued on. In a few feet, I found a memorial for Desmond Henry Fogg, who died a year after Coffey. Next to this tomb, a disintegrated block of concrete held a bronze plaque hanging askance by two rusty bolts. "This cairn was erected by HRH Princess Margriet of the Netherlands and her husband Mr. Pieter van Vollenhoven to commemorate their visit to the Northwest Territories June 15–18, 1978." As I stood there in the cold contemplating the oddity of the site, it occurred to me that whatever urges had compelled the Dutch princess and others to stake their claims on this desolate gravel beach, they smacked of the same hubris that had marked centuries of European exploration in the Arctic and elsewhere.

That evening, we gathered in *Polar Sun*'s cabin. Foul-weather gear hung from hooks on the walls. Tools, charts, and other detritus covered every horizontal surface. Ben tended a lentil stew with carrots, onions, and thin slices of kielbasa that bubbled away on the propane stovetop. Our floating home was a mess, but it was warm, the wind had died down, and the boat felt so still, it could have been parked on land. It had been a heavy day filled with death and thoughts, at least in my mind, about the futility of it all. But for some reason, I felt a deep level of contentment as I sipped on a glass of vodka and fed the woodstove with chunks of the "Franklin coal" I'd found on the beach.

These black rocks differed markedly from the nut coal that we had on board. The hunks were more tarry and they burned hot enough to turn the flue bright red. Later, I learned that this stuff was probably "patent fuel," a mix of coal and pitch hydraulically pressed into bricks. At the time of the Franklin expedition, it was ideal for ships like *Erebus* and *Terror* because it weighed less than regular coal.

As I watched the patent fuel crackle and pop in the woodstove, I pictured *Erebus* and *Terror* frozen in the ice somewhere close to where we currently sat. And I imagined a merry scene of cozy domesticity below deck—perhaps not unlike the one I was currently enjoying. The men would have lain in their bunks or hammocks, or sat together in the ship's common areas, as we did now, reading by candlelight or lanterns and perhaps drinking their daily ration of grog. Personal space would have been minimal, as it was on *Polar Sun*, but *Erebus* and *Terror* had brick furnaces that pumped hot air throughout the ships through square cast-iron pipes that ran down both sides of the hulls.

As the Franklin coal burned down to ashes, it hit me how proud I was at having made it this far. Lying close to 75 degrees, Beechey Island was undoubtedly the farthest north that I'd ever sail. But alongside my pride, I wondered if this adventure, which felt enormous to me, was ultimately misguided, much like the Franklin expedition—just another insignificant monument rotting on the shore.

As I placed another potato-sized chunk of patent fuel into the firebox, I noticed that Jacob's face was set into a hard scowl.

When I caught his eye, he stared back at me with cold intensity.

"That Franklin coal," he said, "we shouldn't be burning it."

"Why not?"

"It's bad luck," he said, "to mess around with dead people's things."

CHAPTER 8
The Hydrographer

Date: August 16, 2022
Position: Peel Sound

I lay diagonally across my bunk, feet pressed against the aft bulkhead, my shoulder jammed at an odd angle into the teak cabinet, where I kept my library. Inches from my head, the sea hissed and burbled through *Polar Sun*'s hull as she accelerated down the face of another wave. After so many months at sea, I could sense the speed of the boat instinctively, and every squeak, sigh, and shudder had become as familiar as my own voice. All around me, the cabin's woodwork creaked and groaned, and above, I heard the telltale clicking of the winch as Rudy trimmed the Whomper up in the cockpit.

He could have just set the sail and left it alone, but Rudy was too good a sailor not to try to squeeze every ounce of juice out of each puff of wind. With every click of the ratchet, the sail tightened, and with each *screech* of the sheet slipping back out from around the drum, it slackened. Rudy knew, like every long-distance voyager does, that quarters and halves of knots, over days and weeks and months, can add up to a lot of miles. And we still had thousands left between us and the finish line in Nome, Alaska.

When the Whomper filled, there was always a moment of lift, as if the boat wanted to take off into the sky. It was a feeling I had come to know as *Polar Sun*'s way of telling me that she was happy. By now my ship felt more like a living, breathing entity than an inanimate object. Sometimes I spoke to her out loud, especially when I was wedged tightly into my bunk. It was down inside the ship more so than topside that I could feel most keenly what kind of a mood she was in, be it happy, sad, annoyed, or—God forbid—downright pissed off. Now, as we rode over, across, and through the waves, I noticed that the moments of lift were growing longer and more forceful. The wind was building, and *Polar Sun* was suddenly sailing very, very fast.

I tried to tell myself that everything was fine, but curiosity and concern got the better of me, as they always did, so I forced myself out of bed and up the companionway. In the cockpit, I found Rudy sitting astride the coaming. He had one leg stretched casually across the deck and the sheet to the Whomper in his hands. The evening sun shining in through the enclosure brightened his auburn-colored beard, and I could tell by the faraway look in his eyes that he'd been having his own tête-à-tête with *Polar Sun*. When he saw me standing there in my long underwear, he flashed a guilty smile, acknowledging that he'd been caught pushing the boat just a touch too hard. "Sorry, Dad," he said. Then, without another word, he went forward to shorten sail by socking the Whomper, and I rolled out the jib in its place.

|||

Beechey Island, which we had departed the day before, lies at a bit of an Arctic crossroads with icy channels radiating outward in every direction. To the east, Lancaster Sound cuts toward Baffin Bay, Davis Strait, and the Atlantic Ocean. To the north runs the Wellington

Channel, which Franklin explored in the late summer of 1845, proving that there was no Open Polar Sea in that direction. To the west, the Parry Channel stretches for hundreds of miles toward the ice-choked McClure Strait and Beaufort Sea.

By the time the Franklin expedition departed England in the spring of 1845, John Barrow had decided that the best chance of finding a Northwest Passage lay in any channels that might be found to the south of Beechey Island. William Parry and John Ross had explored a massive bay to the southeast called Prince Regent Inlet in the 1820s and '30s to no avail, so Barrow ordered Franklin to search for a similar inlet that might be located farther west in the vicinity of Prince of Wales Island. Barrow's hunch proved prescient, and we now know that Franklin did find such a channel in this locale. Today, it is known as Peel Sound. And it was here that *Erebus* and *Terror* sailed in the summer of 1846, and where, 176 years later, we followed in their wake aboard *Polar Sun*.

On the first night out from Beechey, Ben and Renan took the two night watches, battling their way through fog, strong currents, and rogue sheets of pack ice that drove us toward a tadpole-shaped rock called Limestone Island. In the morning, the air cleared, and I caught my first views of Somerset Island to the east and Prince of Wales Island to the west. Wispy patches of mist still clung to the water, and a thick band of clouds hovered above the two shorelines. Prince of Wales presented as a thin dark line, while on Somerset, a range of gently sloping ginger-colored hills glowed in the morning sun.

Jacob sat with me at the helm, glassing the coastline to the east with binoculars. After a few minutes, he pointed toward a low spot on the horizon. "Bellot," he said in his typical terse manner. I had been watching Bellot Strait on the chart plotter all day, knowing that it was a possible bailout back to Pond Inlet if ice blocked our route ahead. This infamous narrow cut, which connects Prince Regent Inlet to Peel Sound, is now the standard route taken by most Northwest

Passage vessels. We had bypassed it by going up and over Somerset Island and following Franklin's and Amundsen's routes down Peel Sound. (Only later did we learn that *Regina* was stranded, with a failed starter, in Depot Bay on the east end of the strait. Without a working engine, Gregor was eventually forced to sail back to Pond Inlet.)

Bellot Strait, sixteen miles long and approximately a mile wide, is bordered on the north by Somerset Island and to the south by the Boothia Peninsula—the northernmost tip of mainland North America. It was discovered in 1852 by one of the many Franklin rescue missions. As the conduit between two massive bodies of water, Bellot Strait is subject to ferocious currents that can carry dangerous masses of ice that crash and grind against one another with deadly force in the strait's many whirlpools and eddies. While searching for the Northwest Passage in the early 1830s, John Ross described a similar phenomenon in the Gulf of Boothia, where "mountains of crystal hurled through a narrow strait by a rapid tide; meeting, as mountains in motion would meet, with the noise of thunder, breaking from each other's precipices huge fragments, or rending each other asunder, till, losing the former equilibrium, they fall over headlong, lifting the sea around in breakers, and whirling it in eddies."

The dangers of Bellot Strait are not just theoretical. In 1942, during the second overall and first west-to-east transit of the Northwest Passage, Henry Larsen nearly lost his ship, *St. Roch*, in Bellot. "As the ice came pouring in behind us, there was nothing else to do but crash into it and attempt to drift through," wrote Larsen. "This we did; the strong current causing large whirlpools in which large cakes of ice spun and gyrated. Many times we thought the ship would crack like a nut under the pressure."

More recently, in 2018, two Argentineans aboard a modern thirty-five-foot aluminum-hulled sailboat called *Anahita* arrived at the eastern end of the strait when it was still filled with pack ice.

While waiting in a small estuary for the strait to clear, they tied their boat to an oval-shaped floe that was sixty-five feet long and twenty-five feet wide. The floe's draft was deeper than the ship's, and the vessel's skipper, Pablo David Saad, a fifty-year-old professional sailor, figured that if the wind blew them toward shore, the floe would ground first, keeping *Anahita* afloat on the outside. While Saad was asleep below deck, his mate, Dario Ramos, a fifty-five-year-old youth hockey coach with minimal sailing experience, failed to notice that the floe had drifted out of the estuary into the ice-choked main channel. When Saad stepped back into the cockpit after a few hours' rest, he looked up just in time to see a fin of ice bearing down on the boat. Before they could react, the missile struck *Anahita*'s starboard side, lifting the ship slightly out of the water and knocking both men off their feet. Ramos remembers the sound of an aluminum can being squeezed and then *pop pop pop*, followed by the gush of water pouring inside the boat.

Saad dashed below to try to locate the breach, but the impact had deformed the hull, making it impossible to remove the floorboards for access to the bilge. He tried to find the hole from the outside, but it must have been deep and big, because within a few minutes, water covered the cabin sole above the tops of his boots. A few minutes earlier, Saad had been asleep in his bunk, all warm and cozy. Now his boat was going down. Fast. He gave the command to abandon ship; then he put out a mayday on their VHF radio and activated the ship's EPIRB, a satellite locator beacon.

As Saad and Ramos hustled to move survival gear—their dinghy, a life raft, flares, a first aid kit, an inReach satellite texter, a handheld VHF radio, warm clothing, food, and water—off *Anahita* onto the ice floe, the boat began to sink. After they took care of the essentials, Ramos filled a bag with some souvenirs he had purchased along the way for his five children: bone earrings, decorative pins, some crayons, and a pair of gloves. He was about to take the bag out to the ice floe when he looked around the interior of the boat, which was Saad's home. Saad had not grabbed a single personal item, and Ramos knew that in a few minutes, it would all be lost. So he left behind the souvenirs out of respect for his friend, and headed out to join him on the ice.

While Saad did his best to ignore *Anahita* as it sat heavily in the water next to the floe, Ramos held a line attached to the ship with which he tried to heel the boat so it would sink more slowly. Eventually, Saad told him to let go, and a few minutes later, they heard a loud *whoosh* as all of the air inside the cabin was forced out through the companionway. The stern rose, the propeller and rudder pointing unnaturally skyward. Then *Anahita* disappeared below the surface and sank to the bottom of Bellot Strait in one hundred fifty feet of water.

Three other private yachts anchored nearby heard *Anahita*'s mayday and tried to come to the rescue, but there was too much ice and they risked losing their own ships if they entered the grinding pack. In Iqaluit, the Canadian Coast Guard, which had received *Anahita*'s EPIRB signal, alerted a cargo ship called *Claude A. Desgagnes*, which had been steaming north up Prince Regent Inlet. The captain immediately altered course for Bellot Strait.

In the meantime, the ice island spun in dizzying circles and crashed against other chunks. Saad communicated via their handheld VHF with the skipper of another Northwest Passage–bound yacht called SV *Atka*, which in turn relayed messages from the stranded sailors to the coast guard icebreaker HMS *Henry Larsen*. When *Claude A. Desgagnes* arrived a few hours later, the captain launched two tugs, which valiantly fought their way into the ice to try to reach the stranded men. Soon Saad and Ramos could hear the tugs' engines and smell burned gasoline. After hours of daring maneuvering through the swirling pack, one of the tugs got within two hundred fifty feet before it was caught in a fierce current in which it nearly suffered the same fate as *Anahita*. "We're sorry," the sailors of the tug relayed via radio, "but it's too dangerous and we have to give up for now."

Hours later, Saad was pacing back and forth across the floe when Ramos pointed and cried out, "Pablo, look." On a nearby floe, about sixty feet away, an adult polar bear was walking across the ice in their direction. In Greenland, the men had read a brochure about polar bear safety, and they'd heard stories about surprise attacks resulting in Inuit being maimed or killed, and about a young girl who had shot a polar bear point-blank—and it hadn't even flinched. Numerous locals had advised the men to buy a gun, but they'd resisted, doubting they would need one and being reluctant to kill a bear in any case. Now they were facing the consequences of their decision to remain unarmed.

To show the bear that they were the owners of their floe, they spread their things around and used the oars to stand their dinghy upright and make it look like a giant person. Then they walked to the edge of floe closest to the bear and waved their arms and jumped up and down. These actions had the desired effect, and the bear stopped moving toward them and instead began to circle, hopping easily from one ice chunk to the next, maintaining a radius of about two hundred feet.

The bear prowled like this for the next two hours until a thick fog rolled in and the visibility dropped to about ten feet. At that moment, Ramos knew that at any second the bear could appear as if from behind a curtain, giving them no time to react. He imagined his death, not in the abstract but viscerally as he pictured the bear tearing him to pieces and eating him alive. He turned to Saad and said, "*Estamos muertos*" ("We are dead").

Saad said that if the bear attacked, he would shoot it with a flare. Ramos replied that if it came to that, he would sacrifice himself. After all, it was his fault that they'd lost the boat and found themselves in this predicament. And something about this irrational suicidal thought snapped him out of his fatalism. He said to himself, *"No es el día de mi muerte"* ("It is not the day of my death").

Using *Atka* as an intermediary, Saad sent out an urgent message: The polar bear was likely to attack any moment and they had nothing with which to defend themselves, save for a few parachute flares and a rope on which they had tied some heavy items to swing over their heads like a mace. The captain of *Henry Larsen* responded that it couldn't get to their position in time, but he would send a helicopter.

Forty-five minutes later, the Argentineans heard the *whop whop whop* of their approaching salvation. The chopper hovered just above the ice without touching down, and they jumped in, leaving all of their equipment behind—but remaining unhurt and alive.

| | |

When *Erebus* and *Terror* sailed past the western entrance of Bellot Strait in August 1846, this remarkable waterway was still unknown. Fifteen years earlier, it had eluded Franklin's friend and predecessor, John Ross, who had spent four years searching for the Northwest Passage in Prince Regent Inlet and the Gulf of Boothia from 1829 to 1833. The irony is that Ross sailed past Bellot Strait, lost his ship, then walked by the strait again, and yet failed to note its existence—despite Inuit telling him it was there.

Ross's first expedition into the Arctic in 1818 resulted in the discovery of a mountain range that didn't actually exist. A year later, while proving this, Parry also determined that the channel he had followed nearly to the edge of the Arctic Ocean was encircled by impenetrable ice to north, south, and west. Upon his return, Parry told the Admiralty that future expeditions should look farther south than the seventy-fourth parallel, which he had followed, and stick to shorelines, where game, driftwood, and antiscorbutic plants could be gathered.

The most obvious place to search next was the bay tucked between Baffin Island and Somerset Island. Parry had briefly explored these waters in 1819 and named the area Prince Regent Inlet. Barrow and others had long suspected that somewhere deep in this bay there might be a strait leading west into the central Arctic, so in 1824 he sent Parry (with Francis Crozier serving as a midshipman) to find out. After wintering in a bay along the west shore of Baffin's Brodeur Peninsula, Parry's two ships, *Fury* and *Hecla*, were pushing deeper into Prince Regent Inlet when they were nipped in the pack ice off Somerset Island. *Hecla* managed to escape, but *Fury* became trapped between pitching six-foot-thick plates of ice that stove in her hull, snapped off her rudder, and drove the broken ship up onto a beach. Before moving the crew onto *Hecla* for the trip back to England, Parry unloaded all of the ship's provisions, including dozens of barrels of flour, biscuit, and salted meats—enough supplies to last a crew for years—onto a cliff-lined spit of land that would henceforth be known as Fury Beach.

No other Europeans would venture into these waters again until 1829, when John Ross and his nephew James Clark Ross, backed by a wealthy British gin distiller named Felix Booth, set off from England with a crew of twenty-one aboard a 165-ton motor sailer called *Victory*. The ship, built in 1826, had originally been commissioned to serve as a ferry between Liverpool and the Isle of Man. She drew only eight feet and was equipped with a side-mounted, steam-powered paddle wheel that could be lifted out of the way when the ship encountered ice.

It was another mild summer in the Arctic, and Ross crossed Baffin Bay and Lancaster Sound in less than two weeks. After stopping at Fury Beach to take on some of the abandoned supplies, Ross and his crew continued south down a wide bay. By the end of September, having covered another hundred eighty miles, Ross brought *Victory* into a small cove on the western shore of the Gulf of Boothia that he named Felix Harbour. As they stripped down the ship for the coming winter, Ross ordered the engine and its paddle wheel dismantled and dumped onto the beach. The contraption had given them no end of grief since leaving England—it would power *Victory* no more. Inuit later salvaged the valuable iron from the engine and used it to make to make knives and spearpoints. And in honor of this bounty, they later named the place Qilanartot, which means "joyful beach." Ross's impression of the area was decidedly less sanguine. In his log he wrote: "The landscape was one indiscriminate surface of white; presenting, together with the solid and craggy sea, all equally whitened by the new snow, the dreariest prospect that it is possible to conceive."

In January, a lookout aboard *Victory* spotted a small group of Inuit on the ice about a mile from the ship—the first people the expedition had encountered since leaving Greenland. Ross and some of his men approached them, calling out *"tina ti ma,"* a friendly Inuktitut greeting. A large party emerged from hiding places in the ice, and Ross counted thirty-one Netsilingmiut Inuit, all males ranging in age from teens up to mid-sixties. Two members of the party sat apart, confined to sledges. One, Ross noted, was a man named Tulluahiu, who had lost his lower leg in a polar bear attack.

After the Inuit called out *"tina"* in return, Ross and his men dropped their guns, and the Inuit reciprocated with their spears and knives. As both parties embraced, Ross stroked the Inuit's fur clothing in admiration, which produced "great delight, expressed, on all hands, by laughing, and clamour, and strange gestures."

The Netsilingmiut Inuit were known as the seal people, and this was their first encounter with white men (*kobluna*). Various accounts of this first meeting with Ross and his men would become enshrined in Inuit lore. According to a woman named Ohotktoo, Aglikuktoq ("he who is unclean") was out sealing when his dog began to strain in its harness. Following its lead, he came upon "some strange men who were walking near a strange kind of house with smoke coming out of it." Aglikuktoq, frightened, ran back to his village "faster than his dog," which caused great mirth amongst his friends. Aglikuktoq shared his story with a shaman who "gathered his charms and his white cloak of Caribou-belly hide," which he hung on a wall and then slipped behind to converse with the spirits. When he emerged, he proclaimed that the strangers "were friendly and would welcome a visit from the Inuit."

James Clark Ross, who had learned some Inuktitut while overwintering with Parry in Igloolik a decade earlier (the same trip on which Crozier got his nickname Aglooka), invited three Inuit men back to the ship. Once aboard, they "showed abundant signs of wonder" at the various items on display—especially the images of themselves reflected in a large mirror. But they "never once showed a desire to possess themselves of anything," wrote John Ross, "receiving, merely, what was offered, with signs of thankfulness that could not be mistaken." While the ship's carpenter, Chimham Thomas, worked on fashioning a wooden peg leg for Tulluahiu, Ross offered his new Inuit friends some preserved meat. One took a bite and said that it was "very good." But when pressed, the man admitted that he didn't actually like it. Ross gave two of the men pieces of iron, but to Aglikuktoq, he offered a knife with a curved blade called an ulu, which he had obtained in Greenland and which he perhaps did not realize was a tool used only by women. Aglikuktoq, a great hunter, was offended. "I am not a woman," he said, pointing to a handsaw that he wanted instead. Ohotktoo later reported that "the eshmuta [chief] became angry and he took the [ulu] back and chased Agliluktoq out of the house emptyhanded."

The next day, John Ross and some of his men followed the Inuit to their camp on a spit of land sticking out into the bay. Here, the

British sailors met a multitude of women, children, and elders. The settlement was made up of a dozen igloos with long, crooked entryways and small side chambers for the dogs. Single-family igloos were ten feet in diameter, two-family structures fifteen feet across. Oval pieces of clear-water ice served as windows, and raised platforms covered in furs took up a third of the interior. Here, the natives stripped naked and slept together, sharing body heat. The snow houses were lit and warmed by seal-oil lamps, with wicks made of moss or cotton grass. Stone bowls hung over these small flames to cook seal, deer, caribou, and fish. "It was for philosophers to interest themselves in speculating on a horde so small, and so secluded, occupying so apparently hopeless a country, so barren, so wild, and so repulsive," wrote Ross, "and-yet enjoying the most perfect vigour, the most well-fed health, and all else that here constitutes, not merely wealth, but the opulence of luxury."

Ross understood that if there was a Northwest Passage leading west from the Gulf of Boothia, his indigenous friends would know its location. After repeated inquiries, he was introduced to a middle-aged Inuk named Ikmallik, a "man of unusual power and stature" who stood five foot ten and commanded universal respect. In his cabin aboard *Victory*, Ross placed a chart in front of Ikmallik, handed him a pencil, and asked him to fill in the blank spaces. Despite the facts that Inuit did not utilize a written language at this time and that Ikmallik had likely never held a writing instrument before, he painstakingly added hundreds of miles of coastline along the Boothia Peninsula with uncanny accuracy. His drawings included numerous bays, inlets, rivers, and lakes, and his knowledge of the area's geography was so extensive that Ross nicknamed him "the Hydrographer." Later, a woman named Tiriksiu added even more details to Ross's map, including many islands as well as camping and hunting spots.

Unfortunately, despite the invaluable information provided by their new friends, the kindness of Ross and his men didn't last. "In return for these services, [the Inuit] were at first caressed and treated with all the urbanity and kindness of the equal," wrote William Light, a steward who kept a detailed journal and later wrote a highly critical unauthorized account of the voyage. "However as their stock of information declined . . . a treatment was adopted towards them, which did not stop at mere unkindness and incivility, but it degenerated at last into downright cruelty and inhumanity" as if "they were but a degree removed from the bears of their native land."

According to Light, a month after Inuit first visited the ships, a group including women cradling babies under their furs showed up on a blustery day when the ambient temperature was −45 Fahrenheit. The women waited in the lee of a snow wall beside the ship for six hours, hoping to be invited aboard. But Ross ignored them and offered not "the slightest sustenance or support" until they finally gave up and headed home. The incident caused Light to note with great disappointment just how little so-called civilization actually humanizes the heart or calls "all the finer charities into active play."

That spring, James Clark Ross, Second Mate Thomas Abernethy, and three other crewmen set off with two small sleds, a lightweight skin umiak, eight dogs, and three weeks' worth of food to explore the lands west of Felix Harbour. Following an old Inuit route, they crossed the Boothia Peninsula and continued across a frozen strait to a landmass that Ross named after King William IV. At Cape Felix, on the northern tip of this land, Ross found masses of ice heaped on the shore "in a most extraordinary and incredible manner: turning up large quantities of shingle before them, and in some places, having travelled as much as a half mile beyond the limits of the highest tide-mark." Though he didn't know it, Ross had come face-to-face with the great ice stream that has poured down the McClintock Channel from the Beaufort Sea since time immemorial.

Continuing southwest along the bleak, ice-heaped coastline, they stopped at a nondescript headland where they constructed a six-foot-high cairn and named the place Victory Point. To the west, Ross spied two distinct headlands that he prophetically named after his friends Sir John and Lady Jane Franklin. He had no way of knowing, of course, that these very promontories would be the closest landmasses to the spot where the Franklin expedition would become hopelessly trapped in the ice sixteen years later.

On the return trip, Ross traced the eastern coastline of King William as they worked southward, hoping to determine if this landmass connected to the mainland. Conditions, though, were deteriorating quickly. Their ship lay two hundred miles away on the opposite side of the Boothia Peninsula, rations were running low, and of the original eight dogs, only two could still pull. When the coastline veered southwest, away from *Victory*, Ross abandoned the reconnoitering and headed back to the boat.

Back aboard, the elder Ross used his nephew's sketches and survey notes as the basis for a chart that would later show King William as a peninsula attached to mainland Canada. Deep in this imaginary backwater, he even drew several headlands and named them after English poets—a cartographic error that would later have fatal consequences for Franklin and his men.

The ice never relented that summer of 1830, and *Victory* moved only four miles before it was sealed in for another winter. In the spring, Inuit left the area in pursuit of better hunting elsewhere, and James Ross set off on another sledging expedition in search of the magnetic north pole. Scientists had understood for centuries that the geographic and magnetic poles were not located in the same place and that the Earth's magnetic field varies in intensity and moves from year to year. Latitude, longitude, and the geographic coordinate system used to define the location of points on the surface of the Earth (including the North Pole) had been figured out by the mid-eighteenth century. But in 1830, geomagnetism was still poorly understood, and until the exact location of the magnetic north pole could be located (and its movement over time measured), no one would know exactly where their compass needles were pointing.

James Ross found his way to the magnetic north pole by using both a compass and an instrument known as a dip circle, a type of magnetometer that measured the angle between the Earth's magnetic field and the horizontal plane. On June 1, 1831, at a random spot on the Boothia Peninsula about eighty-seven miles northwest of Felix Harbour, the needle on Ross's dip circle pointed straight down, while the horizontal needle on his other compasses wouldn't move at all. In celebration, Ross and his men erected a large cairn on the spot, planted a British flag, and claimed the magnetic north pole and its surrounding territory for Great Britain and King William IV. Later, in his official report, Ross wrote that his only regret was that they hadn't been able to construct "a pyramid of more importance, and of strength sufficient to withstand the assaults of time and of the Esquimaux. Had it been a pyramid as large as that of Cheops, I am not quite sure that it would have done more than satisfy our ambition."

Later that summer, *Victory* rode an unusually high tide over a gravel bar into Victoria Harbour, from which it would never escape. By late winter 1832, Ross realized that he had no choice but to abandon ship. In April, he directed the men to begin ferrying loads of supplies along the twisted contours of the frozen coastline leading north to

Fury Beach. There, he reasoned, they could resupply with Parry's stores and make a push for Lancaster Sound, where he hoped they might be spotted by a passing whaler. But after a month of brutal labor, they had established only two supply depots, the farthest a mere twenty miles north of Victoria Harbour. On May 29, 1832, Ross gave the order to hoist the ship's colors and nail them to the mast. As he raised a toast of Booth's gin, the men removed their hats and everyone called out, "Goodbye, *Victory*."

"It was like the last parting with an old friend," wrote Ross.

The mountain of supplies the men schlepped included Ross himself, who rode like a king atop one of the boats that they dragged over the ice. By June 8, bedraggled, footsore, and malnourished, they had advanced only forty miles, and Ross realized that they would have to abandon the longboats and make a desperate all-out push for Fury Beach with three small sledges and whatever they could carry on their backs. Over the next three weeks, they battled blizzards, waded fast-flowing rivers, and tromped through knee-deep water pooled on top of the melting ice. At night, they slept side by side in trenches dug like shallow graves in the snow. By the time they straggled onto Fury Beach on July 1, the half-starved men attacked Parry's old provisions—against Ross's orders. As he put it in his journal, their bowels "suffered smartly for their imprudence."

A month later, the ice in Prince Regent Inlet broke up suddenly, and the men set off aboard three of Parry's longboats, which they had repaired in the interim. Sailing when possible, rowing when not, and dodging rockfall while camped below crumbling cliffs, they worked their way through the pack to the northern tip of Somerset Island. In early September, Ross climbed a promontory from which he gazed out over an unbroken field of ice blanketing Lancaster Sound. They waited another three weeks for the ice to break up before giving up and heading back to Fury Beach, where they spent their fourth winter in a drafty hut built from stones and hunks of ice. They called it Somerset House.

When they finally set off again on July 8, 1833, they found better conditions and reached open water in Lancaster Sound a month later. They were sailing east in Parry's old longboats, near the entrance of Navy Board Inlet, when they spotted a whaling ship. After coming alongside, Ross asked the name of the vessel. "The *Isabella* of Hull," said the mate, "once commanded by Captain Ross." When Ross responded that he was none other, the sailor, "with the usual blunder headedness of men on such occasions," informed the captain that he had been dead for two years.

Ross returned to England a conquering hero, his reputation finally redeemed, the Croker Mountains now (mostly) forgotten. He had not found the Northwest Passage, but he had surveyed unknown lands and his nephew James Clark Ross had discovered King William Island and the magnetic north pole. More important, though, Ross senior had brought his men through four winters in the Arctic and lost only one man, the carpenter Chimham Thomas, who had built that wooden peg leg for Tulluahiu. Thomas died from scurvy during the last winter at Somerset House, shortly after carving a wooden chess set for his mates.

The Admiralty offered the entire crew double pay, even though they had been on a privately financed expedition. And if having geography named after you has value, Felix Booth got his money's worth: To this day the Gulf of Boothia, Boothia Peninsula, Felix Harbour, and Cape Felix bear the name of the gin magnate who financed the mission.

When asked by a parliamentary committee if a Northwest Passage would have any practical value, were it ever found, Ross replied that it would be "utterly useless." But that didn't mean that there was no value in searching for it. "Perhaps the only satisfaction that can ever be derived would be, that there is, on a piece of paper, a black line instead of a blank," wrote Ross. "But of such imaginary joys does human happiness full often consist: and what matter, if even less than this, the anatomy of a fly's toe, or whatever else, will serve to make men happy and proud of themselves?"

One glaring omission in this perspective was the obvious fact that many of these "black lines" were already well-known by Inuit, who had lived and prospered for thousands of years in the lands that the British were attempting to "discover." While exploring the Boothia Peninsula, John Ross and his nephew had gone ashore repeatedly to build cairns and monuments upon which they raised the British flag, claiming the land as far as they could see for the king of England.

During the first winter in Felix Harbour, a group of Inuit had built their igloo a bit too close to *Victory* for Ross's comfort. He confronted them and explained that they were trespassing on crown land. The Inuit, of course, didn't have the vaguest idea what he was talking about. As Berton writes in *The Arctic Grail*, "It had not occurred to [the Inuit] that anybody owned the land any more than anybody owned the sea or the air. . . . No landlord or tax collector ever came to their door. The concept of permanence, of real estate, of tithe, title and deed, was foreign to them."

Of course, Ross and his men wouldn't have survived their four-year epic in the Gulf of Boothia if it hadn't been for the help of the people that Ross called the "rude and savage tribes of the world," who taught him and his men how to survive in the Arctic. One simple example was the immense benefit of eating "Greenland food"—the seafood, seal and whale oils, and fatty meats that helped ward off scurvy in a land devoid of fruits and vegetables. Indeed, after their first winter in Felix Harbour, some of Ross's men showed obvious signs of the disease. But the symptoms—malaise, bleeding gums and hair follicles, rashes, bruising, and the reopening of old wounds—disappeared within a couple weeks of their eating seal blubber and hundreds of pounds of salmon, for which they had traded three knives to the Hydrographer.

Unfortunately for Franklin, the Admiralty failed to heed this lesson and many others. The king and the public lavished Ross with praise and commendations, but Ross's old nemesis wasn't having it. In an essay in the *Quarterly Review*, Barrow called the *Victory* expedition "ill-prepared, ill-conceived . . . and ill-executed" and called his old adversary "utterly incompetent." As Berton puts it in *The Arctic Grail*: "The navy's failure to analyse the very real strengths of his [Ross's] expedition was to cost it dear in the years to come."

|||

Back in Peel Sound, *Polar Sun* skimmed through the ice-free water with ease. We hadn't seen a speck of the frozen stuff all day, and the crew was enjoying a reprieve from constantly scanning the horizon, when, out of nowhere—*WHUMP!*—we hit something in the water. The boat shuddered as Jacob pointed astern at a massive white object that flashed just below the surface.

What the fuck is that?

The thing rolled over, revealing a broad face and a dark eye.

"Beluga," said Jacob.

I scanned the water for blood but couldn't see any, and a few seconds later, the beluga was gone. Jacob lit a cigarette as I sat there muttering to nobody, "We hit a whale. We actually just hit a fucking whale."

Twelve hours later, at dawn of our third day out from Beechey Island, I pulled in a message from a forecaster at the CIS. "As of 13:22z [Zulu], it looks like that area of 9 tenths (with the trace of old) has drifted eastward, nearly pinching the coast at N70 11."

Translation: *You guys are screwed.*

We had been flying down Peel Sound now for two days, gunning for a slot of open water in James Ross Strait between Matty Island and the Boothia Peninsula. This choke point was the crux of the entire Northwest Passage. If we could scrape through the ice here, there was a good chance we'd have Nome in the bag.

But sometimes, the slot never opens at all. Such had been the case the year before. Now, according to our contacts at the CIS, the northwesterly that we had ridden down Peel Sound had also pushed miles of ice directly into our path. Worse yet, a southeast gale was brewing. If we pressed on, we risked pushing into nine-tenths ice, with thirty knots of wind in our teeth.

I, of course, couldn't stop thinking about *Erebus*, *Terror*, and the Argentineans aboard *Anahita*, and my mind filled with visions of *Polar Sun* getting popped like a zit by Ross's "mountains of crystal hurled through a narrow strait by a rapid tide."

ı|ı

Franklin was a voracious reader, and the library aboard *Erebus* featured some twelve hundred books and journals. These included religious texts, novels by Charles Dickens, and issues of the satirical weekly magazine *Punch*. Amongst this floating collection was the *Narrative of a Second Voyage in Search of a North-west Passage*—Ross's account of the *Victory* expedition.

This book included that telltale map showing King William Island as a peninsula connected to the Canadian mainland—rather than the island it actually is—and it's fair to assume that when Franklin sailed out of Peel Sound, he used Ross's chart as his guide. And if he had, we know it showed that the only way to reach Coronation Gulf (which he had explored twenty-five years earlier during his Coppermine River expedition) was by leaving King William Island to port.

On his sledging journey in the spring of 1830, James Clark Ross had discovered an excellent harbor on the northeast shore of King William Island. He reported that its entrance was two miles wide, the bay "divided in the middle by an islet that would effectually cover it from the invasion of heavy ice." The fact that *Erebus* and *Terror* didn't shelter in this harbor, or one like it, suggests that Franklin might have thrown down all his chips as the summer of 1846 drew to a close. Off the northern tip of Cape Felix, they were only about a hundred twenty sailing miles from known territory on King William Island's southwest shore. This final stretch of uncharted waters—the last piece in a geographic puzzle that the British had been working on for centuries—might have proved irresistible to Franklin.

I suspect that as the ships approached the north shore of King William Island, *Terror*'s ice master, Thomas Blanky, who had served with Ross on the *Victory* expedition, climbed the ratlines into the crow's nest, where he spied leads of open water to the southwest. Blanky might also have seen the low, flinty land to the south—a place with which he and his crewmates would soon become bitterly acquainted.

If the wind blew from the southern quadrant, it might have pushed the ice back from shore, opening a corridor of open water. But as we would soon learn, southerlies are often followed by strong northerlies, which might have caught *Erebus* and *Terror* on a lee shore when sheets of marauding pack ice blew down on King William from the McClintock Channel.

When the ice closed around the ships, cementing them in place, on September 12, 1846, everyone aboard would have understood that they were now in a very different position from the one that they had faced the previous winter at Beechey Island. Out in the open ocean, ice behaves differently than it does in a sheltered bay. In Victoria Strait, squeezed between the ice flowing down the McClintock

Channel and the north shore of King William Island, the ice was subject to unimaginable forces that could cause the surface to buckle, split apart or explode into pressure ridges thrusting multiple stories into the air. If one of these had happened to erupt under or beside the ships, *Erebus* and *Terror* could have been swallowed whole, tossed out onto the surface of the ice, or simply cleaved in two.

Blanky, who had already spent four winters in the Gulf of Boothia with John Ross, must have understood, better than anyone, that they would be lucky if the ships weren't crushed like toothpicks during the coming winter.

We are left to wonder, of course, why someone who had barely survived those four winters in the Arctic with John Ross would choose to go back into the Northwest Passage. Perhaps the intervening years had softened the edges of his memories, allowing him to forget the suffering and misery and to focus instead on the beauty and otherworldliness of the place. But Blanky certainly knew the stakes involved. Before departing the Whale Fish Islands, he had written to his wife, Esther, that if he was not home by the spring of 1849, "you may anticipate that we have made the passage, or are likely to do so; and if so, it may be from five to six years,—it might be into a seventh—ere we return."

Blanky must have understood that it could be a one-way trip. Shortly before departing England, he had drafted his last will and testament, which was signed and witnessed by two of his fellow crew members.

|||

After reading the ice report, I called a war council in the cockpit. Should we press on into James Ross Strait and take our chances with the nine-tenths ice and the incoming storm? Or would it be more prudent to duck into a nearby harbor called Pasley Bay, where Henry

Larsen had wintered in 1941–1942, and wait for the storm to blow through?

Apparently, I wasn't the only one having visions of our fiberglass hull being crushed like that of *Anahita*. The decision to bail into Pasley Bay was easy and unanimous.

We anchored a few hours later in the south arm of the two-mile-wide bay. When we cut the motor, the silence was powerful. I paused and gazed out at the clouds reflected on the water like an Impressionist painting. On Beechey Island, cliffs and rolling hills drew the eye, as they had in Baffin too. But here, nearly three hundred miles to the south, the country had flattened out into a low monochrome moonscape. Beechey had felt grim and forbidding, but compared to Pasley Bay, it was a goddamn paradise.

I popped below and dug a bottle of vodka out of the icebox. A few minutes later, the crew gathered in the cockpit for "captain's hour." Renan, Rudy, Jacob, and I sipped spirits from coffee mugs while Ben drank tea. The sun bathed the bay in soft orange light and the ice chunks that bobbed in the water all around us glowed like jack-o'-lanterns.

Renan, who is insatiable when the light is good, spent the next few hours launching one drone flight after another. On the second flight, he passed me the monitor, which showed bergy bits and floes of pack ice dotting Pasley Bay like geese scattered across a pond. Farther west, on the other side of a narrow peninsula separating us from James Ross Strait, larger sheets of pack ice loomed.

I was eyeing the lazy floes that swirled in the water all around us when, without warning, a car-sized hunk spun in from the north and snagged on the anchor chain. It hung up for a minute, pressing our bow down, before popping loose and slamming into the hull.

Ben sat on the other side of the cockpit sipping his tea. When he looked in my direction, I asked him if he was concerned.

"What could happen?" he replied.

He was right. He was always right. We had a fleet of drones. We were holed up in a sheltered bay out of the ice traffic, kind of like if we had pulled over into a rest stop off the highway. The water was dead flat. The floes oscillated around us in our little pond like ice cubes in a vodka glass. *Relax,* I told myself. I looked over at Ben. He smiled placidly. *Everything is fine.*

I went below and fell into a deep sleep. I'm not sure how long I was out before I woke up to Renan shaking me.

"Hey, Mark! MARK! You need to get up," he said. "We're fucked."

PART 3

INTO THE ICE

CHAPTER 9

Franklin's Tomb

Date: July 19, 2022
Position: Gjoa Haven, King William Island

The more I got to know the Northwest Passage, the more I came to realize that it isn't really a passage—or passages—at all. It might be more accurate to call it an archipelago, but even this description is deceptive, conjuring as it does a whiff of the exotic and creating a vision in our minds' eyes of a collection of islets confined neatly in the corner of a map.

More than anything, it's the sheer size of the Canadian Arctic that defies the imagination and makes it almost impossible to fathom its dimensions. The total area, when you combine land and sea, is about two million square miles, equivalent to more than half of the continental United States. Within this vast territory lies a collection of 36,563 islands, ninety-four of which are classified as "major" (which seems a bit arbitrary but means larger than fifty square miles). Three of these—Baffin, Victoria, and Ellesmere—are amongst the ten largest islands in the world. Baffin Island alone is approximately the same size as the UK, Portugal, Austria, Ireland, and Switzerland combined. And the slightly smaller Devon Island is the largest uninhabited island in the world.

Interwoven through it all is a saltwater maze of twisting, ice-choked channels and straits, shallow bays and shoals—all so remote that 90 percent of it has still not been surveyed to modern standards. Explorers in the nineteenth century invariably referred to the confounding geography as a "labyrinth," and even today, the Canadian Arctic is a place that continues to defy those in the south who seek to understand it.

The eastern edge of this region is bound by a long rampart of mountains, known as the Arctic Cordillera, that gradually washes out to the west. By the time one reaches King William Island, where so much of the Franklin story plays out, the landscape is Midwestern: a pancake-flat plain of boggy tundra and gravel-strewn, glacier-strafed badlands across which are scattered thousands of unnamed lakes, ponds, and tarns. In winter, which still spans most of the year in this part of the world, the channels, bays, and inlets between the islands—not to mention all of the fresh water—freeze hard as rock. And when they are blanketed under snow and ice, the distinction between land and sea becomes almost imperceptible—and irrelevant.

As more and more expeditions pushed into this frozen kingdom in the nineteenth century, only to return (or not) with tales of ships crushed in the ice and men driven to the last resort—but also with stories of incredible heroism and rescue—the fabled shortcut to the Far East gradually evolved into something so much more than a trade route for moving expensive goods from one world power to another. By the time the Franklin expedition set off in 1845, the Northwest Passage had become an idea as much as a place, a sort of mythical Holy Grail that gave its antagonists a chance to personify the stuff of Arthurian legend: chivalry, romance, bravery, and duty—all in the service of slaying dragons and creating order out of chaos.

When viewed through this prism, the Northwest Passage reminds me of what Mount Everest represents to many climbers and armchair mountaineers today: a prize that shines so brightly, it can blind its seekers from seeing its truth. Everest, much like the Northwest Passage, isn't a trophy; it's a place of deep spiritual power that has always been worshipped by those who live in its shadow. To the Tibetans, it's Chomolungma, "goddess mother of the world," and to Nepalis, Sagamartha, "goddess of the sky." The whole reason that I had chosen to venture back into the Arctic in the first place was because I wanted to have an experience that might resonate as deeply as the one I'd had on Polar Sun Spire as a young man. But I'd come to realize that there was something alien about moving through these waters aboard a modern sailing yacht with a fixed timetable, requisite fuel stops, and a need to always keep pushing as fast and hard as possible. Ensconced within *Polar Sun*'s canvas enclosure or in my bunk deep inside the ship, I often felt removed from the landscapes, the wildlife, and the people who lived in the communities through which we passed. It was obvious that the best way to travel through this realm was by foot and paddle, as Inuit have for millennia with their dogsleds and kayaks, moving nimbly from one community and hunting ground to the next. And it saddened me that I might be missing the essence of the place that I so badly wanted to experience.

Part of it was also that I kept thinking about an essay I'd found early on when I was researching the feasibility of this voyage. It was written in 2014 by a well-known high-latitude sailor named John Harries and titled "Enough with the Northwest Passage, Already." "This whole transiting the Northwest Passage . . . in a yacht is getting out of hand," he wrote, "and many (maybe most) of the crews and boats trying it shouldn't be anywhere near the place." In Harries's estimation, most Northwest Passage sailors were in it so they could "get the tick"—just like your stereotypical Everest climber. "For these crews it's not a voyage with the associated appreciation and learning about the surrounding lands and seas, it's a mad dash, starting way too late in the season, just to say they did it."

I thought it a bit presumptuous for Harries to imagine that he knew what motivated people in their private hearts, but was I really all that different from the hordes he was describing? I had, in fact, climbed Everest, and now here I was apparently going for the sailing equivalent. What exactly did that say about my motivations? This voyage was a dream come true for me, but it was also expensive, had no intrinsic purpose, and certainly wasn't serving the locals in any way. As my first wife had once so eloquently put it when I returned home from a long climbing expedition: "Who's getting anything out of this other than you?"

And let's not forget that, at least to a certain degree, it all revolved around a search for Franklin's tomb, which was the real reason that I was able to afford to do any of this. National Geographic, of course, wanted to focus its program on the overland search with Tom Gross. And Tom was dead set on conducting his expedition in July, when he knew the weather and conditions would be ideal for a ground search. And since this meant arriving in Gjoa Haven a full month before the ice typically breaks up in James Ross Strait, I did the only thing that was practical under the circumstances. I left *Polar Sun* in Ilulissat, Greenland, with Ben for a few weeks, and while Hampton, Tommy, and Taylor headed off on their next adventures, Renan, Rudy, and I flew to King William Island, leapfrogging hundreds of miles ahead of our sailing itinerary.

So much for an organic sailing voyage fueled by the purity of an explorer's heart.

But who knew? Maybe we'd actually find Franklin. And at the very least, I hoped that stepping off the boat and spending two weeks exploring King William Island via four-wheeler would give me the chance to immerse myself in the landscape in a way that just wasn't possible from the confines of *Polar Sun*.

And this is how, a full month before we made that fateful turn into Pasley Bay, I found myself huddled over a map spread across the kitchen table in a Gjoa Haven Airbnb that Tom had rented as the staging ground for our overland expedition.

"I've been thinking *a lot* about the search," Tom declared, as he drew my attention to a spot on the map labeled Collinson Inlet, "and I have a really good feeling we're going to find the tomb. I've never had this feeling as strong as I have it now."

Collinson Inlet lies just south of Victory Point, where Graham Gore and Charles Des Voeux left that record in 1847. The bay's mouth is framed by Franklin Point to the west and Cape Jane Franklin to the east, both of which James Clark Ross had prophetically named in 1830 during his sledging expedition to Victory Point. From where we stood, it was about eighty roadless miles to the head of the inlet, which lay clear on the other side of the island.

Our plan was to get there by way of all-terrain vehicles, aka ATVs, four-wheelers, or "bikes," as the locals call them—a fleet of which sat outside in the driveway. There were only a couple available for rent in Gjoa, so National Geographic had underwritten the significant cost of purchasing six of these machines and shipping them to Gjoa Haven. (At the end of the expedition, we would donate them to a local nonprofit.) All of our provisions and equipment—plastic crates packed tight with canned goods, pilot biscuits, bags of cashews, summer sausages, peanut butter, candy bars, and precooked kielbasas; plus tents, sleeping bags, cookstoves, and jerry cans of gasoline for the bikes—would be strapped onto two wooden sleds called komatiks that we'd drag along behind us.

Renan had enlisted a veteran cameraman named Matt Irving to assist him with shooting this part of the trip, and Tom had brought his wife, Eileen; his daughter, Pam, the member of the Canadian

Parliament; and her partner, Devon Oniak, a heavy-equipment operator from Cambridge Bay. Tom's old buddy Jacob Nevekitok Keanik, whom I met for the first time in that Airbnb and who would later accompany us on the *Polar Sun*, would serve as our guide. This iteration would be Tom and Jacob's seventh expedition together in search of Franklin's tomb. And what I hadn't realized until now was that Jacob had been in the back seat of that plane on the day Tom and Darcy King had spotted the "stone house" near Collinson Inlet. But when I asked him what he'd seen, he shrugged and didn't say anything. Later, Tom told me that he thought Jacob had been airsick at the time from all the sharp bank turns the plane had made over the site.

Gjoa Haven was similar to many of the small hamlets I had visited in the Canadian Arctic. It's home to about thirteen hundred people, mostly Inuit, many of whom subsist by hunting and fishing. The village is dusty, crisscrossed with gravel roads, and mostly devoid of vegetation or anything else that might act as a windbreak against the northerlies that churn off the polar ice cap. Staring out a paint-splattered window past a cluster of ramshackle homes, I could see a sliver of the harbor after which the village is named. It was here that the Norwegian explorer Roald Amundsen landed in his seventy-foot converted fishing vessel, *Gjøa*, in September 1903. Snug anchorages protected from marauding ice are rare in the Northwest Passage, and Amundsen called this one "the finest little harbor in the world." He liked it so much, in fact, that he stayed for two years, researching the magnetic north pole and learning about the Inuit way of life—knowledge he would later put to use on his expedition to the South Pole in 1911.

While Eileen cooked us a dinner of barbecue chicken wings and Pam and Devon watched a reality TV show in the next room, Tom carried on. "There's only one place it could be, and this is it," he said, tapping his finger on a series of tiny brown hatch marks that looked like eyebrows. According to the map's legend, these features denoted a gravel ridge that in this part of the world is known as an "esker." This one ran from the head of the inlet for several miles into the island's interior. Tom paused and looked up at me to make sure I was ready to grasp the gravity of what he said next. "I'm putting it at 90 percent that we're going to be successful."

Jacob stood beside us with his hands stuffed in the pockets of a pair of insulated baggy green army pants. When I looked at him and raised my eyebrows to gauge his reaction, he stared back at me, his face expressionless. He didn't seem convinced. I, on the other hand, had been totally swept up in Tom's Franklin fever. All we had to do was jump on those bikes and scoot across King William Island, and we'd soon be standing over Franklin's grave, congratulating one another on the archaeological find of the century.

The next morning, I was sitting astride my bike, trying to figure out how to turn it on, when Jacob sidled over. "You ever ridden one of these before?" he asked.

I told him that I hadn't, so he showed me the basics: on and off, the throttle, front and rear brakes, how to shift gears. While giving me the rundown, he noticed my duffel bag flopped over the back of the seat. "Why isn't this strapped down?" he asked.

"Oh, it seems like it's wedged in there pretty good," I said.

Jacob eyes widened; then he started to giggle, which soon built into an unrestrained belly laugh. There was something about the whole clueless-white-guy-in-the-north routine that he found *extremely* amusing. The joke was on me, but his hilarity was contagious, and I started laughing too. When Jacob finally pulled it together, he grabbed two ratchet straps and lashed my duffel to the steel frame on the back of the bike, cranking them so tight, I thought the bag might pop.

"That should hold for a little while," Jacob said, still chuckling as he walked away.

|||

Shards of limestone shingle tinkled like broken china under my bike's tires as I crested a ridge that rose a few stories above the surrounding tundra. After thirteen hours and sixty miles of the bike acting like a bucking bronco and trying to toss me (and during which I had relashed that duffel at least a dozen times), my backside was raw, and it felt like someone had stuck a knife between my shoulder blades. Worse yet, my right thumb was so sore from pressing the throttle lever that I had started riding with my hands crossed on the handlebars. *Please be our camp, please be our camp,* I prayed to any god that would listen as I pulled up beside Jacob, whose bike idled next to a large stone marker known as an inukshuk. Clouds hung low over the wide landscape, breathing a malevolent northeast wind that pierced through my foul-weather gear and two layers of fleece and long underwear. Jacob, kicked back against a duffel bag with his legs resting on his handlebars, glassed a broad valley to the north with his binoculars. Following his line of sight, I gazed out over a pearl necklace of small slate gray lakes stranded together by a string of twinkling streams.

After our party left Gjoa Haven, Jacob had led us along an ancient Inuit hunting route that cut through the center of the island. We tracked northwest but with frequent detours around a maze of shallow lakes and ponds that quilted a hummocky grassland redolent of the American prairie and that smelled like compost or a freshly turned field on a hot summer's day. In the air all around us, bubbles of lake foam wafted across the land, animated by the wind.

About an hour out of town, we encountered our first wildlife: two musk oxen, a mother and her baby, munching on some grass at the crest of a small rise. They looked like shaggy buffalo, with long black fur and curved horns that rose from their wide heads. As we approached, they ran off and disappeared behind a ridge. Later, Jacob spotted a small herd of caribou far off in the distance. Through his binoculars, they looked to me like a pack of brown dogs. Jacob told me that he planned to harvest a caribou on our trip, but not until the end because he didn't want the meat to spoil before he could get it back to his family.

If you've ever watched a flock of snow geese flying northward in a chevron anywhere in the continental United States, there's a decent chance they were heading for King William Island, which is one of the largest goose-breeding grounds in the world. July is molting season and snowy white goose feathers, mixed with the airborne foam, danced in the air all around us like thistledown. Without their plumage, the geese couldn't fly, and we watched as thousands of them ran hither and yon, honking like bike horns, as scraggly black-furred Arctic foxes trotted along behind, picking off the stragglers. And while they didn't pique my appetite, I wondered how many of these birds were harvested by Franklin's men during their desperate march across this island after they abandoned their ships.

According to our map, we had made solid progress, but the first day had been marked by a series of misadventures. Renan was on an older rental bike, which lost a wheel while he was going thirty miles per hour. Miraculously, he emerged unscathed from the ensuing crash. After chasing down the runaway tire, Tom pounded it back onto the mangled axle with a grapefruit-sized rock. A little later, Devon's komatik did a front flip when one of its metal runners came loose and stuck into the ground like a javelin. The sled was demolished, and we lost a few hours waiting for Jacob to run back to town for a replacement.

In the afternoon, with newfound confidence in my bike-handling abilities, I second-guessed Jacob's route-finding at a small stream crossing and quickly learned that there was a reason he had avoided the patch of smooth cranberry-colored ground. When I charged into it, the bike instantly sank to its handlebars in a goopy quagmire of mud soup. My extrication plan was to hop off the bike and gun the throttle with one hand while pushing from behind with the other. The bike didn't budge, but plenty of mud shot out from the tires, spackling me from head to toe like a chocolate-dipped strawberry. Jacob, who had watched the whole thing from the other side of the bog, laughed so hard, he almost fell off his bike. Then he unspooled a cable from the front of his machine and winched me out.

We passed dozens of inukshuks that day but none as prominent as the one Jacob idled beside. He informed me, to my great relief, that this was indeed our camp for the night. And we'd be in good company because according to Jacob it was an old Thule site, eight hundred to a thousand years old, where his ancestors had camped for generations. "You can tell this is an Inuit camp, and not a white man's, because it's on a high place," he said. "We always camp up high because it's where you can see the game." I noticed a few circles of stones scattered around the inukshuk. For a millennium, the same rocks had been used to hold down the edges of sealskin tents. The hill was devoid of vegetation except for a soft carpet of vibrant green moss that blanketed the stony ground inside these ancient tent rings; the moss was fed by minerals leeched from the game slaughtered and consumed here.

We continued north the next day, following a circuitous path between the ponds and lakes and crossing countless streams and rivers that connected them. One waterway in particular was so broad and deep that Tom and Jacob drove up and down the bank, trying to find the best place to ford. When they settled on what appeared to be the shallowest part, Tom went in first and told me to follow. I gave my bike full throttle, leaving a rooster tail of spray in my wake, but when I got about halfway across, I felt the bike's tires lose contact with the river bottom, and suddenly, I was floating downstream like a swan boat. Amazingly, though, there was enough buoyancy in the tires that the machine actually stayed afloat, and when I gunned the engine, the spinning wheels created just enough momentum to crab me across.

What struck me the most during the two days we spent questing across King William Island was the mind-numbing sameness of the landscape. The view changed only by degrees. After a few hours, my mind craved something, anything distinct that might draw the eye. Often this was an inukshuk, and on the vast Arctic plain, we could sometimes see them off on the horizon from as far as twenty miles away. And I know I wasn't the only one who felt drawn to them like a moth to a candle.

Late on the second day, after homing in on another of these ancient monuments, we crested a rise and caught our first view of King William Island's northern shore. The visibility was excellent and directly ahead lay Collinson Inlet, shimmering in the evening sun like a mirage, its waters ice-free and Caribbean blue. On either side of it, long gravel eskers, bleached colorless by eons of time, extended outward into Victoria Strait, where *Erebus* and *Terror* had been caught in the ice on September 12, 1846. Out in the open ocean, pack ice covered the surface for as far as I could see, and oddly, it seemed to hang in the sky too. When I pointed this out to Jacob, he said that it was "iceblink," a phenomenon peculiar to the high latitudes, where sunlight reflects off the frozen ocean, projecting a mirror image onto the underside of the cloudscape like the screen of a drive-in movie.

As we sat there silently, looking out over the place where the Franklin crew had deserted their ships and set off on their doomed march, Jacob turned to me and said, "Whenever I come to the north part of the island, I feel the spirits of dead people."

I knew what he meant, because I felt them too.

|||

If you look in the margins of the original Victory Point Record, you can see James Fitzjames's pinched handwriting from the note's second installment in 1848: "HM Ships Terror and Erebus were deserted on the 22nd April 5 leagues NNW of this . . ."

Now that I'd seen the terrain on the north shore of King William Island with my own eyes, I couldn't stop wondering why Franklin's men would have left the security of their ships to set off on foot across this barren wasteland. Parks Canada divers have found unopened food tins on both ships, and additional intact cans have been found in camps the sailors inhabited during their retreat. Food didn't seem to be the main issue. So *why* did they leave their ships?

In the spring of 1847, after enduring their second winter in the Arctic, Franklin sent Lieutenant Graham Gore, First Mate Charles Des Voeux, and six other seamen on a sledging expedition to explore the coastline of King William Island. Since they left a duplicate copy of the Victory Point Record in a cairn on the west shore of Collinson Inlet, it's generally assumed that they headed west, not east.

In the first entry of the Victory Point Record, dated May 28, 1847, Gore signed off with: "Sir John Franklin commanding the expedition. All well." But if things were "all well" that spring, they clearly weren't anymore eleven months later, when Fitzjames scratched into the margins of the original note: "Sir John Franklin died on the 11th of June 1847 and the total loss by deaths in the Expedition has been to this date 9 officers and 15 men." Twenty-one men, including the commander, had died in less than a year. Dave Woodman notes that "had rescue come at this time, the expedition would still have had the dubious honor of possessing the highest death rate in the history of British Arctic exploration." Many have speculated about the cause of so much death, with theories ranging from botulism and lead poisoning from improperly canned food (mostly debunked at this point), to trichinosis from eating undercooked game, starvation, trench foot, or perhaps illnesses like cholera, typhus, influenza, and tuberculosis. But I believe the most obvious and logical explanation is that scurvy, the scourge of seafarers since ancient times, had begun to rear its hideous face during the Franklin expedition's second winter in the ice.

When this deadly affliction first appeared aboard *Victory* in Felix Harbour in 1830, John Ross and his men survived only when they lucked out and secured hundreds of frozen salmon from the Hydrographer. Ross didn't know that scurvy was caused by a deficiency of vitamin C nor that salmon contained significantly more of it than the dram of rancid lime juice the crew quaffed each day, but he was astute enough to see that Inuit ate a lot of fish and fresh game—and that they seemed to be immune to the disease. So he ordered his men to consume three pounds of fish per day, and within a month, the scorbutic crewmen had fully recovered.

If Franklin and his men were without Inuit help and unable to harvest a significant amount of fresh game or fish, which is likely considering they were stranded in the ice fifteen miles out in Victoria Strait, they would have had no way to fight the disease once it took hold.

For seafarers of that era, the specter of scurvy always lurked in the shadows. Between the sixteenth and eighteenth centuries, the disease is thought to have killed more than two million sailors. These were the years when voyagers began undertaking multiyear sailing expeditions between Europe, the Far East, and the Americas. Since the human body doesn't naturally produce vitamin C, it must be ingested via fresh fruits, vegetables, fish, or organ meats like liver and heart. Without this essential vitamin, the body stops producing collagen, a protein that is a major component of muscles, cartilage, tendons, bones, and skin.

When the Portuguese explorer Vasco da Gama became the first to sail to India from Europe via the Cape of Good Hope in 1499, he lost 116 of his 170 men to scurvy. Two hundred fifty years later, after a four-year circumnavigation, an admiral in the British Royal Navy named George Anson returned to England with 188 men remaining from an initial crew of 1,854—an appalling 90 percent attrition rate, most of which has been attributed to scurvy. "As the scourge invaded the sailors' faces, some of them began to resemble the monsters of their imaginations," writes David Grann in his book *The Wager* about one of Anson's ships. "Their bloodshot eyes bulged. Their teeth fell out, as did their hair," and their breath reeked of what one sailor called "an unwholesome stench, as if death had already come upon them."

In 1747, a Scottish physician named James Lind discovered that scurvy could be cured with citrus, but it took the British Royal Navy another fifty years to mandate that every sailor be administered three-quarters of an ounce of citrus juice per day, usually from lemons or limes. The practice all but eliminated the disease on British warships, but led to U.S. sailors teasing their counterparts, whom they called "limeys."

But for all the progress made by the British Royal Navy to fight this disease, it still plagued Arctic expeditions for the simple reason that the lemon juice carried aboard in five-gallon wooden kegs lost its potency over time. And when expeditions stretched across multiple years and rationing became mandatory, the daily allowance of citrus amounted to only a fraction of the vitamin C needed to ward off "the plague of the sea," as one British captain called it.

As *Erebus*'s and *Terror*'s scorbutic patients' connective tissues likely liquefied, old injuries would have reappeared and scars reopened. Reverend Richard Walter, the chaplain on George Anson's round-the-world voyage, recounted how one sailor, upon "being attacked by the scurvy," had a broken bone, which had been healed for some fifty years, dissolve "as if it had never been consolidated." Perhaps Fitzjames, with his own old war wounds, suffered similarly.

Scurvy also has a mental component, and it's the cerebral aspect of the disease that might have caused *Terror* to live up to its ominous name during that second winter the crew spent trapped in the frozen wastes of Victoria Strait. James Lind wrote that the late stages of the disease are often accompanied with "*affectio hypochondriaca*, or the most confirmed melancholy and despondency of mind." Walter saw plenty of these "strange dejections of the spirits" aboard Anson's flagship, *Centurion*, including "shiverings, tremblings, and a disposition to be seized with the most dreadful terrors on the slightest accident." This gloominess often manifested in scorbutic patients as an overwhelming homesick nostalgia once described as a thing that "fastens upon the breast of its prey, and sucks, vampire-like, the breath of his nostrils."

| | |

On our second night out from Gjoa, we camped at the mouth of a burbling river that drained a chain of several large lakes into Collinson Inlet. The evening was clear and mild; wispy cirrus clouds curled across the troposphere, and a light breeze from the southwest ruffled the Arctic cotton grass that quilted the banks of the river. While a flock of geese tootled nearby and a kaleidoscope of Arctic grayling butterflies flitted amongst floating tufts of goose down, Tom sat on a cooler and opened, what I had dubbed, his "Franklin bible" to a page he referred to often. At the top it read, "The Bayne Map," and below this, in all caps: "REDRAWN FROM MEMORY."

The hand-drawn map showed a dotted coastline with two overlapping ridges and a cairn marked by a black cross near a shoreline bump labeled "Victory Point." A small rectangle denoted a camp, and farther east, a line of graves dotted the lower tier of the ridge. The final detail was a river to the south marked with a squiggly line.

The story behind this legendary map, which purportedly shows the location of Franklin's tomb, begins in the spring of 1866 when Charles Francis Hall was camped along the Gulf of Boothia, about two hundred miles to the south of where the Ross expedition spent those three winters aboard *Victory*. Here, Hall met the Inuit named Supunger, who shared stories about the stone vaults he'd found on the north end of King William Island. Hall told Supunger that he planned to cross the Boothia Peninsula to King William Island to look for evidence of the Franklin expedition, but Supunger warned him that Inuit in that part of the Arctic, known as the Netsilik, were extremely territorial and notorious for killing anyone who ventured into their lands.

When Hall returned to his camp at Repulse Bay (known today as Naujaat, located in the upper-northwestern corner of Hudson Bay), in the fall of 1867, he found two whaling ships out of New Bedford, Massachusetts—*Ansel Gibbs* and *Concordia*—anchored in the harbor. The ships intended to overwinter in Repulse Bay so they could get an early start on the next fishing season. Hall arranged with the two captains to hire five of their men to accompany him on his spring expedition. The contract, worth $500 per man, was set for one year.

Early the next spring, while Hall set off to investigate Inuit reports of strange boot prints they'd found to the northwest on the Melville Peninsula, two of his new recruits, Patrick Coleman of Ireland and Peter Bayne of Nova Scotia, headed out to scout the route ahead and lay in some caches of meat for the upcoming expedition. Near the southern shore of the Gulf of Boothia, Bayne and Coleman met a group of Inuit—a couple with three grown children, and another group of two men and a woman—who were on their way south. Unlike most of the white travelers in the north, Bayne, a Civil War veteran, was an excellent hunter, and he had killed several caribou, which he now shared with the Inuit. Bayne had learned enough Inuktitut to communicate, and he later reported that one of the women spoke some English. As they all camped together and shared meals, the Inuk father relayed an intriguing story.

Years ago, when his son was two years old, the family had traveled to the north end of King William in the summer to hunt seal.

Polar Sun's first mate, Ben Zartman, prepares to haul the dinghy back aboard with one of the spare halyards. He wore this homemade coat, which he called the "Westport Whaler," throughout the voyage.

Left: Sir John Franklin, 1786–1847. In 1845, he led 128 men on an ill-fated expedition to discover a Northwest Passage linking the Atlantic and Pacific Oceans. He and his crew were never seen again.

© Engraved by D J Pound from a drawing by Mathias Negelen. From the book The Drawing Room of Eminent Personages, *vol. 2, published in London 1860/Alamy*

Right: Jane Griffin, later Lady Jane Franklin, at age twenty-four.

© The New York Public Library Digital Collections/Public Domain

Unless otherwise noted, all images © Renan Ozturk

Officers from HMS *Erebus*, plus Francis Crozier, sat for daguerreotype portraits before leaving England in May 1845. In the middle row left is James Fitzjames, with Franklin and Crozier to his right. © *Alamy*

Facing page: The Victory Point Record is the only detailed written note that has been found from the Franklin expedition. It has two installments, written about a year apart. The first note was signed: "Sir John Franklin commanding the expedition. All Well." The second note, written into the margins of the first, states that John Franklin died on June 11, 1847 (just two weeks after the first note had been written), and that the 105 survivors deserted *Erebus* and *Terror* on April 22, 1848. © *Wikimedia Commons*

H. M. S. hips Erebus and Terror

28 of May 1847 { Wintered in the Ice in
Lat. 70°5' N Long. 98°23' W

Having wintered in 1846—7 at Beechey Island
in Lat 74°43'28" N. Long 91°39'15" W After having
ascended Wellington Channel to Lat 77° and returned
by the West side of Cornwallis Island.

~~Commander.~~

Sir John Franklin commanding the Expedition.

All well

WHOEVER finds this paper is requested to forward it to the Secretary of the Admiralty, London, *with a note of the time and place at which it was found*: or, if more convenient, to deliver it for that purpose to the British Consul at the nearest Port.

QUINCONQUE trouvera ce papier est prié d'y marquer le tems et lieu ou il l'aura trouvé, et de le faire parvenir au plutot au Secretaire de l'Amirauté Britannique à Londres.

CUALQUIERA que hallare este Papel, se le suplica de enviarlo al Secretario del Almirantazgo, en Londrés, con una nota del tiempo y del lugar en donde se halló.

EEN ieder die dit Papier mogt vinden, wordt hiermede verzogt, om het zelve, ten spoedigste, te willen zenden aan den Heer Minister van des Marine der Nederlanden in 's Gravenhage, of wel aan den Secretaris der Britsche Admiraliteit, te London, en daar by te voegen eene Nota, inhoudende de tyd en de plaats alwaar dit Papier is gevonden geworden.

FINDEREN af dette Papiir ombedes, naar Leilighed gives, at sende samme til Admiralitets Secretairen i London, eller nœrmeste Embedsmand i Danmark, Norge, eller Sverrig. Tiden og Stœdit hvor dette er fundet önskes venskabeligt paategnet.

WER diesen Zettel findet, wird hier-durch ersucht denselben an den Secretair des Admiralitets in London einzusenden, mit gefälliger angabe an welchen ort und zu welcher zeit er gefundet worden ist.

Party consisting of 2 Officers and 6 Men
left the Ships on Monday 24th May 1847.

Gm. Gore Lieut.

Chas. F. Des Voeux mate.

(Left margin:) 1848. HM Ships Terror and Erebus were deserted on the 22nd April, 5 leagues NNW of this, having been beset since 12th Septr 1846. The officers & crews consisting of 105 souls — under the command of Captain F. R. M. Crozier landed here — in Lat 69°37'42" Long 98°41' This paper was found by Lt. Irving under the cairn supposed to have

(Right margin:) been built by Sir James Ross in 1831 — 4 miles to the Northward — where it had been deposited by the late Commander Gore in June 1847. Sir James Ross' pillar has not however been found and the paper has been transferred to this position which is that in which Sir J. Ross' pillar was erected — Sir John Franklin died on the 11th June 1847 and the total loss

(Top margin, inverted:) by deaths in the Expedition has been to this date 9 Officers & 15 Men.

James Fitzjames Captain HMS Erebus

F R M Crozier Captain & Senior Offr

And start on tomorrow 26th for Backs Fish River

Above: A sketch of Franklin's two ships on the River Thames in 1845, shortly before they departed for the Arctic.

Below: *Polar Sun* stretches her wings under full sail in the Strait of Belle Isle, which lies between Newfoundland and Labrador. Gros Morne National Park is just visible in the background.

Top left: Rudy Lehfeldt-Ehlinger in Davis Strait. He served as both crew and as a member of the production team that documented the voyage for *National Geographic*.

Top right: Hampton Kew, a licensed US Coast Guard captain. She and the author's son, Tommy (age six), joined the expedition for two and a half weeks in Greenland.

Left: Tommy did his first overnight sailing missions as a baby on the coast of Maine, hanging in his car seat from the ceiling of *Polar Sun*'s predecessor, SV *Camelot*. © *Hampton Synnott*

Left: Mr. Dirt, aka Erik Howes, riding a scooter he found in an irrigation ditch in Nuuk, Greenland. If you look closely you can see his motto written on his foul weather pants: "Stay Wild, Never Mild."

Right: Jens Kjeldsen, who sailed his own salvaged boat around the world via the Northwest Passage, befriended the crew of *Polar Sun* in Nuuk, Greenland. The prevailing wisdom is that Northwest Passage–bound boats should have a fuel range of 1,200 miles. To increase its range, *Polar Sun* carried twenty, five-gallon jerry cans on deck. © *Mark Synnott*

Above: *Polar Sun* sails past the "God iceberg" in Disko Bay, Greenland.

Below: Renan Ozturk, a world-renowned professional climber, photographer, and filmmaker, led a small production team that documented the expedition for *National Geographic* magazine and television. © *Rudy Lehfeldt-Ehlinger*

Left: Povl Linnet, a mechanic and former ship's engineer, holds a timber from the *Fox* in his home in Nuuk, which the author describes as a "Greenlandic Aladdin's Cave." © *Mark Synnott*

Right: The crew of *Polar Sun* tied up alongside a whaling ship in Sisimiut, Greenland, which had just harvested a minke whale in Davis Strait. Since 2010, the International Whaling Commission has allowed limited aboriginal whaling in Greenland as long as it's done humanely and sustainably and only for local subsistence and cultural needs.

Below: *Polar Sun* sails into Tasiujaq Sound (formerly Eclipse Sound) en route to Pond Inlet after a seven-day crossing of Baffin Bay.

The author contemplates the Whale Fish Islands anchorage where the Franklin expedition spent eight days provisioning in July 1845 before setting off into the Northwest Passage.

Polar Sun proved a popular attraction in the hamlet of Pond Inlet, where many locals came down to say hello and learn about the crew's voyage into the Northwest Passage.

Above: The author ties the dinghy to a piece of beached ice on the shore of Erebus and Terror Bay on Beechey Island. It was here, at the northernmost point of *Polar Sun*'s Arctic odyssey, that the Franklin expedition spent its first winter and where it lost three crew members to tuberculosis and pneumonia.

Below: When exiting Navy Board Inlet into Lancaster Sound, *Polar Sun* was buffeted by 50-knot katabatic winds blowing down off the glaciers of Bylot Island.

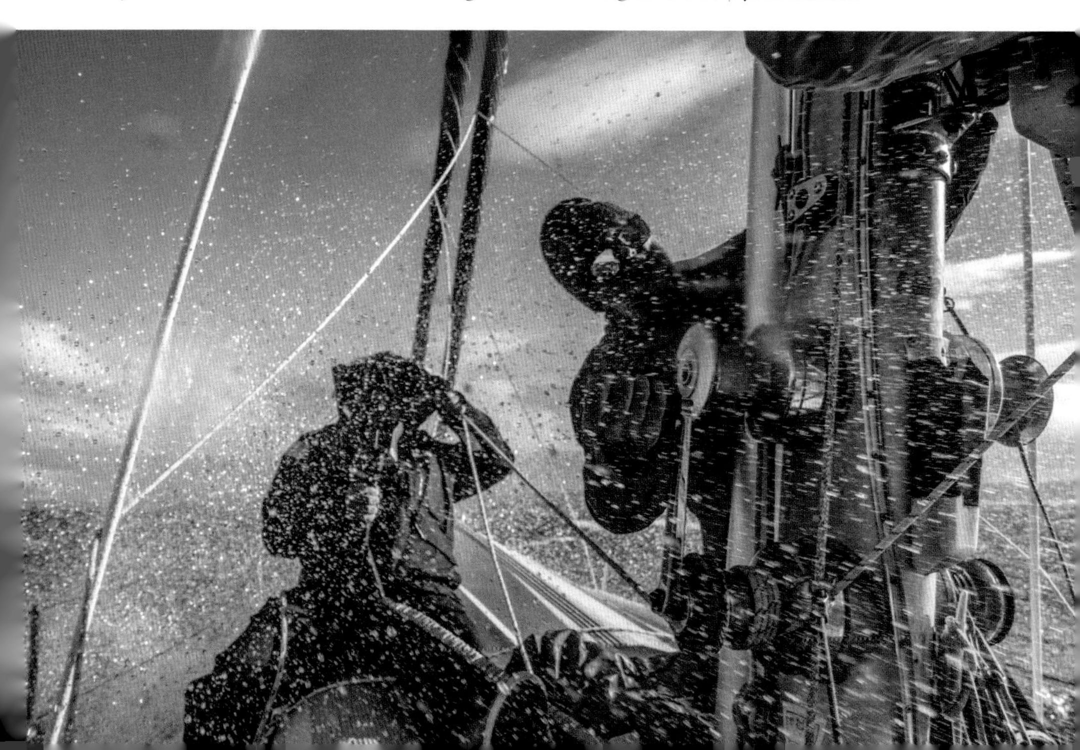

The Inuit call belugas *qilalugaq*, which means "white whale." Early explorers nicknamed them the "canaries of the sea" due to their extraordinary range of vocalizations, which some believe include the ability to sing. In Peel Sound, *Polar Sun* collided with one of these beautiful creatures, which may have been sleeping at the time.

Jacob Keanik, president of the Nattilik Heritage Society and a former wildlife conservation officer, joined the crew of *Polar Sun* in Pond Inlet and also served as a guide on the overland search for John Franklin's tomb on King William Island. Here he takes a turn scanning for ice in Peel Sound as *Polar Sun* is drawn south by their light air gennaker, which they called the Whomper.

As the team approached James Ross Strait, the ice crux of the Northwest Passage, an approaching southeast gale led them to seek shelter in Pasley Bay, which lies about 100 miles south of Bellot Strait on the Boothia Peninsula. Little did they realize that it was a trap that would nearly prove their undoing.

Rudy, Ben, and Jacob passed the time while trapped in Pasley Bay playing Shanghai rummy.

The storm drove miles and miles of dense pack ice into Pasley Bay, forcing *Polar Sun* (below) and SV *Taya* (center) into an ever shrinking pool of open water. The darker patches are multiyear ice, which has declined by 95 percent since 1985.

Right: Alan Cresswell and Annina Barandun, a French/Swiss couple, had sailed into the Northwest Passage aboard *Taya*, a 50-foot aluminum sailboat, after meeting on a dock in Patagonia. They, too, entered Pasley Bay to escape the storm, and found themselves ice bound alongside *Polar Sun*. Through the ordeal of trying to protect the two ships from being crushed in the ice or run aground, the crews became fast friends.
© *Mark Synnott*

Below: After losing all their ground tackle, the crew of *Polar Sun* anchored the vessel to floating chunks of pack ice with screws used for ice climbing. When these chunks threatened to push them up onto the shore, Ozturk donned a dry suit and swam a dock line around a grounded ice boulder. © *Rudy Lehfeldt-Ehlinger*

The inveterate Franklinite and amateur historian Tom Gross has been searching King William Island for the tomb of John Franklin since 1994. In 2015, during an aerial reconnaissance, he spotted a stone house that matches a description of the burial vault from Inuit testimony.

In July, some of the crew took a break from sailing to join Gross and Keanik on a 500-mile overland expedition. In the years since his 2015 purported sighting of the tomb, Gross has narrowed the search zone down to an area covering approximately 30 square miles.

In Collinson Inlet, near the northern tip of King William Island, not far from the place where the Victory Point Record was discovered in 1859, the author found a brass fitting that may have come from one of the steam engines that propelled *Erebus* or *Terror*.

Left: After Keanik, Ozturk, and Lehfeldt-Ehlinger left the expedition, Synnott and Zartman sailed on to Tuktoyaktuk, in the Northwest Territories, as a team of two. Here, they were met by two last-minute stand-ins, including David Thoreson, a veteran high-latitude sailor who is the only American to have sailed the Northwest Passage twice. © *David Thoreson*

Right: Another last-minute recruit was Ben Spiess, a lawyer from Anchorage who was an early climbing partner of the author back in his college years. © *David Thoreson*

Polar Sun's first mate, Ben Zartman (left), and skipper, Mark Synnott. The author describes the pair as "two friends who see the world through radically different lenses and yet somehow found enough common ground, solidarity, and goodwill to succeed at the hardest thing that either of them had ever set out to do."

There they found two large wooden ships icebound off the west coast. The sailors had set up a camp onshore with several tents filled with sick and dying men. A man, whom Inuit called "Aglooka," came ashore and spoke with them. According to Bayne, Inuit told him that "seal were plentiful the first year, and sometimes the white men went with the natives and shot seal with their guns; that ducks and geese were plentiful, and the white men shot many." Most of the dead were buried on a nearby ridge, but one man died aboard the ship, and was "brought ashore and buried on the hill near where the others were buried . . . but in an opening in the rock," which was covered with something that turned to "all the same stone," and "many guns were fired."

This hospital camp was located about a quarter mile in from the beach atop a narrow, flat-topped, southeast-facing ridge marked with "projecting rocks." The camp had "several cemented vaults—one large one, and a number of smaller ones." These Inuit said that the smaller vaults were for the many papers that had been brought ashore by the white men, some of which were buried, while others "blew away in the wind." The account was so detailed that Bayne was able to draw a map.

When Hall returned empty-handed in mid-June from his exploration on the Melville Peninsula, Bayne and Coleman shared the story with him. Details as to what happened next are sketchy, but some of the other whalers later testified that Hall became angry that Bayne and Coleman had interviewed Inuit without his permission. Two days later, Bayne, Coleman, and a Portuguese man returned to camp late from a hunt. Hall confronted Coleman, demanding to know why it had taken him so long to return. Bayne's account of the incident was later published in the *New York Sun*.

Coleman: *I can't do the work of a horse on half a cake and a cup of coffee.*

Hall: *Your conduct is mutinous.*

Coleman: *Well, I can't help it; you impose on me more than you do on the other men. You know you keep us short of bread.*

Hall: *It's a damned lie. Did I propose to give you bread daily?*

Coleman: *Yes, Doctor.*

Hall: *Well, I didn't think you would throw such a damned lie in my face.*

Hall then asked for Bayne's rifle, which he took back to his tent and exchanged for a revolver. When he returned, he asked Coleman, "Are you going to be a dutiful man or not?"

Coleman replied, "How dutiful: what more do you want of me?"

In response, Hall raised the revolver and shot Coleman in the chest. While the other sailors tended to their fallen comrade, Hall went back to his tent, where he cried and said to Bayne, "Oh, why did Pat talk to me like that?"

Coleman hung on for thirteen days, during which time a distraught Hall tried to nurse him back to health. When Coleman finally died, Bayne and the other whalers fled to *Ansel Gibbs* and *Concordia* and sailed to New Bedford, but Hall stayed. Back in the States, Bayne reported the murder to the authorities, but no one could decide in whose jurisdiction the killing had taken place, and Hall was never tried. Hall had friends in high places, including the philanthropist Henry Grinnell and the president Ulysses S. Grant. Instead of going to jail, he was awarded a $50,000 grant from Congress to lead a North Pole expedition aboard the U.S. Navy ship *Polaris*.

Polaris left New York City in late June 1871 and, by the end of the summer, had sailed to 82 degrees 11 minutes—a record for the farthest north reached by a ship. But once again, Hall proved a contentious leader, and several of the crewmen, including Emil Bessels, a German who was the expedition's chief scientist and physician, openly questioned his authority. In October, while the ship wintered in Thank God Harbor in northern Greenland, Hall fell gravely ill after returning from a sledging trip and warming himself with a cup of coffee. Over the next two weeks, he suffered severe stomach pain, seizures, and dementia and became paralyzed on his left side. During moments of clarity, he accused the crew, including Bessels, of poisoning him. Hall rallied briefly, then fell into a coma and died in early November 1871; he was buried on shore nearby.

In 1968, a Dartmouth College professor and biographer named Chauncey Loom exhumed Hall's body. Tests conducted on his tissues, similar to those that would later be done on the Franklin sailors on Beechey Island, revealed that Hall had ingested a large amount of arsenic during the final weeks of his life. More recently, it has come to light that Hall and Bessels had been writing affectionate letters to the same woman, a famous twenty-three-year-old sculptor named Vinnie Ream, who had unveiled a marble statue of Abraham Lincoln in the U.S. Capitol Rotunda six months before the *Polaris* expedition set off. While the evidence strongly suggests that Bessels poisoned Hall, he always denied doing so and no charges were ever brought against him.

Bayne, it seems, never gave up on the idea of finding Franklin's tomb. In 1913, the *Morning Oregonian* reported that he had purchased a schooner named *Duxbury*, in which he would soon set sail for King William Island to find the lost Franklin burial vault. It appears, though, that Bayne never made the trip, because three years later the *Spokesman-Review* carried a similar report. This time, Bayne would have been making the voyage with his new bride, Stella, who was twenty-four years his junior. Again, it's unknown if he ever sailed north, but in 1925 *Duxbury* was beached in Nome, after being wrecked in the ice on the North Slope. By 1920, Bayne was living in San Francisco, and four years later, perhaps in failing health, he moved into a veterans' retirement home in Port Orchard, Washington.

Bayne died there in 1926 at the age of eighty-three, but at some point in those final years, he shared what he knew about Franklin's burial vault with a Seattle man named George Jamme, who'd spent years prospecting for tin in the Arctic. According to Tom Gross, it's likely that the two first met in Nome while Jamme was working for the U.S. Geological Survey. After Bayne's death, Jamme wrote up Bayne's burial-vault story and shared it with a judge in Vancouver named T. W. Jackson. In 1929, Jackson tried to sell the "Jamme document," which included a map of the Franklin vault (Jamme had lost Bayne's original map, so he had redrawn it from memory), to the Canadian government for $25,000. They settled on $1,000, and in February of 1930, an article in the *Evening Post* reported that the Canadian government would launch an expedition to search for Franklin's grave. According to the story, an English-speaking Inuk woman had told Bayne that she had personally witnessed the burial of "the Great Chief, Sir John Franklin."

"She explained to Mr. Bayne," said Jamme, "how men had brought the body ashore from a ship in a coffin, made cement with powdered rock and tar, and sealed the coffin in the rock. The white man placed papers in the coffin before burial. She remembered the salute fired by the guns of the Erebus and Terror."

That summer, Major Lachlan Taylor (L.T.) Burwash landed a float plane at Cape Felix and spent three days searching in the vicinity of Victory Point. Burwash never found Franklin's tomb, but he did find the remnants of an old campsite, including bits of broadcloth, rope, and tents. In 1931, Burwash published his findings (including a verbatim copy of the Jamme document) in a publication the Northwest Territories and Yukon's Department of the Interior called *Canada's Western Arctic: Report on Investigations in 1925–26, 1928–29, and 1930*. Up until this point, the Bayne story, with its contention that Franklin had been buried in a cemented vault near Victory Point, had never been made public. Some historians still question why it took sixty-three years for this intelligence to come to light, while others have wondered why there was no mention of Franklin's burial in the Victory Point Record.

I had taken a seat next to Tom as he flipped through page after page about Bayne in his notebook. "This map has been the basis of my search for the past twenty-eight years," said Tom. "For a long time, I thought the Bayne and Supunger stories were describing the same thing, but I now believe that they're different. I think the Bayne story describes Franklin's vault, which is the one I saw from the air in 2015. And the Supunger story describes a similar vault, which may be that of another Franklin officer. My theory is that they brought the ships into Collinson Inlet in the summer of 1847, and that's when they buried Franklin in the stone house."

"Okay," I replied. "But if that's true, why doesn't the Victory Point Record mention that the ships were brought into shore? Doesn't it say that *Erebus* and *Terror* were deserted five leagues north-northwest of Victory Point?"

"The record is all dates and numbers, right?" said Tom. "And we know that some of those numbers are wrong." He explained how the coordinates for Victory Point are off by nine miles, and that the record states the expedition overwintered on Beechey Island in 1846–1847, when we know that it was actually the previous year. "The men were in a hurry, and their minds may have been addled," he continued. "Since we know they made several errors, who is to say they didn't make more that we don't know about? The note only says that the ships were abandoned five leagues north-northwest of Victory Point. But it doesn't say what happened *in between*."

I knew that longitude was often difficult to pin down in Franklin's day, because unlike latitude, it depended on a precise recording of time, and the clocks aboard *Erebus* and *Terror* might have developed errors during the three years since they'd been calibrated in England. I also knew that Tom was simply trying to make the existing clues fit with what he'd seen from the air in 2015.

"If they brought the ships in here," said Tom, gesturing toward the bay, "it would explain everything. Suddenly, all the Inuit testimony makes sense."

In the morning, Tom led our ATV armada across a vast tidal plain at the head of Collinson Inlet. The ground here was billiard table smooth, and I felt like a character in a Mad Max movie as I ripped across the mud flats at fifty-five miles per hour. After a few miles, we turned north and Tom led us out onto a skinny, hook-shaped peninsula that protruded into the inlet. The water surrounding us was glassy and mostly ice-free, save for the occasional pool-table-sized chunks floating along the shore.

As we traversed the thin finger of land, a ring of limestone boulders—another tent circle—caught my eye. While poking around, I found a scattering of camp items including an old ladle, a rusty fox trap, and a few spent bullet casings. But inside the circle, there was one item that didn't fit the picture of an old Inuit camp: a hunk of metal that looked like it could have been a brass pipe fitting. I picked it up and turned it over in my hand. It felt solid and had four small openings, three of which had hexagonal heads. One hex head had a section of pipe threaded into it that appeared to have been sawed off.

The outside of the fitting was dinged with little indentations, as if someone had hammered it with a nail punch.

Tom arrived, and I showed him what I'd found.

"Will you look at that," he said.

"What do you think it is?" I asked.

"I'd say that it looks like a piece of the *Erebus* or *Terror*'s steam engine," he replied.

A few minutes later, I found a marble-sized metallic ball that looked like it had been burned. Jacob examined it and said he'd never seen anything like it. We later learned through lab analysis that it was iron pyrite, which was used as a flint-and-steel-style fire starter in England in the 1800s.

"Hey, over here," yelled Pam. She'd found an old wooden peg about an inch in diameter that tapered to a point, with a notch in the end for securing a line. Tom pulled a tape measure out of his pocket. The peg was precisely sixteen inches long. Jacob said that Inuit didn't use tent pegs, and when they cut wood, they did it by eye, not to exact measurements.

Tom quickly decided that the brass fitting and the tent peg were likely Franklin artifacts, which he took as confirmation that we had to be close to the tomb. He pulled out his copy of the Bayne map and we all crowded around and tried to match it up with the gravel eskers that shimmered in the sun to the south of us. "The Inuit testified that there were many men in the camp," said Tom. "So they would have needed a reliable source of fresh water. Find that river, and we'll find the camp."

The base map on my Garmin GPS, which was remarkably accurate, showed a couple such rivers flowing into the bay from the southwest. I jumped back on my bike and took off, practically trembling, so excited was I at the prospect that I'd soon be standing over Franklin's tomb. I knew it was totally unfair, since Tom had twenty-eight years invested into the search—and I was only close because he had told me where to look—but secretly, I wanted to be the one to make the actual discovery.

King William Island, it turns out, knows how to hide its secrets. Over the next several days, we scoured every one of those gravel eskers that extend like bony fingers from Collinson Inlet into the interior of the island. The terrain was maddeningly uniform, and even with the track recorded on my GPS, I felt like a dog chasing its own tail.

Eventually, we homed in on one particular ridge that seemed like the best fit. That morning, I had asked Tom, "If someone put a gun to your head and said, 'You have to find the tomb today, or you're dead,' where would you look?" This was that spot, so Renan and Rudy decided it was time to unleash their secret weapon: a four-foot-long remote control foam airplane mounted with a specialized high-resolution camera. We kicked back on our bikes in the blazing sun and watched the tiny plane take off like a helicopter. A hundred feet overhead, it rotated its propellers vertically and then zipped out of view to the south. For forty-five minutes we watched as it flew a carefully choreographed grid pattern, its little electric engine whirring as it made big bank turns overhead, clicking off hundreds of sequential aerial photographs. When it landed, Rudy popped out its SD card and began downloading the images to his laptop, where software would stitch them together into a giant GigaPan-style image of this particular ridge.

As we waited for the photos to download, the wind died and a suffocating cloud of mosquitoes enveloped us. I slipped a bug net over my head and sidled over to where Tom was examining a topo map he had laid on the seat of his bike. He showed me that we were sitting atop the gravel esker he'd pointed out back in Gjoa Haven, the one he'd called "a sure thing" and that I had named "Gun to the

Head." Maybe Rudy's GigaPan, when it was complete in a few hours, would reveal the stone house, but I wasn't holding my breath.

"How did they do it?" asked Tom. "How did they get Franklin from the ship to here?"

"I'm thinking more like 'Why?'" I replied as I looked out over the miles of glistening marshland that lay between our present position and the spit of land where we'd found the brass fitting. "Why in the world would the men drag Franklin all the way out here when they could have buried him closer to shore? Besides, Franklin died in June, when those bogs would have been impassable on foot."

Tom had long held that the men chose to bury Franklin in an esker because the loose gravel would have made it easy to dig a tomb. The sailors could have used a small shallow-draft rowboat to transport Franklin's body up one of the rivers that drain into Collinson Inlet. But the stone on this ridge was flint hard and there was no flowing water nearby—just skeletal claws of lifeless, blanched limestone surrounded by miles of mosquito-infested quagmire.

The next morning dawned cold and windy, the sky draped with dark clouds that threatened rain or possibly snow. The mild southerly that had provided a few days of T-shirt weather had given way to a stiff northeasterly that carried an air mass deep chilled after crossing hundreds of miles of frozen ocean. As I nursed a cup of gritty, lukewarm instant coffee, Eileen told me that Tom hadn't slept a wink—and so neither had she. Rudy's GigaPan had turned out to be a masterpiece: a collage of roughly seventeen hundred images adding up to a file bigger than the hard drive on the average laptop. The night before, Tom and I had spent hours studying it, zooming in and out with the cursor on Rudy's MacBook. But we couldn't find anything resembling a stone house. Tom had been stressing about it ever since, and now, emerging from his tent with red-rimmed eyes, he announced that we were throwing in the towel on Collinson Inlet and relocating to Erebus Bay, forty miles to the southwest, where we'd search for the Supunger vaults.

Jacob led us along the same shoreline where the Franklin expedition had split into sledging parties and begun dragging their longboats on that march to nowhere. After stopping to refuel the bikes at a cache that Jacob had placed via snowmobile earlier that spring, we bumped along until midafternoon, when Tom pulled up alongside a low, sprawling cairn that sat on the crest of a hillock overlooking an ice-covered Victoria Strait. Bolted to a large plate of reddish sandstone that didn't look native to King William Island, a brass plaque read: "To the memory of the crew members of Sir John Franklin's expedition of 1845 who died here in the spring of 1848."

"This is the Boat Place," said Tom. "A lot of Franklin's men died here."

I looked around and saw the now familiar detritus of the island: bits of wood, the remnants of a tin can, and a random scattering of animal bones, including the skull of a musk ox.

In 1992, two Canadian archaeologists, Anne Keenleyside and Margaret Bertulli, spent a summer here. In addition to finding an array of artifacts, they excavated nearly four hundred human bones and bone fragments, representing the remains of at least eleven individuals. Additional excavations in years since have increased this number to twenty-three. The bones were scattered across the hill with broken pieces on one side and more intact skeletons on the other, which suggested that Franklin's men had been driven here to the last resort. When examined under an electron microscope, the bones revealed cut marks that were, according to Keenleyside and Bertulli, "consistent with defleshing, or removal of muscle tissue," and there was evidence of decapitation and the fracturing of long bones to remove the marrow.

These grim findings are consistent with testimony Hall collected in 1869 from an Inuk named In-nook-poo-zhe-jook. At the time, In-nook-poo-zhe-jook and Hall were traveling together on the south shore of King William Island in search of Franklin clues. In-nook-poo-zhe-jook told Hall that on an earlier trip to Erebus Bay, he had discovered a tent and two wooden longboats. Near one of the boats, he'd found a firepit with countless bones stacked like cordwood and a knee-high leather boot in which human flesh had been boiled.

It's likely that In-nook-poo-zhe-jook had been led to the Boat Place after hearing about its discovery by an earlier Inuk visitor named Pooyetak. In late February of that year, Hobson and his commander, Francis Leopold McClintock, had set off on a sledging journey in search of Franklin. From a bay in the eastern portion of Bellot Strait (near the location of *Anahita*'s sinking) where the *Fox* had overwintered, they traveled south down the western shore of the Boothia Peninsula. Not far from where James Clark Ross discovered the magnetic north pole in 1831, they split into two parties. Hobson headed west across a frozen James Ross Strait toward Cape Felix, while McClintock continued south on a clockwise circumnavigation of King William Island.

In April, just two weeks before his discovery of the Victory Point Record, Hobson was working his way along the shore of Erebus Bay when he found a twenty-eight-foot longboat sitting atop a sledge made of heavy oak timbers. The boat contained the remains of two men. The one in the bow was young and slight, and much of the body had been destroyed by animals. The other skeleton—that of a strong, middle-aged man—lay across the stern dressed in clothes and furs. Both were missing the caps of their skulls, with only the lower jaw remaining. Nearby stood two shotguns, loaded and cocked.

When McClintock arrived at the Boat Place two weeks after Hobson, he could hardly believe what he found. In marked contrast to his own lightweight sledges, which were fashioned similarly to the ones used by Inuit, the sled marooned at the Boat Place was a behemoth built atop two twenty-three-foot-long iron-sheathed oak planks that were cross-braced with more stout oak timbers. McClintock estimated that the sledge alone weighed about six hundred fifty pounds and the twenty-eight-foot boat atop it, another seven hundred or eight hundred. And this didn't include the "accumulation of dead weight" inside the boat: silk handkerchiefs, towels, soap, sponges, hair combs, a gun cover, twine, nails, saws, files, rolls of sheet lead, and a number of books, including *The Vicar of Wakefield* and a Bible with handwritten annotations. McClintock noted that other items—cookstoves customized for overland travel and homemade sunglasses—suggested that the men had put considerable time and effort into preparing for their retreat.

Perhaps McClintock's most telling discovery, though, was the fact that the sledge was pointing northeast, back toward Victory Point. "A little reflection led me to satisfy my own mind at least that this boat was *returning to the ships* [italics as in the original]," wrote McClintock. By the time the Franklin sailors had reached Erebus Bay, they had dragged their two longboats approximately sixty-five miles. We don't know how long it took them to get there, but it is not hard to imagine that the retreat had quickly devolved into a battle for survival. If the ocean was still frozen solid (as it was when we arrived in Erebus Bay in late July) and there was no chance for launching the boats, a return to the ships, where some food remained, might have been seen as their only chance to survive the coming winter.

The idea that the ships might have been remanned after the initial desertion in April 1848 has never fit neatly into the classical reconstruction of the Franklin narrative, but there is one clue suggesting it that seems to be incontrovertible. In the summer of 1879, a lieutenant in the United States Army named Frederick Schwatka arrived at Crozier's Landing—the site a few miles south of Victory Point where the Franklin sailors first landed after they deserted the ships. Schwatka's team, which included thirteen Inuit and a journalist with the *New York Herald*, traveled to the site by foot and dogsled from Repulse Bay, hunting for almost all of their food along the way.

On the crest of a gravel ridge about sixty feet from shore, they found an elaborate grave site that consisted of large stones arranged into a sort of sepulcher. The tomb appeared to have been raided and bones were scattered about the area. Inside, they found a telescope lens and brass buttons from an officer's coat marked with the insignia of an anchor and crown. The remains of a skeleton were wrapped in canvas, and beneath the head lay a well-preserved silk handkerchief. The deceased was identified by a silver medallion sitting on or near the grave. One side featured the bust of King George III while the other read, "Second Mathematical Prize Royal Naval College Awarded to John Irving Midsummer 1830."

Third Lieutenant John Irving, a Scot, was thirty-three the year the ships were abandoned. He had joined the navy at thirteen, and with the exception of a few years in Australia trying his hand at sheep farming, he spent his entire career at sea. While stationed in Portsmouth, England, in February 1845, Irving wrote to his sister-in-law Catherine that he was still in suspense whether he would be chosen to join the Franklin Northwest Passage expedition. "I shall be glad to be put off it," he wrote, "as it affects my prospects for the summer very materially, there being some difference between the regions of thick-ribbed ice and perpetual snow, and the green fields I might visit [at Catherine's home in Scotland] if I did not get appointed. . . ." But, of course, Irving was chosen, and now here he lay. Schwatka decided to take his remains home to Edinburgh, where Irving was laid to rest in Dean Cemetery.

Irving's grave is a crucial clue in the Franklin mystery because, as per the Victory Point Record, he was the one who retrieved the note in April 1848 from where it was originally stashed (in 1847, Gore and Des Voeux couldn't find the Victory Point cairn erected by James Clark Ross in 1830, so they built their own somewhere in the vicinity). The fact that Irving had been chosen to collect the record written the year before by Gore and Des Voeux suggests that he must have been one of the strongest members of the party at that point. And we also know that Crozier wrote in the margins of the second installment that they were planning to set off the next day overland for the Back River.

So had Irving died during the brief period between when he retrieved the record and when the group departed the next day? And then, overnight, had his companions built him an elaborate grave? Not likely. McClintock, Schwatka, and Dave Woodman have all suggested an alternative narrative in which Irving died at some later date when the men returned to the ships after it became clear that their overland escape was doomed.

If In-nook-poo-zhe-jook's account of bones piled up by the fire is accurate—and the archaeological work of Keenleyside and Bertulli showing cannibalism suggests that it is—it may be that some of the men, perhaps suffering from scurvy, trench foot, and various other ailments, were too weak to make the return trek when Crozier, who would have taken command of the expedition when Franklin died, ordered everyone back to the ships. Perhaps, when they parted ways, those leaving their sick companions behind promised to sail to Erebus Bay to pick them up if the ice ever released the ships.

In September 2024 we learned from a paper published in the *Journal of Archaeological Science* that one of the men who perished at the Boat Place was James Fitzjames. For many years, a research team led by a University of Waterloo archaeologist named Douglas Stenton (who you'll hear more about later) has been conducting DNA analysis of the bones, teeth, and hair collected in Erebus Bay by Keenleyside and Bertulli in 1993. According to Stenton, DNA from the dental root of a molar extracted from a lower jawbone matches with a British furniture dealer named Nigel Gambier, who is Fitzjames's second cousin, five times removed. Gambier is one of twenty-five living descendants of the Franklin crew who have donated their DNA to the project. Fitzjames is only the second Franklin sailor to be positively identified. The first was an engineer and warrant officer aboard HMS *Erebus* named John Gregory, who was identified in the same way in 2021.

Fitzjames's match was made thanks to a Dutch academic named Fabiënne Tetteroo, who became interested in Fitzjames's story after watching the 2018 AMC miniseries *The Terror*, which is based on the Dan Simmons novel of the same name. While working on a Fitzjames biography, Tetteroo came across a book published in 1924 entitled *The Story of the Gambiers*. In it, she found a family tree that led her to Nigel Gambier.

Like some of the other fragments collected from the Boat Place, Fitzjames's jawbone is etched with cut marks consistent with posthumous dismemberment. Thus, the logical conclusion is that after his death he was cannibalized by some of the other sailors. "This shows that . . . neither rank nor status was the governing principle in the final desperate days of the expedition as they strove to save themselves," says Stenton.

As we stood around that monument marking the spot where so many of Franklin's men met their end, we were all left to imagine what those poor marooned sailors must have experienced as the weaker members fell from starvation, disease, and exposure—and the survivors wrestled with the terrible choice of whether to eat them or starve.

While these dreadful thoughts swirled in my head, that north-easterly, blowing in off the ice in Victoria Strait, flowed over the shallow ponds all around us. The fresh water must have still held some warmth from the recent heat wave because cold smoke formed over the surface, creating an eerie haze in which I half expected to see the ghosts of Franklin's men trudging off into the gloom, back toward the ships. Those who could still walk might have felt like they were the lucky ones. But deep down, they too must have realized that no one was getting out of this alive.

CHAPTER 10

Icebound

Date: August 20, 2022
Position: Pasley Bay, Boothia Peninsula

With Renan's words—"We're fucked!"—echoing in my head, I raced up on deck, where I found Pasley Bay bathed in the crimson light of a never-ending sunset. Jacob stood silhouetted on the bow, holding one of our homemade ice poles. He pointed to the north, where a massive floe of ice, half the size of a football field, was steaming straight for us.

"We need to move. Now!" said Jacob.

But that wasn't possible because another floe, this one about the size of a tractor trailer, had floated down on our anchor chain, which had cut deep into the ice like a chain saw bound inside a tree trunk. I grabbed the other pole and Jacob and I attacked the floe, but it was fifty feet long and must have weighed thousands of pounds. Even with the two of us heaving against it with everything we had, it wouldn't budge. Ben came up, acknowledged our struggle, and then disappeared. A few seconds later, I heard the engine start. Without waiting for it to warm up, Ben put us in gear and drove straight into the floe snagged on the anchor chain, using our bow as a battering ram. The boat crunched against the ice with a sickening thud, nearly knocking Jacob and me off our feet. As the boat rebounded, the chain pulled partway out of the ice. "Again!" I yelled. Ben rammed the chunk even harder, and when the boat recoiled, the chain ripped free.

I stepped on the rubber foot switch for the windlass and slowly reeled in the chain. It came in painfully slowly, and we all stood on deck watching helplessly as the tanker-sized ice sheet spun toward us. With seconds to spare, I felt the anchor break free from the bottom. I called out *"CLEAR!"* and Ben spun us around and motored in the opposite direction with the hook still hanging in the water beneath the ship.

It didn't take long for us to find a patch of ice-free water on the other side of the bay, but when I loosed the anchor for the second time that day, I knew that this spot wasn't any safer than the last. This time, though, I assigned a watch rotation. Jacob drew the first two-hour leg, followed by Renan, Rudy, me, and Ben.

When I was awakened two hours later, it was Jacob standing over my bunk. "You need to come up," he said. "We're stuck again." The sky had darkened to a deep purple. A few stars twinkled faintly overhead. To the northwest, the sunset still simmered on the horizon. We used the same technique as before to ram our way out, and it worked again—but barely. Exactly what it would mean if we couldn't break loose was unknown and not discussed. As I secured the anchor in the bow roller, Jacob pointed toward the other side of the bay where a light shone about forty feet above the water.

"Is that an Inuit camp?" asked Renan. "Hunters maybe?"

"I think it's a ship," said Jacob.

"He's right," I said. "That's a mast light."

I ran below and hailed the vessel with our VHF radio, and to my delight discovered that it was *Taya*, a fifty-foot aluminum sailboat whose crew of two I had met in Pond Inlet. The skipper, Alan Cresswell, was a sixty-seven-year-old Frenchman. His bushy eyebrows and unruly mop of gray hair reminded me of Doc Brown from *Back to the Future*, and fittingly, it turned out he was a retired physics professor from Shippensburg University in Pennsylvania. His partner, Annina Barandun, was in her mid-forties, slim, with an angular, handsome face and long brown hair she wore in a bun. She was Swiss and had the ruddy complexion and crow's-feet of someone who spends a lot of time outdoors.

Alan had purchased *Taya*—his second boat by that name—in Tahiti in August 2018 and had been sailing her nonstop ever since. He and Annina had met in Chile's Strait of Magellan. Alan had just sailed in from French Polynesia and Annina was passing through on her way home from a ski trip in Antarctica. She wasn't a sailor, but she and Alan quickly hit it off. When *Taya* shoved off for Brazil a month later, Alan had a new first mate. Within days, Annina was steering, navigating, taking reefs, and covering 50 percent of the watches.

Neither of them had ever thought much about the Northwest Passage, but they loved the high latitudes and thought it would be cool to see how far north they could get. But then COVID hit, and Canada shut down all travel in the Arctic. As a consolation, they sailed *Taya* to Chesapeake Bay, Bermuda, the Azores, Ireland, the Canaries, and the Cape Verde Islands. In December 2021, they sailed back across the pond to Dominica in the Caribbean. Then they started working their way north into the Arctic.

"Hey, Mark, it's good to hear your voice," said Alan. "Where are you guys?"

Unlike *Taya*, we had not turned on our mast light, so they hadn't known we were also in Pasley Bay. I told Alan that we were across the way and about our tribulations with the ice. He said that they had tucked in behind a protective spit of land, and there was room for us. Ten minutes later, we dropped the hook about two hundred feet south of them, hoping we'd be safer there.

When my alarm went off for my anchor watch at five a.m., I found Rudy alone in the cockpit, listening to music on his headphones. It was eerily still, the water smooth and glassy like a lake in Maine on a windless summer morning. "I haven't seen any ice," said Rudy after pulling out an earbud. "If you want, you could go back to bed." So I did. A few minutes later, Rudy crawled into the bunk next to mine. Within seconds, he passed out and began his best impression of a bear gagging on Jell-O. I'd been dealing with his nocturnal garglings for weeks now, and they often drove me out of the captain's quarters. I grabbed my pillow and sleeping bag and headed out to the settee in the salon.

When I awoke a few hours later, bright sunshine streamed through the open hatch onto my face. Ben was in the galley cooking breakfast, and the thick aroma of frying bacon hung in the air. "Coffee's ready," he said. I poured myself a cup from the French press and stepped up into the cockpit, which was warm like a greenhouse in the morning sun. Jacob was still asleep, but Rudy and Renan lounged in T-shirts, soaking in the rare warmth. The southeast gale that the CIS had warned us about was forecasted to begin midday, but for the time being, the conditions in Pasley were as settled as any we'd seen so far on the trip. As Ben served breakfast, I downloaded the messages that had come in overnight via the satellite modem. Among them was a short note from a contact at the CIS, which I read aloud: "Our latest models are showing that the ice [in James Ross Strait] may move north but could pile into Pasley Bay."

This new intel struck me as immensely significant because if true, it meant that our temporary comfort could disappear quickly, and Pasley Bay could easily turn into a trap. And it wasn't lost on me how close we were to the location where *Erebus* and *Terror* had become beset in the ice—and we all knew how that had worked out for them. But as I looked at my companions to gauge their reactions, everyone just stared back at me blankly as they tucked into their eggs and bacon.

"I'm wondering if maybe we should get the hell out of here while we still can," I said. "If we leave now, we could head back north to Weld Harbour or the Shortland Channel by midafternoon and ride the storm out there."

"If we backtrack, we'll be too far out to make it through the strait during the lull that's supposed to start tomorrow morning," said Ben. "I think our best bet is to stay put." But then he added that if we started backtracking now, I would probably just keep on going—all the way home. Big sailing voyages are like big climbs, he said. As soon as you lose momentum, "it's over."

It annoyed me immensely that Ben thought that I wanted to bail. In fact, I was as determined as ever to finish what we'd started, and I wondered what had caused him to suddenly question my resolve. But before I could reply, Rudy said that he also wanted to take his chances in Pasley Bay.

I looked at Renan. He shrugged and didn't say anything.

Sailing has a long tradition of maintaining a strict chain of command in which the captain always has the last word. This allows for quicker decision-making and avoids the dysfunction that can sometimes come into play when the majority is allowed to rule. But as a career climber, I have always believed that expeditions should be run democratically. And since Renan appeared to be abstaining, Jacob would have to cast the deciding vote. So I went below to wake him up. As I touched his shoulder, he rolled over and gave me a look that said, *I dare you to do that again.* A few minutes later, still scowling, he joined us in the cockpit, and I read him the text.

"I think we should stay," he said without hesitation.

A few hundred feet away, I could see Alan and Annina drinking coffee in their cockpit. I hailed them on the radio, and Alan popped below to take the call. I told him that we had gotten the ice report and that we were debating whether to backtrack to the Shortland Channel, which lay about thirty miles north. After conferring with Annina, he came back on to say that they wanted to stay too. That sealed it. The vote was five to one, with one abstention. I felt frustrated, especially because my gut told me that we were making a poor decision. But I didn't feel comfortable bucking the majority rule, so I resigned myself to the fact that we would take our chances in Pasley Bay.

One silver lining to staying put was that it meant sticking with *Taya*. Sitting alongside a well-appointed expedition boat crewed by deeply experienced world voyagers made Pasley Bay feel a lot less remote and scary. But when I tossed out what I thought was a total no-brainer—that we combine forces with *Taya* for the next leg through James Ross Strait—Ben pounced once again. I suppose I should have realized that "buddy boating," as he called it, runs contrary to everything he holds sacred about sailboat voyaging: namely, self-reliance, traditionalism, and the freedom to sail where one wants when one wants without being beholden to the whims of another ship.

Ben pointed out that they were a faster boat than us, and he thought it would be unfair to ask them to wait if we struggled to keep up. I, on the other hand, felt that we'd have been in a stronger position if we stuck with *Taya* because they had a metal hull, and if we got stuck in heavy pack, they might be able to forge a path through the ice for us.

"I like our chances with *Taya*," said Renan, breaking his neutrality.

"If those were Inuit on that boat, what would you do?" I asked Jacob.

"We'd go together," he said.

Ben didn't say anything more.

ǀǀǀ

The wind turned on at two p.m. In an instant, the breeze grew from a light zephyr to a fifteen-knot southeasterly. Long, wispy tendrils of cloud streaked across the southern sky, reminding me of the old saying I'd first learned while climbing in the Alaska Range as a young man: "Mares' tails and mackerel scales make lofty ships carry low sails."

By midnight, it was blowing a near gale and *Polar Sun* strained at the end of the anchor chain. To ease the stress on the boat, I had attached a nylon rope called a "snubber" between the chain and the bow cleat to act as a sort of shock absorber. And I knew it was doing its job because I could hear it groaning and squeaking where it attached to the ship—right over Renan's bunk. The little hook of land that had protected us from marauding ice when we'd first arrived now formed a lee shore about three hundred fifty feet to the northwest. If the anchor dragged or our ground tackle failed, it would take only a minute before we'd run aground on the inside of the spit.

I again set up camp in the salon, where every clunk, bump, and thump sent me scurrying to the portlights to investigate. The southeasterly was clearing the bay of ice, and on the way out, some of the chunks were colliding with the ship. So far, though, none had snagged on the chain. Twice during the night, I went topside to inspect the chafe guard that I had wrapped around the snubber. Each time, I paid out a little more line to create a new wear point where it made its sharp turn over the edge of the deck. Occasionally, I heard Rudy's wolverine snorts emanating through the closed doors of the executive suite. How anyone could sleep under the current conditions was beyond me, but at least one of us was getting some rest.

At four fifteen a.m., Alan hailed me on the radio. Neither of us had explicitly said, *Hey, man, do you want to buddy-boat this next section?* It ended up just being an unspoken agreement. The forecast called for a twelve- to eighteen-hour lull beginning later that morning, so we made a plan to set off together at six thirty a.m. In a clay-colored world of gray water, gray clouds, and gray land, we cranked in the anchor and followed *Taya*'s canary yellow hull out of Pasley Bay. As we turned the corner and came out from behind the point, the wind and waves hit us dead on, and our speed dropped in half. But we could see at least a mile to the south, and there was no ice visible anywhere, which set off a cheer amongst the crew. "Gjoa Haven, here we come!" I shouted, pumping my fist in the air.

Ben stood next to me in his Westport Whaler, and without thinking, I put my hand on his shoulder. He looked at me with raised eyebrows, and I thought I saw the hint of a *Mona Lisa* smile tugging at the corners of his mouth. My old friend from Yosemite was now a bespectacled middle-aged man with the beginning of a paunch and a shaggy, gray-flecked beard. A painful-looking rash, probably the result of exposure to the elements, covered his peeling cheeks. I held up my hand with my fingers and thumb spread apart, and Ben grabbed it in a GI Joe grip—our standard greeting back when we used to cross paths en route to our neighboring caves in Yosemite.

But that wave of optimism evaporated a few minutes later when *Taya*, motoring two or three boat lengths ahead of us, suddenly disappeared into a dark miasma. Such dense fog could mean only one thing: ice. When the surface of the ocean is colder than the atmosphere above it, water vapor in the air rapidly cools and creates fog. And when it's *much* colder, as it is when covered in ice, that fog can be so thick that it can feel as if you've gone blind. As we pushed into the whiteout, ghostly walls of ice began to coalesce out of the mist right in front of the boat. The floes lay in a tangled heap, with some standing vertically like dominoes as if an invisible force was snowplowing the ice forward. With so much left to the imagination, my mind conjured an image of the ice closing around *Polar Sun* from every direction.

I couldn't see *Taya* anymore, but I tracked her AIS signal on the chart plotter as the boat turned east and followed the ragged edge of the pack toward the Boothia Peninsula. The wind still blew strong out of the southeast, so it made sense that a lead of open water might have formed along the strait's eastern shoreline. But when we got within a few hundred yards of land, the fog thinned and revealed that the ice ran clear up onto the shore, where dead gray floes lay atop the gravel like beached whales.

We turned around and followed *Taya* as Alan and Annina probed the floe edge in the other direction, hoping for a clear channel leading south. In the gloom, I didn't notice that the ice wall had bent our course back north until we motored out of the fog and I saw the entrance to Pasley Bay off to starboard. We continued north for a few miles, and off a point called Andreasen Head, a ray of sunshine broke through the scud, illuminating *Polar Sun*. Renan launched a drone and flew it north through patchy mist until it popped out into a pocket of blue sky. The view on the monitor showed miles of densely packed ice stretching as far as we could see. And a few miles from our position, it ran right into shore, just as it had to the south.

Now we could no longer deny the horrible truth: Pasley Bay was completely sealed off by ice.

|||

Two days later, I opened my eyes on another frozen morning and watched as my breath floated up toward the ice-glazed hatch above my bunk. Tiny crystals of hoarfrost scintillated on the teak woodwork, and all was silent, save for a muffled grunting coming from the other side of the cabin, where Rudy lay facedown, his head buried in a pillow.

It was time to go on anchor watch, so I unzipped my sleeping bag and stepped into a pair of puffy pants, leather seaboots, and a thick

Icelandic sweater. After putting on a kettle, I pushed back the hatch, noting on my way up that the thermometer at the nav station registered 39 degrees Fahrenheit inside the boat. A cold draft whooshed into the cabin as I poked my head out of the companionway and looked around. On this cheerless, foggy morning, the eighty-first since leaving Maine, the wind blew hard from the northwest, carrying sheets of sleet that coated the boat from stem to stern. We'd done our best to tighten down all the running rigging, but the halyards still vibrated inside the mast with a high-pitched, never-ending *ding-ding-ding-ding-ding* that I did my best to ignore. Through the one open panel in the enclosure, I could just make out the floe edge floating ominously in a vaporous haze about two hundred feet away. In the dim visibility, I could see only its lip, but I knew that miles of densely packed ice lurked out there in the murk.

Since our failed attempt to escape two days earlier, we had watched as Pasley Bay slowly filled with ice. Whenever visibility allowed, Renan launched his drone and filmed this ice invasion. From deck height, the ice blended into an amorphous, dread-inducing gray farrago. But from high above, the bay resembled an ever-shifting and metamorphosing kaleidoscope of frosted pieces of white, gray, blue, green, and brown sea glass—or what Pierre Berton described as a "crystalline world of azure and emerald, indigo and alabaster." In places, jumbled piles of multiyear ice jutted skyward from the matrix like miniature floating Alps. And when random floes swirled past, they fizzled like giant Alka-Seltzers as trapped air bubbles were released from their frozen bondage. Jacob told us that the Inuit call Pasley Bay "Kog-lee-ah-rok"—which in Inuktitut means "drum circle"—"because of the way the tides cause ice to dance in circles around the bay."

The first-year ice had a clean white look, like sheets of freshly washed linen, while the old multiyear, which seemed to be mixed in randomly, was often brown and smeared with dirt and sand. In a few places, rocks were embedded in the ice. These floes must have been stuck fast to shore, and when they whirled off into the void, they took pieces of the land with them. The big floes were the most impressive to me. Some were hundreds of feet across, their backs covered in pools of vibrant blue meltwater. These larger floes were always surrounded by little pieces the size of toasters, television sets, and refrigerators. I thought of these as the "babies," and indeed, they always seemed to cling to the mother floe. Each of the thousands of floes in Pasley Bay had a unique signature, and yet the artful way they interconnected in an endless dance made me wonder if there was something more than randomness at play in this isolated and frozen world.

By this point, nine-tenths ice had invaded every square inch of the bay's south arm, save for one little pool of open water where we now sat alongside *Taya* considering what we would do when the ice finally found our little hideout. I, for one, couldn't stop thinking about *Polar Sun* being crushed in the ice or, perhaps even worse, driven up onto the shore, where she'd lie high and dry as a permanent monument to my hubris and folly. Of course, I also thought constantly about my family, and I texted with Hampton (and her father, Alan) multiple times a day. Her description of life at home, including back-to-school shopping with Tommy, who was about to start first grade, only made me even more homesick. With each passing hour, the boat seemed to grow smaller and more cramped.

I shared my concerns with the crew and admitted that I was kicking myself about not listening more to my gut, which of course was code for *I told you so*. Moving forward, I let it be known, they could expect me to be adhering a bit less to democratic rule, which had gotten us into this mess. An awkward silence hung in the air following this proclamation until Rudy said something to the effect that we could end the ordeal whenever we wanted by hitting, what he called, "the Nat Geo button."

And there it was, right out in the open: RESCUE.

We had several different ways to call for an evacuation, but I think Rudy meant that he'd simply text someone at National Geographic, which had taken out a rescue insurance policy on our expedition. All we had to do was tap out, and they'd send in a helicopter to pick us up. This, of course, would mean abandoning *Polar Sun*, and I was starting to fear that if it came down to it, the crew would walk away without looking back. For me, though, it was a lot more complicated.

"You can all hit the Nat Geo button if you want," I said. "But I've made it this far in my life without being rescued, and I don't plan to start now. If we do get frozen in, my plan is to stay with the boat until it's secure in the ice and then get one of Jacob's buddies to come in from Taloyoak and pick me up by snowmobile." Taloyoak was the closest Inuit village, which lay about a hundred miles to the southeast.

Ben acted shocked that I was willing to stay with the boat until freeze-up, which might not happen until mid-November. But it was the only way that I could ensure that *Polar Sun* would at least have a chance of making it through the winter. It was an awful thing to contemplate, mainly because it would mean being away from my family for months longer than I had planned; but I had made this bed and now I had to sleep in it.

To be clear, no one worried that we would actually die in Pasley Bay. We were only a few hundred feet from land, after all, so even if the boat was crushed and sank, we could hop across the ice or use the dinghy—or, hell, even swim—to get ourselves ashore. The area was uninhabited, but there were plenty of places where a helicopter could land. And because I'd known that losing the boat was possible, I had brought a tent, a camp stove, fuel, and a month's worth of freeze-dried meals with which we could establish a base camp. One way or another, we would survive long enough to be rescued. Of that, I was sure.

The real risk was of losing the boat, which would not only be a huge blow for me both financially and emotionally; it would create an environmental disaster when all of the diesel, oils, and other fluids in the engine and tanks spilled into the bay, not to mention the shipwreck itself. Hell, I might even be looking at some hefty fines from the Canadian government. So if I had to stay on the boat alone for months to ensure that the ship survived the winter, I didn't see how I could possibly choose to do otherwise.

Later that evening, I sat at the nav station with my head in my hands, nursing an intense stress headache. I couldn't shake the picture of my crew jumping ship and leaving me behind. The fact that all of them had come right out and said that if the shit hit the fan, they would leave me behind, made it feel like, in a strange way, they already had.

Renan must have realized that I had hit my personal low-water mark of the expedition because when I looked up, he had his camera pointed at me.

"How's it going?" he asked.

|||

I awoke at three thirty a.m. with Ben standing over me. "The sound of ice grinding against the chain woke me up," he told me. "You need to see this."

I followed him up onto the deck. The ship, draped in fog, shook in a cold wind. It took me a minute to get my bearings, but when I did, I saw that we had turned 180 degrees and a hedgerow of jagged ice pressed against our starboard side. The only thing now stopping this implacable wall from driving us up on the shore was our ground tackle, which stretched, taut as a guyline, off the bow, where it disappeared into a tangled mass of ice. When I looked to the north, I saw Alan and Annina out on deck pulling up their anchor. Despite having five crew members to their two, we had missed the signs of incoming ice. They had reacted in time. We hadn't. And now we were even more fucked.

Ben and I stood together on the bow, wondering what to do. Should we cut loose from our anchor and try to push into the ice before it ran us hard aground? Or was it better to hang on in the hopes that *Polar Sun* might somehow prevail in a tug-of-war against a force of nature that has the power to reshape entire landscapes? If the strain on the ship became too great, would the anchor drag or would the chain snap? Or would *Polar Sun*'s deck cleat rip out of the fiberglass, leaving a gaping hole in the deck? One thing was sure: There was no chapter in *The Annapolis Book of Seamanship* covering the situation we now faced.

I decided to hang on as long as we could. To ease the strain on the ship, we attached a three-hundred-foot nylon rope to the chain and transferred the load onto it to introduce some elasticity into the system. As the strain built throughout the morning, a deep moaning sound emanated from the hull as if *Polar Sun* were a wounded animal crying out in pain. To ease her suffering, I uncleated the rope and let a few feet slip around the winch every half hour or so. Each time, the ice greedily snatched up whatever we offered it, and the only thing accomplished was that we found ourselves inching toward the line where *Polar Sun* would run aground, which now lay less than a boat length away.

By early afternoon, the rope was stretched so taut, I was afraid to even go near it. As a climber and mountain-rescue professional, I'd dealt with complex rigging scenarios for decades, but I'd never seen a load as big as the one we put on that one-inch Dacron anchor rope that day. My gut told me that something was about to fail catastrophically, and when I mentioned this to Ben, he said that he felt the same way. The ice had won. At two p.m., I made the call to "slip the cable."

The standard procedure in these situations is to tie a buoy to the end of the anchor rode so that it can be recovered later. But in our case, this float would just freeze into the ice and then drag our ground tackle away if the pack ever blew out of the bay. Instead, I had Rudy hack the chain out of the ice so it could drop to the bottom. Then we detached the Dacron line from the ship and Rudy used the dinghy to bring it to shore and tie it to a rock for future retrieval. We were now disconnected from the anchor, but a strong northwest wind pinned us inside a U-shaped cutout in the floe edge, and it was only after some aggressive bumper-car-style jockeying that we managed to break loose. In the tiny pool of open water, we motored about two hundred feet south toward a little ice alcove that seemed to be custom-made for the boat. As we turned into the opening, nosing the bow into a stack of bobbing refrigerator-sized hunks, I leapt off onto a floe and drilled in several tubular stainless steel ice screws that I attached to our dock lines with carabiners. Ben and Rudy dropped fenders over the side and *Polar Sun* found herself docked to the ice as if she had just landed in a marina.

Our new slip, which sat within a V-shaped slot cut into the edge of a basketball-court-sized floe, protected us from any rogue chunks that might come floating past. We had also increased our real estate by creating a deck of ice on which we could now stretch our legs and take a break from the poky quarters aboard *Polar Sun*. The main floe had a ridge along its back, which sloped down to a fountain-sized lagoon filled with lapis-lazuli-colored water in which two ice mushrooms sat side by side like surrealist sculptures. As I sat on the ridge, taking it all in, there was no hiding from the fact that the ice connected to the shore on both sides of us. Not only was there no way out, but when the tide flowed into the bay every twelve hours, chunks were breaking free from the main pack and piling up in this little basin. If it filled up even a little bit more, it would close off the entrance to our slip—and we'd be officially icebound.

A few hours later, I was in the cabin when I heard Rudy yell,

"Bear!" I bolted up the companionway and out onto the deck to find Jacob and Rudy pointing toward a seven-foot-tall polar bear that stood atop an upturned chunk of ice about a hundred fifty feet away. Jacob declared that it was an adult female and likely the same one we had spotted onshore a few days earlier with her cub. She was close enough that I could see her gleaming black eyes, which looked almost lifeless, like the little balls of coal I had just fed into the woodstove. Her fur was dirty yellow, and the front of her legs had a slight reddish tint. Standing proud atop her version of the *Lion King*'s Pride Rock, she held her head high as she surveyed her domain—in which we were now trespassing.

Ben had been trying to hop across the floes to get ashore when the bear popped out from behind a piece of ice and glared at him. He had left the shotgun aboard and so found himself caught completely off guard. The bear must have been decently well-fed because if it had wanted, it could have easily pounced on him and dragged his body underwater before any of us blinked.

"What does it mean when a bear holds its head up like that?" I asked Jacob.

"It means she's trying to decide if you would make a good meal," he replied.

Jacob squinted and pursed his lips, trying to keep a straight face. But that same running joke, which never seemed to get old, got the better of him, and slowly, the edges of his mouth turned up into a smile.

ǀǀǀ

Alan and Annina, who witnessed our debacle, had reanchored in a tiny pond of open water about a hundred fifty feet to the northwest. I suppose they took pity on us, because that afternoon they called on the radio to invite us over for dinner. When we pulled up in the dinghy that evening, they were both waiting on their stern swim platform. Alan took our line, and as soon as I stepped aboard, I found myself face-to-face with a two-foot-long dead iguana the color of a dried banana skin hanging by its neck from a string attached to a piece of the rigging. Some of the lizard's skin had flaked off, exposing its ribs. When Alan saw me staring at it, he said, "Ahh, a little souvenir from our travels down south."

Alan walked me to the bow and showed me how it had been "rearranged" by a growler in Prince Regent Inlet. The aluminum hull was crumpled like a car that had been in a head-on collision. The yellow paint had peeled off the dented metal, leaving an ugly scar. "I took my eyes off things for about fifteen minutes," he said. "I should have been more careful." The impact launched Annina out of bed and knocked Alan off his feet. When he made it topside, he saw the ice chunk bobbing in the water right in front of them. He reckoned that the growler probably weighed twenty thousand pounds and that they'd been doing about six knots when they hit it. "It wasn't quite like running into a stone wall," said Alan. "It must have given a little bit, or I don't think we'd be here.

"We were lucky," he said, then added, "very, very lucky."

Both of us looked over to *Polar Sun*, where she was tied to her ice dock. And though Alan didn't say it, I knew he was thinking that if *Polar Sun* ever hit something that hard, she'd go straight to the bottom.

Annina, who wore a baggy tan wool sweater, caught my eye and signaled for me to follow her below deck. As I descended a steep wooden stairway into the cabin, a wave of warm air enveloped me, and something about being in a new space suddenly made me feel better than I had in a long time. Annina went straight to the galley to stir something simmering in a big steel pot while I hung my coat on a hook by the ladder and took a quick look around. The ceiling had three massive skylights that filled the salon with light that reflected off the white beadboard and made the interior feel much airier than *Polar Sun*'s. In the center of the salon sat a circular dinette in which two half-circle settees enclosed an oval wooden table holding a bowl of popcorn. Behind this, orange flames flickered in the firebox of a diesel heater that had a wire grate built around it as a drying rack. A shelf running along the back of the settee housed an eclectic mix of books, board games, a box of pinot noir, an orange Frisbee, and a random forked stick covered in Day-Glo yellow lichen.

Alan and Annina stood beaming in the galley, arm in arm, deep into the wildest adventure a couple could possibly imagine. I thought about how much time they must have spent cozied up on either side of the dinette, warming themselves by the heater, playing cards, reading books, and talking as they sailed *Taya* across the great oceans of the world. Adjacent to the companionway was a snug little cabin with a full-sized mattress covered in pillows and big, puffy sleeping bags. Seeing their little nest made me miss Hampton ferociously.

Everyone squeezed in around the table and Rudy poured gin and tonics while Annina dished out Swiss spaetzle with cheese and chunks of sausage. I had just helped myself to seconds when the boat shuddered. I looked at Alan, and we both called out, *"ICE!"* as everyone shot up the companionway. While we'd been eating and relaxing below—and trying to pretend we were anywhere but trapped in Pasley Bay—an ice floe the size of a river barge had broken loose from the main pack and floated down on top of the boat. Unlike us, though, Alan had anticipated this as a possibility and had dropped his anchor into some loose gravel near shore—but *not* to set it. Annina, up on the bow, reeled in as much anchor chain as she could with the windlass; then Alan reversed the boat hard. The anchor dragged across the bottom and had almost slid out from under the ice when it suddenly bit into the mud and the chain came taut. Meanwhile, more ice was piling up, so Annina opened the clutch on the windlass, letting all of the chain fall to bottom. Alan backed away while Ben jumped in our dinghy with their bow line, which he attached to a cleat on *Polar Sun*'s stern. Then he sprinted across the ice floe where he'd seen the bear earlier, and drilled in an ice screw, while I tossed him *Taya*'s stern line.

Taya's anchor was trapped under the ice, but they were still attached to it and now docked at the ice marina lying at a 90-degree angle to *Polar Sun*, bow to stern. The only positive outcome was that they were now tied to the same floe as we were, which meant we could travel freely back and forth between the two ships.

Later that night, when we got back to *Polar Sun*, I plopped down on the settee next to Jacob. He looked at me and then drew his head back in surprise as if I'd grown a second head. "Who's your little friend?" he asked.

"Huh? What are you talking about?"

Jacob reached over, plucked something off the back of my shoulder, and held it up for everyone to see. It was Alan and Annina's lichen-covered stick, which, in Jacob's hand, now took on a whole new persona as a voodoo stick man straight out of *The Blair Witch Project*. I must have leaned back against it on the settee, and the lichen had caught on my wool sweater. The creepy little guy had been riding around on my shoulder for hours.

I was just happy to have an excuse to go back to *Taya* to return my "little friend."

ǀłǀ

"Well, fuck me," said Rudy the next morning when I passed him the latest ice chart. It showed that James Ross Strait was almost entirely ice-free. Even Victoria Strait, where *Erebus* and *Terror* had been abandoned, was open. In fact, the only place in the entire Northwest Passage where there was any ice at all was Pasley Bay. "I wish we'd listened to you, Captain," added Rudy. "Staying here was obviously the wrong call." It was big of him to say it, and I appreciated it far more than he realized.

Rudy passed the iPad to Ben. *Ahh, this is it,* I thought. *This is the moment when Ben is finally going to admit he was wrong.* Earlier in the trip he had told me in all seriousness that he had been wrong only three times in his life. Well, amigo, it was time to add one more. But instead, he looked at me straight-faced and said, "Just think about how lucky we are to be caught in the ice. How many sailors actually get to experience something like this?"

Later that afternoon, we all gathered in the salon to watch a movie—*Black Swan* with Natalie Portman—on Rudy's iPad. The conversation turned, as it always did, to the question of how the hell we were going to get out of our predicament. "I think our best option is to push into the ice," said Ben. "The hull on this boat is a lot stronger than you think, and if we go for it, there's a chance we could get out."

"I agree," said Rudy.

"I think you guys have cabin fever and you just want to get out of here no matter what. And you're willing to take serious risks to do it," I said, my voice rising.

"Don't you want to go home?" asked Ben.

I looked at Renan. He stared back blankly and said nothing.

Jacob, who sat next to Renan at the dinette, looked up from his phone on which he was playing Candy Crush. "I'm not taking that chance," he said.

On the morning of August 24, our eighth day since pulling into Pasley Bay, I awoke to snowfall and a boat glazed in ice. Overnight, a half dozen pool-table-sized floes had floated into the marina and blocked the entrance to our slip. I'd known this moment was coming, but now that it had arrived, I felt a cold knot tighten in my back behind my left shoulder blade. As more floes swirled in by the minute, I dashed below to get Ben. The guy often drove me nuts with his inflexibility, but I still trusted him implicitly and continued to turn to him without hesitation whenever confronted with a big decision. I feared now that he'd vote to stay put and let *Polar Sun* ice in where she sat, especially because there was a certain logic to it. With solid ice between us and shore, getting frozen in would at least prevent us from being run aground, and it would leave open the possibility that we might eventually float free if the ice ever flowed back out of the bay. But thankfully, when Ben saw that it was now or never, he agreed that we should try to force our way into the pool to the north where *Taya* was anchored.

I called for all hands on deck, and while Ben, using the hull to shoulder aside the larger chunks, backed us out, Jacob and I went to work on the smaller chunks with the ice poles—or *tuks* as Inuit call them. After bashing our way through a gauntlet of ice, we tied to the next floe down the line. I had just stepped below to make myself a cup of tea when I heard a loud crack that reverberated through the hull, followed by yelling. Rushing up on deck, I found that our new ice dock had split in half as if cut by a laser. Renan and Jacob were on the far side about a hundred feet away, running as fast as they could toward the crack, which was widening by the second. When they got to the edge, they both leapt across the gap, landing safely on the side to which we were still tied, which was now swirling out into dangerously shallow water. I checked the depth sounder and it registered only a foot of clearance between the keel and the seafloor. We needed to cut loose from the rogue floe immediately, but the bow was now pointing in the wrong direction, and the northwesterly, which hadn't taken a breath in days, pinned us against its edge.

There was too little room and too much wind to do anything with the engine, so I jumped in the dinghy and motored over to the next floe to the north, where I built an ice anchor and attached our stern line. While Ben cranked that line in on the port-side winch, Jacob released the bow line, and using a mountaineering technique called a "load transfer," we turned the boat around in its own length, coming within inches of the bottom in the process.

It went on like this all day. Each and every maneuver just barely worked, our little crew pulling them off only through a combination of desperation, adrenaline, hard work—and luck. A lot of luck. At one point, Ben drove the boat hard up onto a floe, and when it slid back off, the wind caught the bow and spun us 180 degrees. As the boat swung around, out of control, I looked back at Ben, who put his hands up as if to say, *Sorry, man. I'm doing the best I can.* After a half dozen of these white-knuckle maneuvers in as many hours—including one that Ben called "the slow sandwich of sorrow," in which two gigantic floes tried to squeeze *Polar Sun* like a pimple—I noticed open water between us and a mushroom-shaped ice chunk I'd been observing for several days. Unlike the rest of the ice, this block appeared to be stationary, which meant that it was probably hard aground.

We motored over, and when our bow kissed the face of the ice boulder, I jumped off onto its side wearing MICROspikes over my seaboots. But the boulder had the consistency of melted ice cream, and no matter how deep I chopped, I couldn't get down to any ice solid enough to take a screw. I was scratching my head, trying to figure out a way to securely anchor to this mush, when Renan appeared on deck, wearing a baby blue dry suit and holding a giant camera with a waterproof housing.

"I'm going in," he said. "Give me a line and I'll swim it around the boulder."

"You're a fucking legend," I said as he slipped into the freezing water with a dock line coiled mountaineer-style over his shoulder. He waded for a few feet along an ice shelf that extended underwater from beneath the boulder, and then struck out, dog-paddling with one arm, holding the camera in the other. When he got to the back of the boulder, he threw me the end of the rope; I pulled it tight and passed it up to Ben, who tied it to one of the bow cleats. After securing the stern to another floe with a couple ice screws, I shot bearings to some rocks onshore to confirm our position before I headed below deck to check the level on the vodka bottle.

ı|ı

A day later, our bearings hadn't changed, and Jacob and I dared to believe that it was probably safe for us to go ashore to stretch our legs and collect some driftwood to burn in the stove. He took the gun and headed for a gravel ridge because it offered a better vantage on our surroundings—and on any polar bears lurking in the area. The driftwood, though, was along the shoreline, so we split up: Jacob took the high road while I ambled along next to the ice, scooping up pieces of broken komatiks, bony little twigs, and the occasional waterlogged branch that had floated in from some place thousands of miles away where trees actually grew.

By now we had traveled through and across hundreds of miles of the central Arctic, but I had yet to experience a landscape as daunting and woeful as this one. On Baffin, Beechey, and King William islands, if you looked closely, you could always find a tuft of grass, a little flower, or at least some moss. But here, the land was so sterile and lifeless that we might as well have been on the moon. The potato-sized stones that now crunched under our boots looked like little bundles of razor blades, as if eons of −100 degrees Fahrenheit winters had vaporized anything resembling organic material and caused the rocks to implode. No matter how carefully I looked, I couldn't find even the tiniest flake of lichen, let alone an actual plant.

By the time Jacob and I met up, the haze had thinned and patches of blue sky flashed overhead. Suddenly, a shaft of light slanted down through the clouds, bathing us in the first sunshine we'd felt in more than a week. Instinctively, we both turned our faces upward and let the life-giving solar rays soak into our souls. Jacob sat down on a square chunk of rock, and I plopped down next to him, settling carefully onto the stone blades so as not to slice open my pants. We sat there silently staring out over the bay, listening to the floes as they rasped against one another like stalks of summer wheat rustling in the wind.

Earlier, I'd found a perfectly intact fox skull with pointy, gleaming-white teeth and dried sinew still holding the jaw in place. Thinking about Tommy and feeling homesick, I grabbed the jaw and moved it up and down, pretending to "talk" the skull—a game that had never failed to make my kids laugh. "Hello, Jacob," I said, doing my best impersonation of Kermit the Frog. "How do you like Pasley Bay?"

Jacob looked at me askance. Then he started to snicker. I hadn't seen him laugh since he'd made the crack about the bear eating one of us. It felt good to let loose and be silly. There hadn't been much of that of late.

We started to walk again. A half hour later, we arrived at the tip of the peninsula that forms Pasley's Bays western shoreline, and we gazed out into James Ross Strait. An unbroken, undulating field of pack ice unrolled to the west and as far as we could see to the north and south. The ice reminded me of the Neil Young album *Rust Never Sleeps.* It never stopped moving, especially out here in the strait where the pack was exposed to the full brunt of the twenty-knot northwesterly that had been piping down the McClintock Channel for days. The swirling floes in the bay were scary enough, but in the swell, the plates of ice heaved and slammed and ground against one another in a way that left no doubt about what would happen to

Polar Sun if we ever found ourselves caught in the pack in the open ocean.

Ben and Rudy had continued to hassle me about trying to bash our way out of the bay, but neither of them had yet hiked out here to see the pack ice roiling in a five-foot swell.

"Am I being a wimp, Jacob?" I asked.

"There's no way," he said. "Those guys are crazy."

׀׀׀

Early on the morning of August 26—our tenth day in Pasley Bay—Jacob panned his binoculars over the undulating carpet of blue-green pack ice that surrounded *Polar Sun*, and I knew he was searching for the polar bear that had been stalking us for the past few days. After a few minutes, he put the glasses down and turned to me. "Winter is coming," he said. Jacob had never seen *Game of Thrones* and was likely unaware of the phrase's reference to the show's menacing hordes of ice zombies. But it seemed a fitting statement, considering that the small amount of open water between the floes had frozen overnight, like the skim of ice that forms on ponds in New Hampshire after the first hard freeze in December.

As it washed over me that *Polar Sun* was probably spending the winter right where she currently sat, I felt a sudden, almost manic urge to get off the boat. Jacob and I had gone ashore together most days since we'd arrived in Kog-lee-ah-rok, but for some reason, he didn't want to join me that morning. Maybe it was too early or maybe he simply needed a break from me. The polar bear was out there somewhere, and the thought of facing it alone, without Jacob, was terrifying. But so was the idea of another day cooped up aboard. So I loaded up my pack with spare slugs, bear spray, and flares, and after I climbed into the dinghy, Jacob handed me the 12-gauge.

"Look back over your shoulder as you're walking," he said. "The bears will try to sneak up on you from downwind, or they might even come up out of the water. Be careful."

The dinghy didn't want to float through the thin ice, so I raised the oars and chopped downward with each stroke to break through the water's gelid surface. I pulled hard and the aluminum-bottomed inflatable made a strange crinkling sound as it crunched through the ice, leaving a trail of open water filled with broken windowpanes in my wake.

After pulling the dinghy up on the shore, I set off toward the point with the shotgun slung over my shoulder. Every few minutes, I looked back to check for bears. Jacob had told me that if approached by a bear, I was to fire a warning shot or two and not to actually aim at the animal until I was certain of its deadly intent. Too many trigger-happy white people from the south who know nothing about polar bears had killed these magnificent creatures unnecessarily. But if it came down to it, I was to aim for the bear's chest. I figured I'd be lucky to hit the broad side of a barn, let alone a specific body part on an enraged polar bear charging at thirty miles per hour.

At the point, I gazed out over the same field of ice that Ben and Rudy wanted to bash their way through. Nothing had changed except that the northwesterly had finally lain down and the air was still. The sun pulsed in the cobalt blue sky, and it was eerily silent, save for a faint susurration of waves lapping against the gravel shoreline.

I turned south and walked for miles along the spine of a steep gravel esker that rose above the ice-lined shore. I passed mounds of rotting kelp and the occasional dirty gray floe that had gouged out a deep trough in the limestone shingle when it was driven by the pack high above the tide line. As I strolled along, I collected small pieces of driftwood and set waypoints on my GPS for the bigger hunks that wouldn't fit in my pack. If I was truly going to spend the next few months in Pasley Bay, I figured I'd need every scrap of wood available to stay warm.

If Polar Sun *really was beset, how many days did I have before my crew would hit the Nat Geo button?* I wondered. I pictured a helicopter landing on the beach and everyone but me climbing aboard. How would that feel? Scary, for sure, but there was also a part of me that kind of looked forward to it. If I was alone on the boat, frozen into Pasley Bay, my existence would be pared down to pure survival for myself and the ship. First, I'd have to strip *Polar Sun* down for the winter. Maybe I could get the guys to help me unbend the sails before they left. We'd want to hoist them on a sunny day to dry them before we folded them up and stashed them in one of the empty bunks. The rest of the winterization, I'd have to do myself.

One big question would be when to decommission the engine. Once I flushed its raw-water jacket with coolant and put it to bed, I'd lose access to the truck heater and hot water. And without a way to keep the temperature in the boat above freezing—and there was no way the woodstove could do it alone—I'd also have to decommission the freshwater system and the desalinator. At that point, I'd be without plumbing, which would mean setting up some sort of outhouse on the ice and finding a source for drinking water.

Working through the technical aspects of surviving the winter on the boat seemed to calm my mind and it gave me a concrete list of tasks upon which I could focus my thoughts and fears. It also kept me from thinking about my real concern—that it could be months before I saw Hampton and my children again.

By the time I got back to the boat, a blazing midday sun had lit a fuse in the mass of ice covering the bay. Every few minutes, the air rang with the sound of melting chunks shattering and crashing into the water.

"I'm worried about the ice boulder," said Ben, who stood shirtless on an adjacent floe, holding a *tuk* in one hand.

Water poured off the boulder and a distinct crack now ran clear across its base. If the boulder toppled onto the ship, it could take down our mast. As if on cue, a huge piece of it sloughed off, spawning a wave that rocked the boat. Alan and Annina, who'd also been ashore, rowed up in their green fiberglass dinghy. Alan took one look at the ice boulder and declared that they were moving out.

"We need to go too," said Ben.

I hesitated.

"Come on, man!" he cried, his face turning red. "For once, can't you just make a decision?"

It was the closest we'd come yet to open hostility, and I sensed that the cord that ran between us, already relatively frayed over the last few weeks, was about to snap.

But I didn't know what to do. If we left the security of the ice boulder, where would we go? *Polar Sun*'s fiberglass hull wasn't made for smashing through heavy ice, and the forecast was for the wind to turn northeast. If it did and we got caught between the ice and the shore, we might finally be driven aground. On the other hand, Ben was right that it wasn't safe where we were anymore. As we argued, Alan and Annina fired up their engine, untethered from us, and, without further ado, began ramming open a path through the floes that surrounded us. It wasn't clear where they were going, and I'm sure they didn't know either.

As I watched them leaving us behind, Jacob approached me on the deck.

"It's time to go," he said, holding my gaze for a bit longer than was his wont.

I didn't respond, but he must have seen something in my eyes because he nodded as if I'd given him the go-ahead. Then he jumped down onto the ice and began removing the ice screws and dock lines. We were going, whether I liked it or not, and the next hours would decide whether *Polar Sun* would "float at freedom on the seas," as John Ross once put it, or end up lying on the bottom of the ocean—along with *Victory*, *Erebus*, *Terror*, *Anahita*, and so many others.

Jacob and I perched on the bow, *tuks* at the ready, while Ben drove us into a basin of open water the size of a swimming pool. On *Taya*, Alan and Annina had rammed their way out of this one just fifteen minutes earlier, but my hesitation had cost us dearly: The opening they'd created had already closed up. On four sides, we were surrounded by impenetrable walls of ice, but on the north end, I noticed that the barrier was thinner.

"This way," I yelled, pointing at the spot. Ben gunned the engine, but as *Polar Sun* accelerated, it was immediately clear that we were going *way* too fast. Jacob and I turned and yelled in unison, "SLOW DOWN!" But Ben didn't hear us—or didn't care to. The boat hit the ice with a crunch that lifted the bow out of the water. As Jacob and I hung on to the lifelines, she tipped on her side, and then all thirty-four thousand pounds of her slid backward into the basin, leaving a black streak of paint on the ice. I was so angry, I could barely speak. Aggressively ramming our way out of the ice was what Ben had been advocating all along, and in my mind, a part of why he was doing it now was to prove that *Polar Sun* could take far more abuse from the ice than I realized. And this realization triggered me even more. I had taken two steps back toward the helm to let him have it when Jacob called out, "Look!"

Using *Polar Sun* like a battering ram had worked—a shipping-container-sized chunk had broken loose, opening a narrow lead.

I let it go, realizing that what he'd done might have been necessary, and we motored into another basin of open water, passing boxcar-sized floes that had been shoved twenty feet up the beach. Ahead, *Taya* disappeared around the spit that we'd been hiding behind since we first arrived in Pasley Bay. Earlier in the day, the ice had been hard against the land here, but now someone was smiling on us because a westerly breeze had sprung up and pushed the floes twenty feet out from shore, opening a tiny channel. As we squeezed between the ice and the shoreline, we all held our breath as the numbers on the depth sounder registered the amount of water in feet below the keel—2.1, 1.5, 0.9, 0.6, 0.3. To this day, I still can't believe that our keel skimmed within inches of the seafloor and that we'd been so close to the land that I could have stepped off the rail onto dry ground.

As we rounded the corner, I took the helm. *Taya* was a few hundred yards ahead and relayed instructions and depths as we followed it through the ice maze. In a few spots, when they ventured into water too shallow for *Polar Sun*, I headed farther out into the pack and forged our own path, ramming ice chunks out of way, scraping through narrow openings and occasionally leaving stripes of paint on the edges of the floes.

Ironically, Renan had used up all of his drone batteries that morning shooting photos of *Polar Sun* in the ice, so we had lost our secret weapon for scouting the route ahead just when we needed it the most. Taking the old-fashioned route, Renan climbed up to the first set of horizontal spreaders on the mast. Suspended thirty feet above the water, he could see for miles in the clear air. As he called down directions, Ben relayed them back to me in the cockpit.

Because we had a scout in the crow's nest, it made sense for us to take point, so I passed *Taya* and entered a lead that led southwest for a half mile. I kept our speed below four knots so that I could stop the boat quickly if needed. The depth increased to sixteen feet, and the water was suddenly so clear that I could see dark boulders and kelp on the bottom. Slowly, the gaps between the floes grew wider. We found a rhythm as Jacob and Rudy shoved bergy bits to the side with the *tuks*, Ben relayed directions from Renan, and I did my best to maneuver through the icy obstacle course without hitting anything.

After passing through a section of brash ice, I heard Renan yelling over the drone of the motor: "I can see open water. We're gonna do this. We're gonna get out!"

"Renan's parting the ice like Moses so we can get through!" exclaimed Ben, who held his arms open as if he were the prophet himself.

We pushed through a final barrier of growlers into a large pool of open water, and as *Polar Sun* rocked in the gentle swell, I realized that the azure blue line stretching across the horizon before us was the ice-free water of James Ross Strait.

"We're in open water, boys," I called out, barely able to believe my own voice.

CHAPTER 11

Gjoa (*Jo-Uh*), Come Again

Date: August 27, 2022
Position: Gjoa Haven, King William Island

Twenty-four hours later, we motored into Gjoa Haven, where a cluster of weather-beaten homes clung to a bony hillside. In the front yard of one of them, perched right above some riprap, dried strips of Arctic char hung on a wooden rack swung in the breeze. In the small outer harbor, two men, bundled up in thick parkas, tinkered with an outboard motor on a red-sided aluminum boat. Jacob stood on the foredeck scanning his hometown through his trusty binoculars, and I wondered if he was hoping that a family member or friend might walk down to the shore to welcome him home. But the only one who seemed remotely interested in our arrival was an old husky that sat on the rock-strewn beach and turned its head to track *Polar Sun* as we motored past.

Though he didn't say it, I'm sure Jacob was relieved to trade the cramped and tense quarters aboard *Polar Sun* for some time with his family. The only complaint he'd made during the weeks he'd spent with us was about the lack of fresh game, which he called "country food." He'd never found that caribou on our overland expedition, and now that he was finally home, I knew it wouldn't be long before he set off into the interior of King William Island on his end-of-the-summer hunt. The rest of us, weary and near broken, were only halfway through the Northwest Passage—a fact I tried not to think about too much. It was August 27, four weeks since we had flown from Gjoa Haven back to the boat in Greenland, but so much had happened in the interim that our last visit felt like it had been a lifetime ago.

In the inner harbor, more aluminum boats were beached on the muddy, slate-colored shore, where a few people were fishing, and I decided that Amundsen must have been scarred by his own voyage to get here when he'd called this place "the finest little harbor in the world." Jacob pointed toward a barge tied to a crumbling wooden wharf. If we rafted to it, we wouldn't need to dinghy to shore, so we slid alongside and got our dock lines ready. When we were about ten feet away, a white man wearing a baseball hat and a forest green button-down shirt stepped from inside a shipping container on the barge's deck. When he saw us, he grimaced and conspicuously didn't offer a friendly word of welcome.

After shutting down the engine, I climbed over a metal railing onto the barge and introduced myself to the man, who, it turned out, was a dive tech for Parks Canada. He said that a few others and he were prepping for a season of archaeological work on HMS *Erebus*. The shipping container door was open, revealing an interior packed with dive tanks, hoses, tools, and all manner of mechanical apparatus. While we chatted, Jacob walked silently past with a backpack slung over his shoulder. His form crested the hill and disappeared silently into Gjoa Haven. Jacob knew that he was welcome to continue on with us, because I had told him as much, but it made sense for him to hop off in his hometown, and this had always been the plan. His wife and children and grandchildren needed his attention, and his freezer wasn't going to stock itself with caribou meat to sustain his family over the coming winter.

The last time I saw him was a day later at a little house we rented in town for a couple nights. He stopped by late in the afternoon, looking all clean-cut with a fresh shave and his bangs neatly combed. It took me a minute, as he stood there staring at his feet with his hands in his pockets, to realize that he'd come to say goodbye. I wanted to hug him, but I sensed it wasn't his thing, so instead I just thanked him sincerely for all his wisdom and guidance and for having my back when we were trapped in the ice. What I didn't say, but I hoped he understood, was that spending time with him in the Northwest Passage had given me a tiny window into his culture and a way of life centered on a connection to nature that we have lost in the south. And in this way, I felt that I had come a little bit closer to understanding how Inuit not only survived but thrived—with lives full of love, art, philosophy, and spiritual gravity—in the same country where the members of the Franklin expedition had starved and eaten one another. After we shook hands and he walked silently out the door, I found myself feeling quite emotional. For a guy who didn't say much, Jacob had an outsized presence, and *Polar Sun* would not be the same without him.

I wanted to follow Jacob into town to stretch my legs too, but shortly before we landed, Renan had pulled me aside and told me we needed to have a team meeting before everyone disappeared. When I asked what this was about, he said that Ben had suggested to Rudy and him that finishing the passage before the end of the season might not be in the cards. This came as a shock to me because I had already done quick back-of-the-napkin calculations and decided we still had time to get this done. And when I'd mentioned this to the crew, no one had pushed back. So as far as I knew, we were still all in.

I followed Renan below deck and found Ben sitting in his go-to seat by the galley, looking unusually serious. He wore the brown wool sweater his daughter Antigone had knit for him; it was now so well loved that the neckline sagged down his back. Rudy manned a camera on a tripod at the base of the companionway, and Renan gestured for me to take a seat on the starboard settee, upon which someone had draped a towel decorated with butterflies and wildflowers that I'd never seen before. *Fuck,* I thought. *They're filming. This can't be good.*

After an awkward moment while we waited for Rudy to start rolling, I spelled out my plan for the remainder of the expedition. If we spent two full days in Gjoa Haven, we could leave on August 30 and head straight for Terror Bay, where I wanted to make a stop to tie up a loose end from the overland expedition. From there, it was only another day to Cambridge Bay, then six more to the hamlet of Tuktoyaktuk ("Tuk"), near the mouth of the Mackenzie River. At this point, we would technically be out of the passage part of the Northwest Passage and into the Beaufort Sea, where the twelve-hundred-mile sweep of the North Slope of Alaska, punctuated by Nuvuk (formerly known as Point Barrow), stood as the last obstacle between us and Nome. This leg would probably be the most consequential of them all because we'd be running down a tiny strip of open water between the North Slope of Alaska and an immense sheet of pack ice that covered most of the Arctic Ocean all the way to the North Pole. Currently, the lane of open water was only about fifty miles wide, and all it might take was one northerly gale to close it out and effectively seal off our escape route. The scariest part of all was that there was a section of more than a thousand miles across which there wasn't a single decent harbor in which *Polar Sun* could hide if the ice did invade.

In fact, this exact scenario had played out on September 7, 1871, when forty American whaling ships, including twenty-two from

New Bedford, Massachusetts, were hunting bowheads to the south of Point Barrow (Nuvuk) when a high-pressure system over Siberia began pumping strong northerlies toward the Alaskan coast. When the ice closed in, seven vessels managed to escape, but the rest were all caught in the pack. A week later, all thirty-three ships were abandoned and 1,219 people, including wives and children of some of the sailors, set off on a seventy-mile trek to rendezvous with the ships that had escaped the ice. Amazingly, not a single life was lost, but that winter, every ship save one was destroyed in the ice.

In 1905, after his two winters spent in Gjoa Haven, Amundsen was caught in a similar situation near the mouth of the Mackenzie River, where he was forced to spend a third winter in the ice. *Polar Sun* might be floating free for the moment, but there were no guarantees that she wouldn't find herself frozen fast in the ice once more.

The main thing working against us at this point, apart from low morale, was that we had lost ten days in Pasley Bay, and as Jacob had pointed out, winter *was* coming. I'd been warned by numerous people to be out of the Arctic by the first week of September—no matter what. And now here we were, still only about halfway through the passage, and it was already the end of August. Nome was still twenty-three hundred miles away—the equivalent of sailing across the Atlantic Ocean. Even with perfect conditions, which we knew not to expect, it would take us at least another three or four weeks to complete our escape.

"According to my calculations," I continued, "in a perfect scenario, we could be in Nome by September 15."

Of course, there was no denying that September 15 was ten days later than it could have been, but it also happened to be our original target. I knew that if we stuck together and dug deep, we could finish the passage that season. But something felt off. And as the cameras rolled, I wondered if our team had lost more than just time while we were trapped in the ice.

I looked over at Renan. His face looked puffy, and he had purple bruises under his eyes. If anyone had slept less than me on this trip, it was him. He had struggled with seasickness, and throughout the voyage, he had been torn between his obligation to serve as a fully contributing hand aboard the ship and his professional duties as a filmmaker. It was impossible to do both jobs well, and the moments I called for all hands on deck were the exact times when Renan knew that he had to be shooting. It was hard for me to fully appreciate the pressure he'd been under, but when I looked into his eyes, I could see that his fire for this trip had burned out.

The conversation ended without resolution, but my gut told me that we had peered over the abyss and that it would be hard to regain the energy and commitment we needed to finish the trip. I'd learned this lesson years ago in the mountains, and my partners and I had always had an unwritten rule that bailing was not to be discussed until there was absolutely no way forward. I know that sounds cavalier, but without that level of grit and determination, serious climbs—or Arctic sailing passages—probably won't get done.

Later that night, we were sitting around the dining table in that shabby rental home normally used to house temporary construction workers, when Renan announced that he was flying home from Gjoa Haven in two days. "I'm really sorry, man," he said. "I made the mistake of suggesting that I might be done with the expedition, and Taylor went ahead and booked me a flight."

I tried not to show it, but it stung a bit that he had pulled the plug without at least discussing doing so with me first. I guess that's what the conversation earlier that day had been, and I questioned when exactly this die had been cast. The worst part of it was that he and Rudy were a package deal. Rudy then said he'd do one more leg with Ben and me to Cambridge Bay, and then he was going home too.

And as if to put the final nail in the coffin, Ben now told me that he wasn't comfortable carrying on with only two people. I wondered if that was even more true when one of the two people was me. But at this point in the game, I really couldn't blame him. As much as Ben sometimes drove me nuts with his moral superiority, I knew that it must have been a two-way street, and I wished I'd done a better job of channeling some of Leopold McClintock's legendary stoicism when we were trapped in Pasley Bay.

"We can wait and see if it's possible to haul the boat out of the water in Cambridge Bay," said Ben, "but I've been gone for a long time, and I'd really like to get home in time for my daughter Emily's birthday. I've never missed one." Ben is a tough dude, and he almost always exuded a positive outlook on things, but his eyes got misty now. He'd never spent more than two weeks away from his wife and daughters before, and here we were almost three months in with many more weeks of stressful, dangerous sailing still lying ahead of us.

I felt it too. In two days, Tommy was set to start first grade at the local grammar school. He'd done kindergarten at a private, experiential outdoor program, so this would be his first time attending "real" school. Hampton told me he was putting on a brave face, but she could tell that he was nervous. Tommy has always been apprehensive about the unknown. Of course, as certainly as I knew how hard it would be to find my way out of the Northwest Passage, I knew that Tommy would be stronger and better able to face that unknown if he had *both* parents backing him up. It pained me deeply that I wasn't going to be there on such an important day in my little man's life.

The next day, I was down below alone, writing in my journal, when I heard a knock on the hull and someone say, "Ahoy." I popped up to find a man and woman standing on the edge of the dive barge. "Are you Mark?" said a thirtysomething woman with long, curly dark hair and a friendly face. She introduced herself as Monica, an employee of Parks Canada. Her colleague, David, who was tall and thin with angular features and a short-cropped salt-and-pepper beard, shook my hand warmly and said that he was a producer with the National Film Board of Canada. They were working on a documentary about the Franklin wrecks, and they'd heard about our trip and wanted to stop by to check in.

We sat down in the cockpit, and I told them a bit about *Polar Sun*'s voyage from Maine, our overland expedition, and those harrowing days in Pasley Bay. Monica and David had recently flown into Gjoa Haven to rendezvous with a Parks Canada research vessel called the *David Thompson*, which was supposed to tow the dive barge to which we were still tied out to Wilmot and Crampton Bay, where they'd anchor it beside *Erebus* and spend a couple weeks making dives on the wreck. They had hoped to already be underway, but the *David Thompson* had run into thick ice in James Ross Strait. Apparently, it had shut down for business within hours of us sneaking through. The *David Thompson* was currently hove to, waiting for a Canadian Coast Guard icebreaker to cut it a path through the ice.

Monica and David explained how the film had an Inuit director and would focus heavily on the indigenous people's rôle in the discovery of *Erebus* and *Terror*. David pointed out how neither of the ships would have been found if the Inuit hadn't pointed searchers in the right direction. In the case of *Erebus*, the first clue in this treasure hunt had been unearthed by Leopold McClintock in 1859.

That spring, during the historic sledging expedition that would later result in the discovery of the Victory Point Record, McClintock met a dozen Inuit living in an igloo village out on the ice near the entrance of Pasley Bay. These Inuit had in their possession snow shovels made out of mahogany, meat tins, and a three-foot-long wooden box with brass hinges, all of which McClintock assumed had been salvaged from the Franklin ships. An old Inuk man named Oo-na-lee had testified that years earlier Inuit had found two ships wrecked off the coast of King William Island. One had sunk in deep water and "nothing was obtained from her, a circumstance at which they expressed much regret," while the other had been driven aground by the ice near a place called "Oot-loo-lik." Per Oo-na-lee's recounting, after this ship was destroyed, "the white people went away to the 'large river,' taking a boat or boats with them, and that in the following winter their bones were found there."

McClintock wasn't quite sure where Oot-loo-lik (more commonly referred to as Oot-joo-lik and Utjulik) was located, but Charles Francis Hall later determined that it lay near O'Reilly Island in Wilmot and Crampton Bay, west of the Adelaide Peninsula and about fifty miles due south of Terror Bay. Another Inuk told Hall that the Oot-joo-lik ship had four boats hanging from its side and a plank extending from the gunwale down to the ice. A woman named Koo-nik recounted that some hunters boarded the vessel and "found a very large white man who was dead. . . . The place smelt very bad. His clothes all on. Found dead on the floor—not in a sleeping place or birth [*sic*] . . ." with teeth that were "as long as an Inuit's finger."

Hall's informants probably hadn't been aboard the ship at Oot-joo-lik but rather had heard these accounts from other Inuit. But in 1879, Frederick Schwatka interviewed a man named Puhtoorak, who testified that he had personally boarded the vessel.

"When his people saw the ship so long without anyone around, they used to go on board and steal pieces of wood and iron," wrote Schwatka.

> *They did not know how to get inside by the doors, and cut a hole in the side of the ship, on a level with the ice, so that when the ice broke up during the following summer the ship filled and sank. No tracks were seen in the salt water ice or on the*

ship, which was also covered with snow, but they saw scrapings and sweepings alongside, which seemed to have been brushed off by people who had been living on board. They found some red cans of fresh meat, with plenty of what looked like tallow mixed with it. A great many had been opened, and four were still unopened.

According to Puhtoorak, the ship sank about eight miles off Grant Point in water so shallow, her masts stuck above the surface. Testimony regarding the existence of this ship—collected by McClintock, Hall, Schwatka, and the Greenlandic-Danish polar explorer Knud Rasmussen—is remarkably consistent. Over a span of more than sixty years, including testimony from a dozen different informants, the location was repeatedly placed in the same area.

And yet, for as long as people have been looking for answers to the Franklin enigma, Inuit testimony relating to the expedition has been criticized for being, in Russell Potter's words, "unreliable, garbled, inconsistent, or fraught with potential inaccuracies due to hearsay or poor translation." But as Potter points out, this is grossly unfair considering how accurate it has proven, time and again.

David now pointed out what I already knew: that when the Parks Canada archaeologists discovered *Erebus* in Wilmot and Crampton Bay in 2014, the location—roughly ten miles northeast of O'Reilly Island and the same distance southwest of Grant Point—was more or less exactly where Inuit had always said the ship had gone down.

Monica explained that due to COVID-19, there hadn't been any dives on the wrecks for the past two seasons. Now that the Arctic was finally back open for business, the archaeologists had decided to focus their efforts on *Erebus* because it was in much shallower water than *Terror*—thirty-three feet versus eighty—and storms and waves had been breaking up the ship. It was critical, she said, that they study and salvage what they could from it "before it was too late."

The *David Thompson* arrived later that day and anchored in the bay. The ship was nearly a hundred feet long, with a dark green steel hull and a multistory superstructure framed by two red-and-white antennae that stuck forty feet into the air on either side of the bridge. A yellow crane used for launching its dinghy, which was bigger than most of the vessels in the harbor, hung over the stern. We were told that *Polar Sun* and *Taya*, which had pulled into Gjoa Haven about two hours after us, needed to decouple from the dive barge so that the Parks Canada team could finish their preparations to get underway the next day.

Alan and Annina had decided to leave that night. Their plan was to push as hard as they could to complete the Northwest Passage that season. "I just don't feel good about leaving *Taya* in any of the communities up here," said Alan. "Even if we could get her out of the water, I'm really worried about what might happen over the winter. If anything broke or got damaged, just think about how hard it would be to get spare parts or any work done.

"You should leave with us," said Alan. "We can do this together." I looked at Annina. She smiled and nodded her head in agreement.

I told them I'd think about it; then I gave each of them a hug, like I'd done with Renan earlier that day when he left for the airport.

Rudy, Ben, and I followed *Taya* about twelve hours later, and the boat felt strangely empty without Renan pointing a camera in our faces. We had timed our departure to half tide on the flood to catch a fair current up Simpson Strait, the rock-strewn inlet that separates King William Island from the Adelaide Peninsula. The mainland side of this notorious channel is riddled with islands and shoals, and the navigable section, at its narrowest, is less than a mile across and prone to strong tide rips, back eddies, and currents that can run up to seven knots. Many vessels have come to grief in these waters, including a four-hundred-foot passenger ship called *Hanseatic*, which ran aground in 1996 after honoring a green navigational buoy that, unbeknownst to the captain, had been carried six hundred feet from its proper position by ice. I was nervous and a bit unsure if we had timed it right until I looked back and saw the *David Thompson*, with the dive barge in tow, following us out of the harbor.

As the Parks Canada archaeologists and production crew steamed for *Erebus* in Wilmot and Crampton Bay, I set a waypoint for Terror Bay, where her sister ship rested peacefully on the seabed in eighty feet of water.

At the end of the overland expedition, Tom Gross had dangled one last clue—and I would now use *Polar Sun* to chase it down.

CHAPTER 12

Terror

Date: August 31, 2022
Position: Simpson Strait

At the end of our first day out from Gjoa Haven, *Polar Sun* skimmed across a glassy Simpson Strait on the calmest night I'd ever spent at sea. High pressure had settled over the central Arctic, and the dark vault of the night sky sparkled with a million pinpoints of light. Ahead, to the west northwest, Jupiter pulsed near a V-shaped constellation, while behind us, Mars sparkled like a ruby cast amongst a field of diamonds. A few fingers above the invisible horizon, three satellites zipped in the direction of the North Pole, where a strange miniature cloud illuminated from within hovered just below Polaris. As I watched, mesmerized, this apparition began to slowly descend until it disappeared into the inky abyss. Untethered from the sea, *Polar Sun* glided through the ether like a spaceship, as if there were no boundary between Earth and the heavens. The only thing marking our passage—and keeping me from totally losing my shit—was the low hum of our trusty Ford Lehman motor and a glowing green streak of phosphorescence we left in our wake.

Knowing that Transport Canada was monitoring *Polar Sun*'s

sometimes inaccurate AIS tracking signal—and that the fine for violating the restricted area around *Terror* was supposedly a half million dollars—I took a few photos of the chart plotter as we approached to prove, just in case, that we were indeed skirting *around*, not going *through*, the national historic site. As the sun poked above the horizon to starboard and the low-lying land of King William Island slowly took shape from the gloom, I scanned for the massive inukshuk I'd seen a month earlier as the overland team bounced along this same coastline on our bikes on the way back to Gjoa Haven. Several miles off to the south, that cairn had pierced the sky like a missile silo, screaming, *I'm important. Come here.* But when I had asked Tom Gross if we could detour to check it out, he said it sat on a tiny islet called Fitzjames Island near the entrance of Terror Bay. And since we didn't have a boat on our overland trip, there was no way to get there. He also tossed out, rather casually I thought, that Fitzjames Island was rumored to be a place where an Inuk elder had found graves and some musket balls about a hundred years ago. And as far as he knew, no one had ever gone there to investigate.

Ever since we'd been back on the boat, I'd kept this little tidbit in my back pocket. Fitzjames Island was only about ten miles off the rhumb line to Cambridge Bay, and I knew that if I didn't stop to check it out, I'd always wonder whether that huge monument might mark the grave of Franklin or some other important member of his crew.

ı|ı

A month before we sailed into Terror Bay, Tom, Jacob, and the rest of us had spent the last few days of the overland expedition searching in a cold, unremitting drizzle for the Supunger vaults in the vicinity of the Boat Place. As evening fell on our last night in Erebus Bay, the rain had soaked through my foul-weather gear, and upon seeing my trembling blue lips, Jacob drove out to the shoreline and piled a mass of driftwood onto the back of his bike. That night, at our camp a few miles in from shore, we huddled around a raging fire stoked with a little encouragement by some spare gasoline. As I stared mindlessly into the flames trying to shake the chill, Tom opened his Franklin bible and told us another story.

This one was about an Inuk man named Kok-lee-arng-nun. In 1866, he met Charles Francis Hall, and told him that years earlier he'd been invited aboard a ship frozen into the ice off King William Island. According to Hall's account, Kok-lee-arng-nun described the chief of the ship as "an old man broad shoulders, thick . . . with gray hair, full face, bald head" and referred to him as Too-loo-ark, which means "raven" in Inuktitut. "He was a very cheerful man, always laughing; everybody liked him—all the kob-lu-nas and all the Innuits." Too-loo-ark wore spectacles and was always offering food, but he was "quite lame" and appeared to be sick.

"Was Too-loo-ark Franklin?" I asked.

"We don't know," Tom replied, "but Hall certainly thought he was." Tom flipped to the next page in his bible and showed me the daguerreotype of Franklin taken shortly before the expedition set off from England. In his pointed black bicorn hat and long, dark overcoat, "ravenlike" seemed a fair description for the British captain.

The ship, according to Kok-lee-arng-nun, was anchored in a bay where "a great many, many men on the ice had guns + many had knives with long handles," and they "killed a great many" caribou, so many in fact that the bloody carcasses made a line across the bay.

"What would the Inuit do if they came to hunt on King William Island and found white men killing all the game?" Tom asked. He was looking at Jacob, but his friend had his head down and was staring at the steam rising from his insulated boot liners, which he had placed next to the fire. Having lived among Inuit for most of his life,

Tom was used to such silences, and so he answered his own question. "Inuit shamans would have put a curse on Franklin's men," he said. "I'm convinced that the Inuit may have once known where Franklin's tomb is located, but they didn't want it to be found because it was cursed."

Kok-lee-arng-nun told Hall that after this first winter when he and his wife visited the Franklin ships, one of the vessels had been crushed in heavy ice the next spring. Hall, quoting Kok-lee-arng-nun, wrote that "the men all worked for their lives in getting out provisions, but before they could save much, the ice turned the vessel down on its side, crushing the masts and breaking a hole in her bottom. She sank at once, and has never been seen again. Several men at work in her could not get out in time, and were carried down with her and drowned."

After that, Kok-lee-arng-nun and his wife didn't see Too-loo-ark anymore, and "Ag-looka was Esh-e-mūta [chief]." We know that Franklin died on June 11, 1847, and if he was indeed Too-loo-ark, this would explain why he was no longer around. And you'll recall that Aglooka was the nickname given to Francis Crozier in Igloolik in 1821. Crozier wasn't the only sailor called Aglooka. "He who takes long strides" was a common sobriquet for white men traveling through Inuit lands. But Hall, who spent years collecting testimony from Inuit, always believed that Crozier and Aglooka were one and the same.

Kok-lee-arng-nun's story bears similarities to another piece of Inuit testimony from a woman named Ookbarloo, who told Hall of a lone Inuk man who visited a ship by dogsled when it was frozen in the ice. The captain of the ship, a white man, was very kind. He took the Inuk down into his cabin and gave him food and drink. The Inuk man visited the ship two more times, once in the spring or summer, when "the sun was high." On his third visit to the ships, "a great many men—black men—came right up out of the hatch-way + the first thing he (the Innuit) knew, he couldn't get away. These men who were then all around him, had black faces, black hands, black clothes on—were black all over!"

The Inuk was frightened and even more so when the men erupted in three loud rounds of cheering. The noise roused the captain, who came up on deck, told the men to stop, and sent them below. The captain now pointed to a tent on the land and told the Inuk that "black men such as he had just seen, lived there + he (this Innuit) nor any of his people must ever go there."

Ookbarloo's story is striking for its detail, and it's hard to imagine that she fabricated it out of whole cloth. And, of course, it begs the obvious question: Who were these black men? In his book *Finding Franklin*, Russell Potter suggests that the Franklin sailors might have been celebrating Guy Fawkes Day, a British holiday (celebrated on November 5) that marks the failed seventeenth-century attempt by Fawkes to blow up the British Parliament and assassinate King James I. McClintock recorded a similar occasion aboard the *Fox* during which the sailors, with "blackened faces, extravagant costumes, flaring torches and savage yells," marched around the ship playing drums before they burned an effigy of Fawkes.

For Tom, the crucial clue in Ookbarloo's story is that the captain pointed out the tent from the deck of the ship. If, at the time, the ships had been trapped in the ice fifteen miles from shore in Victoria Strait, there's no way the land—let alone a tent—would have been visible. Tom has always believed that the Bayne, Kok-lee-arng-nun, and Ookbarloo stories recount visits to the ships when they were in Collinson Inlet—or some other bay on the north end of King William—*before* the desertion.

"But if the ships were brought into Collinson at some point," I asked, "why didn't Fitzjames mention this in the Victory Point Record?" Tom said that he had wrestled with this contradiction for years before deciding that the only way he'd ever solve the mystery would be to start thinking outside the box. "What if we set all the European knowledge of this tragedy aside," he said, "and we pretended that the Victory Point Record doesn't exist? If we only listened to the Inuit, where would that lead us?"

Tom's obsession with finding Franklin's tomb, I had come to realize, began and ended with the stone house, and since he knew it lay somewhere in the vicinity of Collinson Inlet, he had focused most of his efforts on searching in that area. But since 2016, we've known that *Terror* lies on the bottom of Terror Bay. And since the location of this ship is one of the biggest clues in this entire mystery, it's only logical that all of the Inuit testimony regarding visits to the Franklin ships should be recalibrated. *What if,* I wondered, *Tom is wrong, and the Bayne, Kok-lee-arng-nun, and Ookbarloo stories aren't from Collinson Inlet but from Terror Bay after the ships were remanned?* I didn't know when or if I would ever return to King William Island, so at the end of the overland expedition, I asked Tom if we could go to Terror Bay and look around. He was reluctant at first, saying that it wasn't included on our permit. But I kept pestering him and he eventually agreed that if we swung through for some "sightseeing" on our way back to Gjoa, it would probably be okay.

And so, on the second-to-last day of the overland expedition, after a long drive south along the shore of a milky-colored inlet that bisects the Graham Gore Peninsula, we popped out onto broad tidal flats stretching across the head of Terror Bay. With a veritable freeway open before us, everyone took off in different directions, reveling in the rare freedom of ripping at full throttle across the land. I headed south to the end of a peninsula jutting into the bay, and then followed the shoreline around into a narrow cove. There I found the rest of the team gathered outside an abandoned HBC trading post where two old wooden buildings were slowly disintegrating. One, which appeared to be a machine shop, had collapsed entirely. Its metal roof lay in a crumpled heap atop a scattered assortment of rusted tools, bits of pipe, and engine parts. Lying amongst the detritus was a wooden board printed with the faded words "Hudson's Bay Company Cambridge Bay."

The main building was missing its doors and windows, but a still upright chimney poked from the center of the roof, and the west wall, facing the inlet, was still sheathed in weathered gray clapboards. Broken glass crunched underfoot as I stepped inside, where Devon pointed out a claw mark dug a half inch into the wall. "Polar bear," he said. "They find joy in wrecking stuff." The main room had a rusty barrel stove in one corner and a few pieces of furniture, including a cupboard and a dining table covered in algae. What immediately drew our attention, though, were the moldy pages from the February 1936 issue of *Esquire* magazine that papered the walls. A story called "Modern Engineering Beats the Killers of New Guinea" sat alongside an ad for Evan Williams Shampoo, which "keeps the hair young," and a column entitled "This Petting Problem" with some sage advice on the question of "what privileges a girl should allow to a boy who takes her out."

While the rest of the team explored the surrounding area, beachcombing and sifting through limestone fossils, animal bones, and a time capsule of trash from the 1930s, I wandered up a small hill where I found two graves marked with thick pieces of tin into which someone had used a punch to make the inscriptions. N. P. Klengenberg had "past away" on August 4, 1943; Jane Klengenberg, buried next to N.P., had died a year later, at the age of eighteen. I'm still not sure of N.P.'s identity, but Jane was the granddaughter of "Patsy" Klengenberg, who had run this outpost back in the 1920s, '30s, and '40s. Patsy was one of nine children born into the family of a controversial Danish whaler, trader, and explorer named Christian Klengenberg.

In 1930, when this trading post was in its heyday, the mining engineer L. T. Burwash, who was investigating the Jamme document, flew to Terror Bay after searching for Franklin's tomb near Victory Point. He landed his plane on a patch of open water near the mud flats that we had torn across on our bikes, and after a brief search of the coastline, he found a grave that he had reason to believe might have contained more than one body. In his report in *Canada's Western Arctic*, Burwash reported that he had carefully replaced the stones covering the grave site. But he didn't record the location, and the site has subsequently been lost in obscurity. What's interesting, though, is that Burwash had a copy of the Bayne map in hand, and he wrote that the location where he'd found the grave site was "much similar in structure to that examined at Victory Point."

The only other clue relating to this grave surfaced seven years later in 1937. In an HBC publication called the *Beaver*, an essay on the Franklin mystery written by a former RCMP officer turned HBC director named William Gibson includes two photographs taken by Patsy in 1931. One shows Patsy's wife, Mary Yakulan, and his daughter sitting atop a glacial erratic beside a crumpled copper tank identified in the caption as "a relic of the Franklin expedition; obviously the remains of a water tank from one of the life boats." The second photo is of a large cairn supposedly rebuilt by Patsy and identified as the grave of a Franklin sailor. The photographs were taken somewhere in the vicinity of Patsy's trading post, but to date, no one has been able to find the tank or the grave.

Standing outside the Klengenberg site, I looked south over the emerald blue waters of Terror Bay, knowing that the eponymous ship rested on the seabed somewhere nearby. Wondering *exactly* where it might be, I opened an electronic chart of the bay on my phone and tried to orient myself to its possible location. A mile and a half from where I stood, a red box was marked on the chart as a "Restricted Area." "No person shall enter the Wrecks of HMS *Erebus* and HMS

Terror National Historic Site of Canada without written authorization," it said. Zooming out, I was able to see where Terror Bay lies relative to Victoria Strait. As the crow flies, the distance from where the Franklin ships were abandoned is about a hundred miles—more, of course, if you follow the coastline, as a ship would do. What struck me powerfully, though, was how perfect an anchorage this was and how there really wasn't anywhere else a ship could find decent shelter on this side of King William Island.

If Lieutenant Graham Gore had made it this far during the exploratory sledging expedition he led in the spring of 1847—the one during which he penned the first installment of the Victory Point Record—he would have known about this bay and reported it to Franklin and Crozier when he returned to the ships. The obvious question, then, is whether *Terror* drifted here or, as Woodman and others have proposed as being more likely, was she sailed into Terror Bay? Woodman told me that he had consulted with a hydrologist at the Canadian Hydrographic Service who confirmed that it's difficult to imagine, based on what is known about currents and the flow of ice in and around King William Island, how a ship could drift from Victoria Strait, continue down through Alexandra Strait (which lies between King William Island and the Royal Geographical Society Islands off its west coast), and then hook a turn back north into Terror Bay. The standard reconstruction of the Franklin expedition mystery has never allowed for the possibility that the sailors retook the ships after the initial abandonment, but the more I studied the clues—and saw the terrain with my own eyes—the more convinced I became that *Erebus* and *Terror* must have been remanned and set to sailing again.

Inuit consistently recounted stories of two shipwrecks, but when it comes to geography, the accounts relating to the first vessel, which sank quickly in deep water, are as vague as the Oot-joo-lik (*Erebus*) stories are detailed. But since we now definitively know that the ship that sank quickly in deep water must have been *Terror* (the wreck has been positively identified by Parks Canada divers and was found in eighty feet of water while *Erebus* was found in thirty-three), and since McClintock, Hall, Schwatka, and Rasmussen all collected testimony pointing to Franklin survivors having a land presence in Terror Bay, we can, for the first time, connect the dots in a new way.

According to McClintock, Oo-na-lee, whom he met near Pasley Bay, had drawn him a sketch of the ship's location using a spear with which he traced lines upon the snow. McClintock later wrote that he could "make nothing out of this rude chart," but perhaps it's not a total coincidence that he named this place Terror Bay when he traversed its shores a few weeks later in the spring of 1859. Of course, McClintock had no idea at the time that he was within a few miles of the ship, but Terror Bay had a long history as a prolific Arctic char fishery and seasonal Inuit camp, so perhaps someone shared something about the bay that caused McClintock to make this prophetic connection.

Ten years later, when Hall finally made it to King William Island, he met a group of Inuit who reported visiting a hospital camp in Terror Bay. Here they had found a scattering of musket balls, three shallow graves, and two large canvas tents. As best Hall could ascertain, these visits had probably taken place in 1849. He'd already heard accounts of this camp years earlier from Ookbarloo, a chilling tale that David Woodman calls "one of the most powerful of all Inuit remembrances." As the story goes, an elder Inuk woman and her husband found "a frozen mass of human bodies" in one of the tents. Some were intact while others had been "mutilated by some of the starving companions, who had cut off much of the flesh with their knives and hatchets and eaten it." One of the corpses, partially embedded in ice, had a silver chain around its neck. Assuming the chain was attached to a watch buried inside the man's clothing, the woman carefully chipped away the ice with a sharp rock until she and her husband were able to free the body. According to Hall's recounting, the man seemed "to have been the last that died, and his face was just as though he was only asleep."

Hall later identified the spot in his field notes as "Fitzjames' Islet," not to be confused with Fitzjames *Island*, which appears on modern charts near the western entrance of the bay. A decade later, Schwatka learned from Inuit that this camp had lain near the tide line where years of wind, storm surge, and the gouging of the shore by pans of sea ice had washed away any trace of it.

Before the discovery of *Terror* in 2016, Woodman believed that the Boat Place in Erebus Bay had such a large human footprint because the men had sailed one or both of the ships there after remanning them in the summer of 1848. His theory was that later, after one of the ships sank, they deserted the remaining "much broken" ship and crossed the Graham Gore Peninsula to Terror Bay, where they established the hospital camp that Inuit reported to Hall and Schwatka. But when *Terror* was found in Terror Bay, not Erebus Bay, Woodman realized that the reason so many Franklin sailors had perished at the Boat Place was probably because it marked the terminus of the 1848 abandonment.

Let us imagine, then, that Woodman is right: Crozier, surrounded by dead and dying men at the Boat Place—including Fitzjames, his second-in-command—realizes that he has vastly overestimated their rate of travel and that they are doomed to end the season stranded somewhere along that forlorn coast with no shelter, no food, and essentially zero chance of surviving the coming winter. He orders those who can still walk to return to the ships, where they have left a stash of food. Later that summer or perhaps the next, the ships finally break loose from the ice, and the survivors sail west in a narrow strip of water between the shore and the pack ice in Victoria Strait. Perhaps along the way, they stop and pick up survivors in the camp at the Boat Place. They continue on, past the western end of King William Island—now named Cape Crozier—and then south into Queen Maud Gulf. Here, they run into more ice, which forces them to shelter in Terror Bay, an anchorage mercifully in the lee of the bitter north winds and the unrelenting crush of ice spilling from the McClintock Channel. With the ships safe for the time being, a hospital camp is established ashore where the men can spread out and at least attempt to secure fresh game with which to ward off the scurvy that was undoubtedly picking them off one by one.

Continuing this hypothetical, I opened my mind to the possibility that the story from the whaler Peter Bayne—in which "one man died on the ships and was brought ashore" and buried in an opening in the rock that was covered with something that "after a while was all the same stone"—refers not to Collinson Inlet, as Tom believes, but to Terror Bay. Since we know that Franklin passed away before the ships were abandoned, in this new scenario, the man who dies aboard the ship in the Bayne story would have to be some other high-ranking officer.

On the second-to-last day of our overland expedition as I stared out at Terror Bay from the hill above Patsy's trading post, the wind gently riffled the water and the sun felt warm on my face. The rest of the party was off doing something else, so I sat quietly for a time turning over the clues in this great mystery, trying to fit them together in a way that made sense. Back in Collinson Inlet, I had taken a photo of the Bayne map in Tom's notebook, and I now pulled it up on my phone. And as I traced the contours of those overlapping ridges, that bump in the shoreline, the line of graves, and the squiggly river, it hit me how closely it resembled the topography of the spot where I currently sat.

By the time we had pulled into Gjoa Haven on July 29, we'd covered more than five hundred miles on the four-wheelers, and I decided that if I never saw one of those machines again, I'd be a lucky man. While we waited for our flight back to Greenland, I stopped at the town office, where I bumped into a guy named Jimmy Pauloosie, who introduced himself as "the Guardian coordinator for the Nattilik Heritage Society in partnership with Parks Canada," almost as if he'd practiced saying the words in front of a mirror. The Guardians are Inuit caretakers who set up camp each summer near the Franklin wrecks to guard them from looters and to support the work of underwater archaeologists. Jimmy—who had thick black hair, close-set eyes, and a large dome-shaped mole on the bridge of his nose—quickly added that he was "supposedly a descendant of Roald Amundsen."

"So what are *you* doing here?" he asked.

I told him about the overland expedition, and when I got to the part about sailing the Northwest Passage and how I hoped to be back through in a few weeks aboard *Polar Sun*, Jimmy's eyes opened wide. "Boy, would I love to be on a boat like that," he exclaimed. "That must be *sooo* cool."

There was something about his manner and the earnest look on his face that instantly won me over. The next thing I knew, I was asking him if maybe he could join our crew when (if) we got to Gjoa Haven aboard *Polar Sun*. "We could take you to Terror Bay," I offered, "and you could check in on the Guardians."

"Are you serious?" he asked.

When I said that I was, he took off his glasses and used a tissue from his pocket to wipe away the fog in the corners, leaving tiny white balls of lint on the lenses. When he looked up, his eyes had a faraway look as if I'd just handed him a DNA test proving that

Amundsen was indeed his great-grandfather. Jimmy and I ended up talking for a long time that day. He told me that the town was already getting calls from people from the south who wanted to hire guides to visit the wreck sites. They weren't ready to give tours yet, mainly because the archaeological work was still ongoing and the divers didn't want people getting in their way. But in a few years, he saw huge potential for Franklin-related tourism to create jobs and give the economy in Gjoa Haven a much-needed boost.

But he also mentioned that not everyone in Gjoa supported the work now underway at the *Erebus* and *Terror* wreck sites. Many in the community felt that no good would come from messing around with long-lost dead explorers. Four years earlier, during a two-week period in August 2018, six Gjoa Haven residents had died unexpectedly. First, Jacob's brother and nephew drowned in a boating accident. Then two other men were killed in an ATV rollover, followed by the death of an elder and a staff member at a local school. Earlier that same year, Louie Kamookak, a friend of Tom's and a legendary historian and Franklinite who'd spent more time than any other Inuit investigating the Franklin mystery, died at age fifty-eight from cancer. "It seemed like there was a lot of supernatural stuff happening in the community, and the elders were associating all that death and bad stuff with HMS *Erebus* and HMS *Terror*," said Jimmy. "Their belief is that we should not be disturbing the belongings of a dead person, that we should just leave all that stuff alone."

To help allay concerns of a Franklin "curse," community leader Fred Pedersen told Gjoa residents that the only things being removed from the wrecks were artifacts. "If any bodies are found, they will be left in place," he said. "[We] will not bring up or disturb human remains."

In 2015, Jacob and Kamookak participated in a ceremony in which local elders blessed some sand from Gjoa Haven that was then carried by boat out to Wilmot and Crampton Bay and sprinkled over

Erebus. After the deaths in 2018, the Guardians suggested that a similar ceremony be done at the *Terror* wreck site, and it took place about a month later. Jacob was there, and his old pal Louie was too, although in spirit only.

When I asked Jimmy if the Franklin curse was real, he squeezed his eyes together and dabbed at the sweat on his face with that same tissue he'd used earlier. Several times he looked up at the ceiling, then caught my eye and sighed deeply, as if to say that he *really* wished I hadn't asked. Finally, he locked me in an unnerving gaze and said, "We, as Inuit people, believe that there are spirits out there that are looking after us. I really believe myself that there is a spirit that sometimes guides and helps me through hardship." A month earlier, the spirit of a dead neighbor had led him to an abandoned camp where he found a tool that he badly needed.

But, of course, not all spirits are benevolent. According to Jimmy, when people die in great pain or with meanness or anger in their hearts, their spirits can be malevolent, and if you speak ill of them, "it's almost like inviting them to come back to curse you and to make you sick too." This, he explained, was why many Inuit wished that Franklin and his men had never come to King William Island in the first place. More than a hundred of these white men died hard in their lands, and their spirits—what Jimmy now described as "the shadow people"—were still roaming the landscape, trying to find their way home. Jimmy told me that for this reason, it was hard for him to hang on to local people who worked as Guardians. Most of them did it for a season and then left the program because when they had been near the wrecks, they sensed that the shadow people were watching.

It was different with Jimmy, though, and that was why he was willing to tell me all of this—even if it meant grinding that tissue into dust in his hand. When Jimmy was a boy, his mom had told him a story about the shadow people. If you knew how to look, you could see them faintly, a mist or a darkness where it shouldn't be. One day, when his mother was a young girl, she was playing outside the family tent in the dirt with her friend, and a shadow person came around. An elder woman saw the spirit hovering over the children. She called to it, scolded it, and told it to leave the girls alone. And the shadow disappeared. Ever since that day, his mother had known that she needn't fear the shadow people, and she passed this on to Jimmy. "I've always been comfortable, you know, by myself," said Jimmy, "because of my belief that you can scold the spirits and they'll leave you alone."

But the key, he said, was "not to dwell on stuff like that."

ǀǀǀ

A month after all of this took place, as we anchored *Polar Sun* next to Fitzjames Island, I scanned the surrounding shoreline of Terror Bay, picturing that Franklin hospital camp and feeling more convinced than ever that this place held the keys to solving this mystery. I thought back to our entrapment in Pasley Bay and how certain it had seemed on that last morning that *Polar Sun* would be frozen in for the winter. And yet a few hours later, we found ourselves sailing free in open water once again. In the spring of 1848, Crozier, Fitzjames, and the 103 surviving crewmen must have felt that *Erebus* and *Terror* were similarly doomed, or they never would have left the ships. But those ships somehow ended up more than a hundred miles away from where they had originally been abandoned, and all of the evidence suggests that they were remanned and sailed to the places where they finally sank. One of those locations was currently only a few miles away, and it was more than a little intriguing to imagine that the answers to what really happened might, at that very moment, lie within the pages of Crozier's logbook stashed inside the desk in his cabin, where it had been hidden from the prying eyes of hungry Franklinites for nearly two centuries.

Parks Canada underwater archaeologists had probably already arrived at *Erebus* and anchored their dive barge over the wreck. They had taken some of the Guardians with them, but another group, led by Jimmy, was spending the summer watching over *Terror* to protect it from treasure hunters—or amateur sleuths like myself. I wondered about Jimmy and whether he might be watching us from that Guardian camp, wherever it might be, at that very moment. We'd stayed in close contact ever since that day in Gjoa Haven, and the plan for him to join our crew seemed all but set when, for reasons I still don't understand, he abruptly stopped communicating. I probably shouldn't speculate as to why, but if he had joined our crew and been willing to stay on until Nome, it would have given me that all important third man, whom I needed to complete the passage. Also, he was an interesting man, and I would have enjoyed getting to know him better.

Having been up all night, I headed for my bunk and fell asleep immediately. I woke at ten a.m., ate some breakfast, and loaded my pack for a day out on land. Rudy and Ben had no interest in this outing, so I grabbed the shotgun and dinghied over to a sandy cove on the north shore of Fitzjames Island. My goal was to investigate that huge monument I'd seen from Patsy's trading post a month earlier. Maybe, if I was lucky, I'd find a Franklin expedition grave, some of those musket balls, and possibly even the Supunger vaults. It was a balmy day fanned by a gentle southerly breeze and crowned with gauzy clouds so thin you could see right through them to the robin's egg sky above. It felt good to be alone and back on the Franklin hunt as I scrambled through some boulders and made a beeline for the monument that rose from the center of the tiny island, where it could be seen from miles away in every direction.

The base was about three and a half feet across, and the overall structure was pyramidal and so tall and solidly built that I could have climbed it and squatted on top without the slightest worry of toppling it. On the peak, which stood about eight feet above the ground, sat a stack of rectangular capstones topped with a perfectly round block the size of a human head. Indeed, from a distance, the cairn bore an uncanny resemblance to a giant human being, and I assumed this effect was intentional. Vibrant orange lichen speckled the rocks, and when I drew in for a closer look, I noticed a hollow interior space. I stuck my hand through an opening and groped around but found nothing. *Why had this inukshuk, of all the ones I'd seen, been built with such care and precision?* I wondered. Had Patsy erected it as a lighthouse of sorts? Was it hollow so a lantern could be lit and placed inside to direct wayward ships to his trading post? Or was this a monument to one or more of Franklin's men who'd made their last stand at that long-lost Terror Bay hospital camp?

A few feet north, the ground was covered in thick soil and moss. Throughout our journey, we'd seen these mossy beds, and Jacob had explained more than once how they marked traditional Inuit camps. Blood and guts and sinew from animals slaughtered in these camps leeched into the ground, allowing soil to take hold in the gravel. Usually, these carpets, which always seemed to lie adjacent to inukshuks, were about the size of a two-man tent, but here, the patch of dark, loamy earth covered about four hundred square feet. The plot was rectangular in shape and, around its perimeter, had trenches about a foot deep that looked similar to irrigation ditches that I've dug around tent sites to divert rainwater. Stranger still were the dozens of pieces of wood sticking out of the ground like little punji sticks.

On the western side of the island, I found more cairns and, on a low patch of ground near shore, extensive evidence that someone had camped here recently: a full sheet of rotting three-eighths-inch plywood, empty fuel cans, bits of copper piping, shell casings, a tin cup, a busted-up Sanyo radio, and a small motor made in Hartford, Connecticut. Amongst this refuse, I found a few brass screws and a single rusted can that looked identical to the ones I'd seen on Beechey Island. Down by the tide line, a rusted fifty-gallon drum lay on its side in a tarry pool of congealed oil or diesel fuel. Nearby, I picked up two items that looked much older: a rusted shackle the size of my hand and a foot-long piece of S-shaped steel on which I could see the layers where the sheets of metal had been sandwiched together by a blacksmith.

I continued to stroll along the beach looking for more artifacts when I came upon a set of fresh polar bear prints in the wet sand. Each one was a foot long, with distinct indentations from the five toes and streaks where the claws had dragged across the ground. Ben had told me that there were no bears in the area and that I didn't need to bother bringing the gun. Fortunately, I'd known that he had no idea what he was talking about, and I'd ignored him.

Keeping a wary eye behind me, I scooted back to the dinghy, leaving Fitzjames Island exactly as I had found it, save for one souvenir: a wood-handled knife with a wide, stout blade that I had dug out of the sand down by the tide line. As we motored south toward Cambridge Bay under a herringbone sky, I sat down at the nav station and opened my computer. It took a few minutes before I found the file I was looking for: the William Gibson article from the *Beaver* with the photo of the mysterious Franklin grave that Patsy Klengenberg had rebuilt in 1931. I compared it to my own photo of the Fitzjames Island monument, and there were some striking similarities, including the overall size and shape and the way the top was layered with flat, horizontal paving-stone-style blocks. The Fitzjames Island monument was no stone house, but it was *very* close to where *Terror* lay. And I had learned long ago that if a coincidence seems meaningful, it probably is.

I was about to close up my computer when I noticed something that had previously escaped my attention. Gibson's article opens with a verse written by Alfred, Lord Tennyson that is inscribed on the Franklin monument at Westminster Abbey:

Not here: the white north has thy bones;
and thou, heroic sailor-soul, art passing on thine
happier voyage now toward no earthly pole.

PART 4

TOWARD NO EARTHLY POLE

CHAPTER 13

Not Great but Good Enough

Date: Summer 1849?
Position: Terror Bay, King William Island

The Victory Point Record tells us that 105 men abandoned the ships in April 1848. We know that four of them were buried between Cape Felix and Erebus Bay and that another twenty-three perished at the Boat Place. No bodies have been found to date in Terror Bay, but the Inuit reported that as many as thirty or forty men died in the washed-away hospital camp there, based on the number of bodies found. All told, human remains attributed to the Franklin expedition, buried and unburied, have been found or reported at thirty-five different locations across King William Island and the Adelaide Peninsula. Archaeologists have only been able to confirm a little more than half of those sites, which means that dozens of Franklin sailors are still unaccounted for. In the absence of a complete record, we still don't know where the vast majority of Franklin's men made their final stand.

But if you trace the locations of casualty sites across the south coast of King William Island, a picture emerges that some of Franklin's men attempted to escape Terror Bay on foot. Inuit testimony and modern-day underwater archaeology both indicate that *Erebus* and *Terror* were moored in Terror Bay for a period of time between 1848 and 1850. Logic dictates that if *Erebus* remained intact as her sister ship was crushed in the ice, a number of the men would have stayed aboard the flagship in hopes that the vessel would be released to sail once again. Their companions residing in the shoreside hospital camp, too ill to press on, would have faced their end with the same stoic resolve as their comrades had the previous year at the Boat Place in Erebus Bay.

All of these men must have been part of the group that had retreated to and remanned the boats in Victoria Strait in the late spring or summer of 1848. I imagine that the ones who set off from Terror Bay must have reasoned that on land they'd have more opportunity to secure fresh game or, even better, find a group of friendly Inuit who might have food they could share. And it's probable that they framed their retreat to their dying companions left behind in the hospital camp as a rescue mission rather than outright abandonment.

According to testimony collected by Hall in 1869, this group did meet some Inuit only fifteen miles to the east in Washington Bay. In the summer of what was probably 1849, three Inuit men named Tetqataq, Ukuararsuuk, and Mangaq were seal hunting on the south shore of Ki-ki-tuk (King William Island) with their families. They were breaking camp on a clear morning with the sun high in the sky when they observed something white out on the ice. Initially, they hoped it was a bear, but as the shape drew closer, they could see that it was a group of men dragging a boat with a sail that "shook in the wind."

The Inuit walked to the edge of a large crack in the ice and waited for two of the white men to arrive on the other side, one of whom held a gun in his arms; the other put his hand to his mouth and moved it down to his stomach to show that he was hungry. The *kobluna* found a place to cross the crack, and when the two parties met face-to-face, the man named Aglooka, the one without the gun, spoke to them, saying, "*Man-nik-too-me*" ("hungry" in Inuktitut) over and over while he stroked the chests of the Inuk men's caribou jackets. Aglooka then pointed south and east and said, "I-wil-ik," (the Inuit name for Repulse Bay, now known as Naujaat), but the Inuit weren't sure if he was asking for directions or simply telling them that this was his destination. After this, Aglooka pointed north, and "drawing his hand & arm from that direction, he slowly moved his body in a falling direction and all at once dropped his head sideways into his hand, at the same time making a kind of combination of whirring, buzzing, & wind blowing noise. This was taken as a pantomimic representation of ships being crushed in the ice."

Several different accounts of this meeting were collected by Hall, Schwatka, and Rasmussen, and one consistency between them is that this party of Inuit tried to help Aglooka and his men. They traded some seal meat for a knife and some of the sailors reportedly slept in Mangaq's tent. According to Hall's account, the Inuit broke camp early the next morning, and on their way out, they passed Aglooka's tent. He was standing outside at the time and "put his hand to his mouth and spoke the word 'Netchuk' or 'Nest-chuk' (seal). But the Innuits were in a hurry—did not know the men were starving."

It was after this encounter in Washington Bay, as Aglooka's men continued trekking southeast along King William's shoreline, that they "fell down and died as they walked along," as one Inuk woman described these final days to McClintock. And indeed, when McClintock sledged through this country on his circumnavigation of King William Island, he found a bleached human skeleton, with brown hair still clinging to the skull, lying face down on a windswept peninsula called Gladman Point. The remains were partly buried in the snow, but bits of tattered clothing stuck out and a hairbrush lay on the ground nearby. McClintock carefully rolled the skeleton.

Much of the sailor's uniform was intact, and inside one of the pockets he found a leather wallet that contained several papers including the seaman's certificate for Harry Peglar, a copy of Lloyd's "Weekly Summary of Maritime Casualties" dated April 6, 1845, a doggerel version of a popular nineteenth-century ode entitled "The Sea," and a series of enigmatic letters written on thin sheets of vellum in what appeared to be German.

The seaman's certificate seemed to offer a definitive identification of the individual, but the blue jacket "with slashed sleeves and braided edging, and the pilot-cloth great coat with plain covered buttons" was not the type of uniform that would have been worn by Peglar, who was the captain of the foretop aboard *Terror.* The uniform was more akin to those worn by stewards, of which there were two—Thomas Armitage and William Gibson (no relation to the William Gibson who wrote the article in the *Beaver)*—who had served with Peglar on previous voyages. Today, most Franklin historians presume that it was Armitage or Gibson who took the wallet upon their shipmate's passing.

What initially appeared to be German turned out to be backward writing. No one knows why the author would have gone to such lengths, but the obvious explanation was that its purpose was to create a code, like pig Latin, to obscure the meaning from other sailors. And yet there is nothing in the letters that appears to be secretive or proprietary, so perhaps it was simply a game.

The original "Peglar Papers" are housed in the Caird Library at the National Maritime Museum in Greenwich, England, which holds many other Franklin artifacts. Some of the pages still contain bits of red sealing wax on their edges as well as addresses on the outside corners, suggesting that they were intended to be posted in Alaska—if they ever made it there. Almost all of the writing is garbled. Every single page has water damage, and a lot of the words were lost when an archivist attempted to darken the ink.

What is clear even to the uninitiated, though, is that much of the material was written as verse. And the overall tone is fun and lively:

The C the C the open C
It grew so fresh the ever free the ever free the ever free
without it without it
guard it will run to Earth above the regions round
I love the C I love the C

"Who, if dying of exposure, scurvy, or lead poisoning," writes Russell Potter, who conducted his own exhaustive analysis of the Peglar Papers, "would compose ditties about dogs and sea turtles, accounts of by-gone parties, and idle references to land-bound matters such as 'the grog shop opporsite'"—or to happy times on previous voyages to the Caribbean and Venezuela?

Richard Cyriax—the godfather of all Franklinites, whose 1939 magnum opus, *Sir John Franklin's Last Arctic Expedition*, is still considered by many to be the definitive analysis of the Franklin mystery—believed that the following passage was written near the time of the ship's original abandonment: "We will have his new boots in the middel watch . . . as we have got some very hard ground to heave . . . shall want some grog to wet houer wissel . . . r now clozes should lay and furst mend 21st night a gread."

Cyriax speculates that "new boots" referred to those found by McClintock at the Boat Place; they had been modified with hobnails, a type of crampon used by early climbers in the Alps for traction on icy "hard ground to heave." And might not the "21st night a gread" refer to the final abandonment of *Erebus* and *Terror* on April 22, 1848, which perhaps got postponed by one day? Other installments, however—"brekfest to be short rations," "the terror camp ["is" or "be"] clear" and "[o] death wheare is thy sting the grave at comfort cove"—might have been written closer to the end.

Details as to what happened after the survivors passed Gladman Point get sketchy. At some point, likely in August or September of that same year, the motley crew must have split up. With Simpson Strait finally iced out, a small group of a half dozen men might have used the boat with the sail that "shook in the wind" to cross over to the mainland, while the bulk of the party continued trekking eastward. A scattering of graves that would later be found by Hall at Booth Point and the Todd Islets (the southernmost point of King William Island, about twenty miles southwest of Gjoa Haven) suggests that most of the main party must have died later that fall or winter.

As these Franklin survivors faced the prospect of a fifth or sixth winter in the Arctic, this time without the shelter of their ships to protect them from the elements, they had no way of knowing that the British Admiralty had launched three separate expeditions to rescue them.

The first Franklin search expedition, led by James Clark Ross aboard HMS *Enterprise* with Leopold McClintock and Robert McClure as lieutenants, came in from the east via Lancaster Sound and searched Somerset Island in the spring of 1848. When *Enterprise* returned to England empty-handed that fall, the Admiralty quickly organized another expedition that approached the Northwest Passage from the west via the Bering Strait. It too found not the slightest trace of the missing sailors, and one of its ships, HMS *Investigator* under the command of McClure, would itself become lost in the Northwest Passage for several years. A third expedition came in overland via the Mackenzie River and spent three years between 1848 and 1851 searching the area around Coronation Gulf and Victoria Island. This party, led by a remarkable Scottish surgeon, HBC fur trader, and explorer named John Rae, came the closest to solving the Franklin mystery.

Late in the summer of 1851, Rae found two pieces of wood in a bay on the east end of Victoria Island, only eighty miles west of Terror Bay. One, which he believed to be the base of a flagstaff, was marked with the initials "S.C." and had a piece of white line with a red inlay nailed to the wood with copper tacks stamped with the Royal Navy's broad arrow. Previous to this discovery, Rae had made several attempts to cross Alexandra Strait over to King William Island. Had he succeeded, he might have found the remains of the hospital camp in Terror Bay.

A year earlier, in part to appease the British public, which was increasingly consumed with Franklin mania, the Admiralty issued a £20,000 reward "to any Party or Parties, who . . . shall discover and effectually relieve . . . the Crews of Her Majesty's Ships 'Erebus' and 'Terror,'" or half that sum to anyone who brought back intelligence that would lead to the crew's relief—or simply ascertain what had happened to them.

Rae eventually claimed the reward after discoveries he made on a subsequent expedition. Ironically, if the Franklin crew had followed his tactics, some of them might have escaped their tragic fate. During his first Franklin search, Rae covered nearly fourteen hundred miles aboard a small boat and more than a thousand miles on foot. He and his two companions survived by adopting Inuit ways of travel and survival on the land—living in igloos and hunting for almost all of their food. On their trek across the south coast of Victoria Island, the team used snowshoes of Rae's own design and averaged more than twenty-five miles per day for thirty-nine days straight. Granted, the surviving members of the Franklin expedition were undoubtedly malnourished and scorbutic by the time they set off overland, but had any of them been able to match even half of Rae's pace, they might have made it out.

In 1854, Rae returned to the Arctic with a mandate from the HBC to fill in the crucial blank in the Northwest Passage map that lay between the Boothia Peninsula and Simpson Strait (the same area where John Ross fatefully fabricated "Poet's Bay" and that land bridge connecting King William Island to the mainland). Rae set off from Repulse Bay on March 31, 1854, with four men and an Inuit interpreter. Over the next three weeks, they made slow progress through storms and heavy snow as they man-hauled their sledges across the frozen tundra. Near Pelly Bay, they met a group of seventeen Inuit who strongly cautioned against traveling any farther west. This territory, they said, was inhabited by hostile Inuit who would murder anyone who ventured into their lands. Rae deduced that this was a ruse to discourage his team and him from heading in the direction where these people had cached meat from recent hunts. Shortly thereafter, Rae met In-nook-poo-zhe-jook, who would prove over the years ahead to be one of the most reliable and articulate providers of testimony related to the Franklin expedition. In-nook-poo-zhe-jook arrived from the west wearing a Royal Navy gold hatband and driving a dogsled piled high with musk ox meat.

In a letter written to the governor of the HBC, Rae explained how he learned from In-nook-poo-zhe-jook "that a portion, if not all, of the then survivors of the long-lost and unfortunate party under Sir John Franklin had met with a fate as melancholy and dreadful as it is possible to imagine."

Much of In-nook-poo-zhe-jook's account matched the details of the Washington Bay encounter that would later be shared with Hall, Schwatka, and Rasmussen, which is not surprising considering he was Tetqataq's brother-in-law. But In-nook-poo-zhe-jook also included information about what happened subsequently. According to his testimony, after meeting the starving sailors on King William Island, Inuit later found the corpses of approximately thirty people on the mainland near the mouth of the Back River and another five bodies on an island a day's journey to the northwest. Rae wrote to his superiors that "some of the bodies were in a tent or tents, others were under the boat, which had been turned over to form a shelter, and some lay scattered about in different directions." The grimmest detail, according to Rae's account, was that "from the mutilated state of many of the bodies, and the contents of the kettles, it is evident that our wretched countrymen had been driven to the last dread alternative as a means of sustaining life." One of the dead appeared to be the group's leader. He had died lying on top of a double-barreled gun with a telescope worn over his shoulder.

We don't know who this man might have been, but it was likely one of the officers. The expedition had begun with two dozen, but as recorded in the second installment of the Victory Point Record, nine of them had died by the spring of 1848. For some reason, the officer casualty rate was much higher than that of the enlisted men, a fact that has long intrigued historians. We know that the officers took on more demanding and riskier jobs, like leading the exploratory sledging expedition in the spring of 1847, and the responsibilities of command might have led to increased psychological stress. It's also likely that the officers ate different food, although why that would have led to greater malnourishment is unknown. If their higher rate of attrition had continued, and we can only assume that it did, many more officers would have died in the months leading up to this last stand. We can assume that the men who finally succumbed at this mass casualty site were probably members of the group that met Tetqataq, Ukuararsuuk, and Mangaq in Washington Bay. But as to who, when, and how, until those logbooks are found, we will probably never know more.

On March 31, 1854, the same day that Rae set off on his expedition from Repulse Bay, the Admiralty announced in its gazette that Franklin and all of his men were officially deceased and to be struck from the Royal Navy muster rolls. Barrow had been dead for six years, Great Britain was now at war with Russia in Crimea, and the lords of the Admiralty had decided that they had sacrificed too many men and ships in search of the lost expedition. Franklin's wife, Lady Jane, was told that as far as the British government was concerned, she was now a widow. And yet she refused to give up. "It would be acting a falsehood & a gross hypocrisy on my part to put on mourning when I have not yet given up all hope," she wrote to Franklin's sister. "Still less would I do so in that month & day that suits the Admiralty's financial convenience."

When Rae returned to England in October of that year, Lady Jane was in no mood for more bad news. The Admiralty, on the other hand, must have seen Rae's report as corroboration of their decision to call off the search, which might be why they quickly released a copy to the *Times*. When the news broke that the polar heroes "had been driven to the last dread alternative," the public outcry, led by Lady Jane, was instant and sharp. Charles Dickens, a close friend of the Franklins, went on the attack. In his illustrated weekly magazine *Household Words*, he wrote that "the noble conduct and example of such men, and of their own great leader himself . . . outweighs by the weight of the whole universe the chatter of a gross handful of uncivilised people, with domesticity of blood and blubber." Dickens went on to describe Inuit as "covetous, treacherous, and cruel" savages. The power of his voice and those of others turned back the discourse such that many Britons soon believed that it was Inuit—and not the elements, bad luck, or unpreparedness—that had killed Franklin's men.

Twenty-five years later, Frederick Schwatka located the mass casualty site on the Adelaide Peninsula described by In-nook-poo-zhe-jook; it appeared to be the place where the final Franklin survivors had made their last stand. Unlike in other spots on King William Island where the dead had been buried by their comrades, here the bones lay scattered across the low, muddy ground—a sign, Schwatka thought, that there had been no one left to bury them. He named the place "Starvation Cove," and Henry Klutschak, a member of his party, later wrote, "Mother Nature could probably not produce a more desolate spot on this wide Earth than that where the last survivors of the Franklin expedition found their end."

It's reasonable to infer, based on the location, that this small group of Franklin survivors was trying to reach the Back River (as Crozier had indicated in the Victory Point Record), which they could have followed five hundred miles south to the nearest HBC trading post on Great Slave Lake. By this point, though, suffering from exposure, starvation, malnutrition, scurvy, trench foot, and God knows what else, these men would have understood that their quest was all but hopeless, for this route was against the current on a river with no less than eighty-three sets of rapids and waterfalls. On top of that, it coursed through those same "barren lands" where many years earlier their deceased commander had eaten his own boots.

ılı

On September 1, about twenty-four hours after departing Fitzjames Island, *Polar Sun* slid into the calm waters of Cambridge Bay and tied up to a scrappy public wharf alongside a steel fishing trawler called the *Martin Bergmann*. There was something familiar about the name and the huge polar bear logo painted on the side of its baby blue hull, but I couldn't place it.

On the dusty waterfront, a row of modest single-story ranch-style homes rested on the edge of the bay alongside a radio tower, an Anglican church, a pharmacy, a school, and the Arctic Islands Lodge—home of the only restaurant in Cambridge Bay. The houses along the waterfront had big picture windows that looked south over the bay and most had satellite dishes mounted on their roofs. When I went out for a stroll to stretch my legs, I could see the flicker of televisions playing in the living rooms, but no one seemed to be around or to have taken any notice of our arrival. I'd had high hopes for Cambridge Bay being a bona fide metropolis. With a population of 1,760, it's the biggest hamlet in the central Arctic. But you wouldn't have known that from walking around town.

Rudy had decided to end his trip here, which left *Polar Sun* with only two remaining crew members—Ben and me. And as Ben had made it clear that he wasn't willing to sail on with a crew of two, it was his preference that we end this year's expedition in Cambridge Bay. But for that to happen, we needed to find a way to get *Polar Sun* hauled out of the water for the winter. A part of me hoped it wouldn't be possible because then I'd have no choice but to press on. After our epic in Pasley Bay, I'd decided that no matter what, I would NOT freeze the boat in for the winter unless forced to do so. On the other hand, as weary, stressed-out, and downright afraid as I was of getting caught by a Bering Sea superstorm off the North Slope of Alaska, it was hard not to dream about ending the trip and going home to my family. The thought of sleeping in my own bed was almost more than I could handle.

The next morning, I found my way to the town garage, where I introduced myself to a man named Dana Langille. Through the grapevine, I'd learned that a number of years ago, Dana had hauled a Northwest Passage–bound sailboat here, and if anyone could get *Polar Sun* out of the water, he was the guy. After hearing me out, he told me to hop into his pickup truck, and we set off for the tank farm on the other side of town.

Dana was in his fifties, with broad shoulders, a rosy face, and thick dirty blond hair. A younger, curly-haired guy named Travis sat in the passenger seat across from him. Dana's phone beeped and rang nonstop. In brief moments between calls, he explained that he ran a construction crew that had several projects underway. But the season was woefully short, presently coming to an end, and he was severely shorthanded crew-wise—a predicament with which I could relate.

He had just poured a foundation for a garage at a place he called "the Pit," but the temperature was a bit too cold and he worried the concrete wouldn't cure properly.

"So what do you do?" I asked Travis as Dana took another call.

"I actually don't know," he said. He had just flown in from southern Canada that morning to work as a heavy-equipment operator, but Dana was so busy, he hadn't had time to tell him yet what he'd be doing. All he knew was that the job was six weeks on, two weeks off—and this was his first day.

A half mile outside of town, we pulled up next to a row of four-story-tall whitewashed industrial storage tanks filled with diesel to fuel the town's electrical generators. About eight years earlier, when the Northwest Passage–bound boat similar to ours had gotten stuck for the winter in Cambridge Bay, there happened to be an industrial crane in town. Dana had helped arrange for the crane to lift the ship out of the water and set it down on a custom-made wooden cradle. The derelict cradle was still there, but, unfortunately, the crane was long gone. According to Dana, without it, the only option for getting us out of the water now was to build our own new cradle out of steel, push it out into the water, motor *Polar Sun* onto it, and then use heavy equipment to drag it up onto land.

Dana took me down to a spot where a massive boat trailer sat in the mud next to a ramp made out of four-by-eight-foot steel plates that ran in parallel tracks into the water. The trailer was about sixty feet long and twenty feet wide, with fifteen-foot-high steel uprights running along the sides. Lengths of thick chain hung from these posts, and the entire apparatus rode on sets of massive truck wheels. For a second, I thought my ship had come in, but Dana quickly clarified that this trailer, which would have been perfect for *Polar Sun*, belonged to the owners of the *Martin Bergmann*, who would soon be using it to haul out for the winter.

Nearby in the mud sat a rusty, dented skidder that had once been used to transport a house. Dana said that he owned it, and if we could find some steel I beams, maybe we could weld on some uprights to turn it into a cradle. Unfortunately, he had no time whatsoever to help me figure it out, but I might be in luck as a team of "real crackerjacks" from Newfoundland was flying in to help haul the *Martin Bergmann* later in the week. Dana was confident they could help.

When I got back to the wharf, there were two guys using a small crane to lift a motorboat off the deck of the *Martin Bergmann* onto a trailer. I hovered for a bit, and when they were done, I walked over and introduced myself. Tom Surian, in his late twenties, wore overalls with big hip pockets into which he had stuffed his meaty hands. He had farm-boyish good looks and thick dark hair and stood well over six feet tall. His colleague, Gerry Chidley, was stocky and older, with a round weathered face that was partially hidden under the hood of a dark blue sweatshirt. He seemed a bit surly, and before he even opened his mouth, the scowl on his face told me that he was the type who didn't suffer fools gladly.

I pointed over at *Polar Sun* and told them how I'd just talked to Dana about some guys from Newfoundland who might be helping me to haul out for the winter in Cambridge Bay. The way Dana had volunteered the Newfoundlanders' services, I'd assumed they were his guys. Nope. Tom and Gerry were the "crackerjacks" to whom Dana had referred, and as I now learned, they worked for an outfit called the Arctic Research Foundation, which owned the *Martin Bergmann*.

When they heard how they'd soon be working their asses off welding a cradle for me and then hauling my boat and parking it for the winter in front of theirs, Gerry looked at me as if I had just told him I was dating his mother. "Well, hold on a second there, fella," he said with a thick Newfoundland accent. "That's all our equipment over there and you will not be parking your boat in front of ours and jamming up our whole scene." And even if they were willing to help, which was a big if, I clearly had no idea how hard it was to build a cradle for a thirty-four-thousand-pound sailboat.

"This isn't something that you can just randomly slap together," said Tom. "You have to *engineer* it, and you might be surprised how much something like that would cost."

"Ka-ching," said Gerry, rubbing his thumb against the tips of his fingers.

Tom looked at me now with an expression that said, *Go ahead, ask how much.* When I didn't take the bait, he told me anyway. "Our crew bills out at seven thousand a day," he said. "And it could easily take a week to build the cradle."

"That's fifty thousand dollars," said Gerry, whose scowl had turned into a smile.

"Well, if that's the case, then I guess we should just end the conversation right now," I said, "'cause I definitely don't have that much money to throw at this."

At my admission, Gerry finally softened up a bit. "No, of course not," he said. "We get that. Look, we actually do want to help you and we're willing to donate some hours to make this happen. If you can find all the steel and get it over to the tank farm, we can do the welding after work. In the meantime, send us some photos of what your boat looks like underwater so we can start thinking about how the cradle would need to be built, okay?"

That night, while I searched online for drawings of the Stevens 47 underwater profile, I typed "Arctic Research Foundation Martin Bergmann" into Google, and suddenly, I remembered why I knew the name of the ship. Tom and Gerry hadn't mentioned it, but the *Martin Bergmann* is none other than the vessel that found the wreck of *Terror* on September 3, 2016.

|||

The story of this historic discovery begins in 2008 with a Canadian businessman and philanthropist named Jim Balsillie, who had been sucked into the Franklin vortex. A year earlier, Balsillie had been the chairman and chief executive of the company that invented the BlackBerry. Research in Motion had all but owned the nascent smartphone market for nearly a decade, but by the time Apple introduced the iPhone in 2007, BlackBerry was in a tailspin, and Balsillie, who had already made billions, was on his way out. By the fall of 2010, he had left the company and spent three summers in a row in the Arctic with his friend Marty Bergmann, a scientist and tireless Arctic advocate who headed Canada's Polar Continental Shelf Program.

When Stephen Harper was elected prime minister in 2006, he immediately announced plans for a new deepwater port in the Arctic and a fleet of Polar-class coast guard patrol ships as well as funding for the search for *Erebus* and *Terror.* In addition, he declared the wrecks as national historic sites—despite the fact that no one yet knew where they were. "Canada has a choice when it comes to defending our sovereignty over the Arctic," he pronounced in July 2007. "We either use it or lose it"—a declaration that seemed prescient when Russia planted a titanium flag on the seabed beneath the North Pole a month later. That summer, when the Northwest Passage became ice-free for the first time in history, global superpowers including the United States, Russia, and China found themselves jockeying for control of the potential shipping lanes and some of the largest untapped reserves of oil, gas, and minerals in the world.

But despite big promises, Parks Canada's effort to find the Franklin ships made little in the way of progress. Balsillie, frustrated that more wasn't being done and worried that some other nation might beat Canada to the discovery of the ships, decided to put his ample resources to the task. With support from a fellow philanthropist named Tim MacDonald, he formed a nonprofit called the Arctic Research Foundation (ARF) and in Newfoundland purchased a sixty-four-foot steel-hulled fishing trawler named *Ocean Alliance*, which was refitted into a scientific research vessel. In the summer of 2011, *Ocean Alliance* was en route to Cambridge Bay when the crew received news that Marty Bergmann had died in a plane crash in Resolute Bay. Shortly thereafter, the ship was rechristened the RV (research vessel) *Martin Bergmann*. That winter, having not yet built their crazy trailer, the crew froze the ship into the ice in Cambridge Bay.

Harper had been making annual pilgrimages to the Arctic every summer to meet with Inuit officials and to make sure his Arctic strategy was moving in the right direction. On his trip north in 2013, he was accompanied by John Geiger, the CEO of the Royal Canadian Geographical Society and coauthor of *Frozen in Time*. Harper, who was a bit of a Franklinite himself, was familiar with *Frozen in Time*, and Geiger and he spoke at length about the Franklin search during that trip. The book opens with an introduction by Margaret Atwood, in which the acclaimed Canadian poet and novelist explores the reasons why Franklin mythology has become so deeply embedded into the Canadian identity and psyche. She attributes the Franklin myth's power to the old saying in china shops—"if you break it, you own it." "Canada's North broke Franklin," she writes, "a fact that appears to have conferred an ownership title of sorts."

Harper clearly felt a bit of ownership himself. In that year's Speech from the Throne (Canada's version of the State of the Union address), the governor general told Canadians that "the story of the north is the story of Canada." With the nation's one hundred and fiftieth anniversary coming up in 2017, he promised to double down on the effort to solve "one of the most enduring mysteries" in Canadian history by working with "renewed determination and an expanded team of partners to discover the fate of Sir John Franklin's lost Arctic expedition."

Harper backed up the rhetoric with funding for what came to be known as the 2014 Victoria Strait Expedition, a massive collaborative effort between government entities (Parks Canada, the navy, the coast guard, the Hydrographic Service, the Canadian Ice Service, the Canadian Space Agency, Defence Research and Development Canada, and the Government of Nunavut) and private partners (the ARF, One Ocean Expeditions, the Garfield Weston Foundation, Shell Canada, and the Royal Canadian Geographical Society). After years of searching in the vicinity of Oot-joo-lik, the consensus was that it was time to start looking elsewhere, and the most obvious place to start was Victoria Strait, where the ships had originally been abandoned.

But heavy ice prevented the small armada of ships from getting anywhere near Victoria Strait that season, so they fell back to the Oot-joo-lik search zone in Wilmot and Crampton Bay, where the *Martin Bergmann* and Parks Canada had already spent years ruling out territory and slowly closing the noose around the missing ship. While setting up a GPS station on a small island, a Coast Guard helicopter pilot named Andrew Stirling went for a walkabout on which he spotted a piece of rusted metal on the shoreline. At the time, two archaeologists employed by the Government of Nunavut, Doug Stenton (who would later identify the remains of James Fitzjames) and Bob Parks, were investigating nearby tent circles. They couldn't identify the item, but there was no mistaking that its base was embossed with the British Royal Navy's broad arrow. Later, back aboard *Sir Wilfred Laurier*, the item was identified by Captain Bill Noon and Jonathan Moore of Parks Canada as a davit pintle perfectly matching the ones used on the Franklin ships to raise and lower their longboats.

Knowing that they might be close, the search team adjusted the grid to an area upwind of the island. The next morning, a small sonar-equipped "day launch" called *Investigator* picked up the ghostly outline of a ship sitting on the bottom in thirty-three feet of water.

The crew immediately launched a remotely operated vehicle (ROV) equipped with a powerful camera that captured high-definition video of the wreck. The images showed a shipwreck festooned with kelp and other marine overgrowth. In the murk, it was hard to make out details, but the initial results matched closely enough with the engineering plans of *Erebus* and *Terror* that Ryan Harris, the head of the underwater archaeology team, confirmed what he had known in his gut from the moment he'd seen the first sonar trace—they had found one of the Franklin ships. The only question that remained was: Which one? They didn't have any dive gear aboard, though, and there was pressure from the government to get to Ottawa for a press conference and photo op at which Harper would announce the discovery. So the *Laurier* steamed back to Gjoa Haven, and Harris caught a flight south.

"This is truly a historic moment for Canada," said Harper at the press conference as Harris stood beside him in his Parks Canada uniform. "Franklin's ships are an important part of Canadian history given that his expeditions, which took place nearly two hundred years ago, laid the foundations of Canada's Arctic sovereignty."

At the time, the *Martin Bergmann*, which had done more than any other vessel to narrow the search for *Erebus*, was still combing the waters of Queen Maud Gulf. Due to government protocols, the crew didn't learn of the discovery until they heard about it on the news.

The next day, Harris flew back to Gjoa Haven with his dive gear, and soon he was back in Wilmot and Crampton Bay for what he later called the best dive of his life. He had personally been searching for *Erebus* and *Terror* for six years, but nothing could have prepared him for the moment that he found himself face-to-face with one of the most famous missing ships in the world.

By the end of that first dive, Harris had identified several anchors sitting on the deck and a large, algae-coated brass bell with the

British Royal Navy's broad arrow insignia clearly visible on its side. While Harris and others dove repeatedly on the wreck, the Hydrographic Service scanned the area with high-resolution multibeam sonar. When they returned to Ottawa later that month, the sonar data was used to create a stunningly detailed image of the wreck that showed the locations of hatches and masts and the exact length and breadth of the ship. This information led to the positive identification of HMS *Erebus*, which was announced by Harper on October 1.

Terror, of course, was still out there somewhere, and Harper's press release granted that, "Finding the first vessel will no doubt provide the momentum—or wind in our sails—necessary to locate its sister ship." Two years later, Parks Canada dispatched three vessels to Victoria Strait: the *Laurier*, the Royal Canadian Navy minesweeper HMCS *Shawinigan*, and the *Martin Bergmann*.

In Gjoa Haven, the crew of the *Bergmann* picked up a local Inuk man named Sammy Kogvik. Kogvik, forty-nine, was a member of the Canadian Rangers, a unit of mostly Inuit army reservists stationed in parts of Canada where it isn't practical to have conventional military bases. Parks Canada had hired him as a guide to assist in the search. As the ship steamed up Simpson Strait, Kogvik sat on the bridge with Adrian Schimnowski, the Arctic Research Foundation's operations director. Kogvik was pointing out various landmarks to Schimnowski when suddenly he turned to him and said, "I have a really important story to tell you. Is it okay if I tell you this story?"

| | |

In the spring of 2009, Kogvik and his friend James Klungnatuk, aka Uncle James, were snowmobiling across Terror Bay en route to a favorite fishing spot when they caught a glimpse of a dark object on the ice. They detoured to it and discovered that it was an old wooden pole sticking up six feet. Kogvik thought it might be the top of a mast. For years, he'd heard stories from hunters about a ship visible under the clear waters of Terror Bay, and from pilots who'd spotted vague shapes from the air. As they studied the strange object in the ice, Kogvik handed his digital camera to Klungnatuk, who took a photo of his friend bear-hugging the pole.

Kogvik was excited to share news of his discovery, but when he got back to town, he reached into his pocket for the camera, and it was gone. It must have fallen out when his snowmobile hit a bump. Was this just bad luck? Or was it an omen? Everyone in Gjoa Haven knew about the stories passed down over the generations of hunters, when they were out on the land, hearing voices and singing in a language they didn't recognize; and of Inuit looking out the doors of their tents and seeing mysterious figures walking off into the mist. Perhaps the restless spirits of Franklin's doomed men who'd been wandering King William Island for decades didn't want that ship to be found. When Klungnatuk plunged through the ice of a local lake and drowned a year later, Kogvik became even more convinced that the shipwreck was cursed.

But now, as an official member of a Parks Canada team tasked with finding *Terror*, he decided it was time to break his silence. Kogvik was not the first to speak of a wreck in Terror Bay. In fact, stories about a sunken ship in that area had been swirling in Gjoa Haven for decades. Inuit had even mentioned it to Parks Canada, which had included Terror Bay in their permits each year as a backup search location in case ice prevented them from reaching Wilmot and Crampton Bay. So when Schimnowski heard Kogvik's tale of what might be a mast sticking out of the ice, he decided that it was worth investigating. The *Martin Bergmann* arrived in Terror Bay on September 3 at four a.m. and eased its way in through the uncharted waters, keeping a close eye on the depth sounder. The crew eventually anchored in the north part of the bay in about fifty feet of water and quickly launched their dinghy, a sixteen-foot aluminum skiff outfitted with twin twenty-horsepower outboards. Using transom-mounted side-scan sonar, they looped around the area while Kogvik tried to orient himself and remember where he'd been when he'd seen that wooden pole. After a couple hours of fruitless searching, they gave up and headed back to the *Bergmann* for breakfast.

By eight thirty a.m., David McIsaac, who served alongside Gerry Chidley as cocaptain, had gotten them back underway and set a course about four hundred yards west of where they came in. Since so much of the Northwest Passage is still uncharted and the *Bergmann* spends more time than any other ship exploring these waters, they almost always collect sonar data for future reference. And besides, there was always the chance they might randomly steam over one of the lost ships. Schimnowski, Kogvik, and Chidley were in the mess hall tucking in to some omelets with sides of bacon when the boatswain Matt Briggs yelled down from above, "Adrian, you better come up. We found something." The three of them practically ran up to the bridge, where the monitor for the sonar showed a grainy but unmistakable crimson-colored outline of what could only have been a ship. "We knew right away what it was," says Schimnowski.

Kogvik, who is described by colleagues as soft-spoken and a man of few words, had tears streaming down his cheeks. "We found it" was all he said as everyone started hooting and hollering, hugging and backslapping one another.

According to Merriam-Webster, the term "synchronicity" refers to "the coincidental occurrence of events . . . that seem related but are not explained by conventional mechanisms of causality." The region in which people had been searching for *Terror* since it first disappeared more than 175 years ago is approximately the size of New England, while the range of the sonar deployed aboard the *Martin Bergmann* was a few hundred yards at most. Even on a micro scale, Terror Bay covers an area of about twenty-five square miles. And yet, somehow, the *Bergmann* had passed directly over the lost ship. Had McIssac chosen a route even a hundred feet to either side, they would have blown right past the wreck without ever knowing it was there. And, of course, the only reason they had detoured to Terror Bay in the first place was because of Kogvik's story.

Schimnowski ordered the crew to deploy several GoPros mounted inside a wire cage rigged with LED lights and an underwater laser system for taking measurements. As the *Bergmann* slowly steamed over the ship dragging the cage at the end of an umbilical cord, the cameras recorded video of the wreck. When the cage was reeled in to swap batteries, the crew pulled out the SD cards and got their first view of *Terror.* To their disbelief, the ship appeared stunningly intact. Pieces of the masts still rose above the deck, and a twenty-foot-long bowsprit projected forward. Curiously, an anchor hung from a hawsehole alongside the bow, suggesting that the ship might not have been anchored when it sank. On the third deployment, Schimnowski realized that the cage was dangling dangerously close to the bowsprit. He had just ordered a crewman to crank it in when suddenly it detached from the umbilical. Some part of it must have snagged on the wreck, triggering its quick-release mechanism. Not only had they lost all their equipment; they had also lost the footage. A bit like Tom Gross in the plane over the stone house and Kogvik with his lost camera, the crew had either forgotten or failed to record the video on the GoPros. They'd all seen it with their own eyes, but now the proof of their discovery was sitting somewhere on the seabed. "It was this weird thing," says Schimnowski. "The spirit of the ships—it's still there. And for some reason, it was not ready to release the images." They had no backup cameras, so they motored straight to Cambridge Bay to collect more equipment.

According to a source who is familiar with the details, the *Bergmann* did not have a permit to search in Terror Bay that season, and they turned their AIS off before entering the area. Upon exiting the bay after their discovery, they turned their AIS back on, at which point they were contacted by the *Laurier* and asked about their whereabouts. According to the source, the *Bergmann* reported that they were diverting to Cambridge Bay because of a "mechanical issue." No mention was made of the discovery. A few days later, when they departed Cambridge Bay, it was assumed by others in the fleet that they would rejoin the search effort. Instead, the *Bergmann* diverted back to Terror Bay. When asked why, they told the *Laurier* that they were doing so to escape bad weather. But there were no storms in the area at the time.

Back in Terror Bay, the *Bergmann* immediately launched an ROV and captured breathtaking footage of the wreck. The hull, sitting perfectly flat on its keel, its bow reinforced with iron sheathing, rose three stories above the seafloor. The planking appeared surprisingly undamaged—an indication, they believed, that the sinking had not been overly violent. The helm had a double steering wheel, like *Terror*, and nearby, amidst a pile of loose timbers, lay the ship's bell and a cannon. Forward of the wheel, the engine's algae-coated smokestack jutted several feet above the deck, and a fifteen-foot stub of one of the masts still rose upright from the deck. It's top and the other two spars lay over the starboard rail, still attached to the rigging. None, though, was tall enough to stick above the surface of the water—not even close actually—which makes Kogvik's sighting of a "mast" all the more baffling.

A technician piloted the ROV into the interior of the ship through the main hatch, where it recorded ghostly video of the communal mess hall, a food pantry, and the private cabins of some of the officers. Footage captured on a later dive, available on YouTube, shows stacks of blue patterned china plates, upright glass jars stashed in cabinets, a box that might possibly contain a daguerreotype camera, chamber pots, and a set of rifles hanging on a wall. In the pantry, two empty wine bottles lie on their sides on a shelf, begging the obvious question as to how they could have avoided rolling off if the ship had indeed drifted, unattended, from Victoria Strait all the way to Terror Bay. In Crozier's captain's quarters in the stern, many of the glass panes in the transom windows are still in place, and an upright armchair sits beside a large desk in the center of the room. The desk's drawers are firmly sealed shut, and as of this writing, no divers have entered the cabin to find out if Crozier might have left his logbook inside.

Perhaps the most compelling discovery, though, occurred three years after the initial find, when a group of Parks Canada divers confirmed that *Terror*'s propellor was engaged in its operating position. Both *Erebus* and *Terror* were designed with retractable propellors that could be drawn inside the hulls when the ships were put into winter quarters. Finding *Terror* with its prop deployed is one of the strongest clues yet that it had indeed been remanned and sailed—*and motored*—to its final resting place in Terror Bay.

By September 11, 2016, Schimnowski and the crew of the *Bergmann* were certain that the ship they'd found was *Terror*. But they still hadn't alerted Parks Canada about the discovery. When I asked Schimnowski why in December 2023, he declined to comment and said that the best source for this part of the story could be found in the book *Ice Ghosts* by Paul Watson. I'd read this book as well as Watson's exclusive on the story of the discovery, which had been published in the British daily the *Guardian* on September 12, 2016. The details as to exactly how and why it all went down the way it did are a bit murky, but Parks Canada and the Government of Nunavut (GN) learned of the *Bergmann*'s electrifying discovery at the same time as everyone else in the world when they read about it in the *Guardian*. By the time the *Bergmann* got back to Gjoa Haven, news of the find was making headlines around the world—and Kogvik suddenly found himself an international celebrity.

Parks Canada visited the wreck shortly thereafter, confirmed the discovery, and published their own press release on September 14. It made no mention of Kogvik, the *Martin Bergmann*, or the Arctic

Research Foundation. This snub did not go unnoticed in Gjoa Haven, where rumors swirled that Parks Canada was miffed that the *Bergmann* had made an unauthorized detour to Terror Bay and then kept its discovery secret for eight days.

While the discovery was celebrated around the world, the GN quietly asked the RCMP to investigate if the *Martin Bergmann*'s discovery was illegal. Through an access-to-information request, a journalist with *Nunatsiaq News* named Steve Ducharme obtained a trove of emails and letters between GN and Parks Canada revealing that the Department of Culture and Heritage had specifically prohibited any of the ships operating under the auspices of Parks Canada from entering Terror Bay on their search. "There is no historical, oral historical or archaeological evidence suggesting Terror Bay as a possible location for the Franklin wrecks," stated the Heritage Department in its response to Parks Canada's application.

During the two-month investigation that followed, lawyers representing the ARF claimed that they had never seen a copy of the permit and therefore had not been aware that Terror Bay was off-limits. The GN eventually decided not to bring any charges against the ARF. The question as to why the ARF didn't alert Parks Canada about the find before going to the press with the story has never been adequately answered, but it appeared to many that Balisille and the ARF violated the spirit, if not the rules, of their collaboration with their government partners. More than likely, the reason he didn't tell anyone other than Paul Watson and *The Guardian* about *Terror*'s discovery is because he simply wanted to make sure that he and his team got *all* of the credit for the historic discovery.

|||

The morning after I met the crew of the *Martin Bergmann*, Rudy, Ben, and I sat in *Polar Sun*'s salon drinking coffee and debating whether it was really possible to build a cradle for *Polar Sun* in Cambridge Bay. Overnight, Ben had mocked up a drawing, which now sat on the table in front of us. Building the cradle looked feasible, but there was something about Ben's absolute certainty that it would work that made me deeply uneasy. When I said as much, he turned to me and said that my decision-making process was "agonizingly painful," and that if I'd been more decisive, "we'd be on our way to Nome right now."

You can stretch a rubber band only so far, and at that moment, the one that had been holding Ben and me together finally snapped.

"The only reason we're not on our way to Nome right now is because you decided that you didn't want to finish the trip," I shot back. "All of this is happening because of you. Not me."

So much for peanut butter and jelly. Recriminations and pent-up frustrations from months of being cooped up together in a damp, moldy boat poured out of both of us until Ben, red in the face, abruptly stood up and said he was going into town to find some breakfast. A second later, he disappeared through the companionway, the Westport Whaler trailing behind him like a cape.

When he showed up an hour later, I told him that we needed to "take a walk."

"Why?" he wanted to know.

"Because we need to talk, and I don't want to do it here in front of Rudy."

Ben said that he didn't want to talk about my "feelings."

"As the captain of the ship, I'm officially asking you to step off for a conference," I replied.

This apparently was the right tack, and reluctantly, he agreed. We walked down to the end of the wharf to get out of earshot of Rudy and the crew of the *Bergmann*, and there, beside a small aluminum skiff that someone had pulled up onto the gravelly shore, we had it out. I told Ben that I didn't think he respected me, and the feeling of constantly being looked down on was ruining the trip for me.

A long silence followed as his blue eyes burned into mine, and a little muscle in his cheek began to twitch. When he finally spoke, he said that my feelings didn't matter and that I was acting like a "little girl" and I needed to "toughen up." In his worldview, we had a job to do, and that job was to get *Polar Sun* from point A to point B. His role in the enterprise was to sail the boat, to stand watch, and to cook and clean. And that was it. Worrying about someone else's feelings was not part of his mandate. And all the tension that was ruining the trip for me? It was in my head. It had to be—because he didn't feel it.

"Well, you must feel something," I said, "because you seem pretty annoyed with me."

At this, Ben admitted that it drove him nuts the way I always reached out to people to collect information and opinions when faced with a big decision. "You've been sailing for a long time and you're good at it," he said. "Why don't you just decide what you think for yourself and then go with it?"

As soon as the words left his mouth, it hit me that he was the person I'd reached out to the most for advice about seafaring over the past twenty years. But I had leaned on Ben only because I thought he fully embraced the role of being my sailing mentor. Now here he was telling me that answering my endless questions was a waste of his time. What he found especially exasperating was that I would ask him for advice but then apply my own logic to the situation. "Why call and ask someone for advice," he asked, "if you're then going to question it?"

This, it seemed, was the essence of his problem with me: my inability to recognize that he was always right. Whenever I disagreed with him, it was more than a shortcoming in my captaincy of *Polar Sun*—in his worldview, it revealed deeper flaws in my character. As an example, he referenced my habit of sometimes orienting the chart plotter head up, rather than north up. I did this on occasion when entering a tricky channel or harbor. I'd always thought that I did it for a change of perspective to make sure I was seeing the marks and topographical features in their proper orientation relative to the boat's heading. No, he said, I did it because it was how I saw the world—with me at the center of it.

By this point, I'd heard enough. "The bottom line, Ben, is that I don't want to do this with you anymore, and I think you should leave the expedition."

Instantly, Ben's eyebrows shot up in surprise as if he'd not realized, until that very moment, that I had truly reached the end of my rope. And perhaps it did matter how I felt. Maybe it even mattered a lot.

"Are you kicking me off the boat?" he asked.

"I'm not kicking you off," I replied, "but if we can't sort this out, I do think it would be best if you left with Rudy."

At that moment, it was still unknown if we could get the boat out of the water in Cambridge Bay. And if not, my plan was to continue on at least to Tuktoyaktuk. The expedition wasn't over. But I had just driven another nail into its coffin.

Ben looked away, and I imagined that he was picturing himself on that plane, looking out the window down at *Polar Sun* tied to that scruffy dock in the middle of the Northwest Passage, with me standing alone out on deck.

When he looked back at me, the muscle in his face still twitching, he said, "I don't want to leave yet. I wouldn't do that to you. And come on, man, you're one of my oldest friends. Isn't our friendship stronger than all of this?"

It was a good question. And simply by him asking it, I felt something shift inside of me. When I was a kid, the one thing in life I had always wanted more than anything but never got was approval from my father. His was a school of tough love, and no matter how hard I tried, it was never quite good enough. But when he died in 2010, one of his old friends told me at his funeral that he was always bragging about how great I was. I couldn't believe it. Later, when no one was around, I cried, thinking about why a father would hold out on telling a son that he was proud of him. Ben wasn't my father, of course, but because he was my mentor, I'd always looked up to him in a similar way; and like with my dad, the harder I tried to get his approval, the more elusive it seemed to be.

I held Ben's gaze, looking into his eyes, which bored into mine. For a minute, maybe two, I said nothing as I tried to weigh the strength of our friendship against all the bad blood. And I wondered about the underlying cause of the tension that had built up between us. When it came to sailing *Polar Sun* through the Northwest Passage, which on the surface seemed to be the primary thing we were doing together, we almost always saw eye to eye. Invariably, if one or the other of us thought it was time to reef, to change course, to charge the batteries, or to make fresh water, the other tended to agree. But living as we had been in such close proximity for so many months, just about every other topic under the sun managed to bubble up at one time or another. Ben is an incredibly well-read and learned man, and he's thought through most subjects and come to well-informed opinions that he does not hold lightly. And when it came to politics, climate change, and whether alcohol has any place on a boat—we found our fault lines. It all simmered below the surface, but the real source of the tension between us might have been nothing more than ego and a nagging, unspoken competition for who would be the top dog on *Polar Sun*.

And that was when I saw Ben. I mean, I really saw him. I saw an old friend who looked hurt and on the verge of tears. And I saw a partner without whom I never would have dared to undertake this grand adventure. But more than anything, I saw a guy who refused to walk away, a guy who'd had my back from the very beginning—and who still had it now. And that, I decided, was the measure of a true friend.

"We *are* stronger than this," I finally said, reaching out and grabbing his hand in that old GI Joe grip. For a second, I almost pulled him in for one of those climber-bro-half-hug-chest-bump things. But Ben's not much of a hugger. And truth be told, neither am I.

| | |

By the time we got back to the boat, we had committed to building that cradle. One way or another, we'd find a way to get *Polar Sun* out of the water so she could spend the winter on the hard in Cambridge Bay. I'd never really cared how long it might take to complete the Northwest Passage, and Ben said that he'd come back with me in a year's time to finish the trip. We could set off from Cambridge Bay as soon as the ice melted, and finish the passage at our leisure—and during a time of year better suited to traversing the North Slope of Alaska. With a firm decision finally made, we went straight to work removing all the sails and running rigging so *Polar Sun* could be put to bed for the winter.

After the down-rig, I scoured the town for steel, eventually locating a few I-beams in a rack behind a random warehouse. With a plan and materials close to hand, I headed out to the tank farm, where I found Tom and Gerry working on their trailer. "We were just about to come find you," said Tom. "Here's what we can do." He pointed to Dana's skidder and suggested that we add a V-shaped cradle to one end to hold the bow and that we weld two vertical uprights in the back to support the stern. The final step would be to weld some steel plate onto the bottom and curl the front up to turn it into a sled.

"You'll have to get some heavy equipment and push it out into the water," said Gerry, "then drive the boat up onto it until it loses buoyancy, block it, and pray like hell you built the thing right and it sits properly on its keel." All of this would have to take place far enough away from their ramp that if anything went wrong, we wouldn't end up in their way.

"You gotta understand," said Gerry, "that if your boat falls over when we're pulling it out, she's toast, because there is no equipment in this town that could get her back on her keel. And if that happens, you've got an environmental disaster on your hands, and you'll be lucky if you don't get fined by the town for the cleanup."

My only recourse at that point would be to hire a heavy-equipment operator to chop *Polar Sun* up into little pieces and then truck them all out to the town dump. The best part was that I'd have to pay someone, presumably these guys, $350 an hour to turn a member of my family into scrap. Tom and Gerry stood there staring at me with their hands in their pockets.

"How does that sound to you?" asked Gerry.

I told him that it sounded bad. Very bad.

Gerry nodded, then proceeded to catalog all the damage that had been done to the *Bergmann* by vandals over the years: windows and portlights smashed with rocks, the radar dome crushed, wires and lines cut, anything not bolted down—gone. The miscreants would always find a way into the boat, even if it meant using an ax to chop through a hatch.

"They're good people here in Cambridge Bay," said Gerry, "but you only need one bad apple. A bored kid with a pocketknife could do twenty-five thousand dollars in damage in a matter of minutes."

Then there was the problem of getting the boat back into the water next summer. Hauling it out would be the easy part because we'd be pulling. When it had to go back in, they'd have to push it, and there was always the chance that the skidder would lodge itself in the mud and refuse to budge. That had happened one year with the *Martin Bergmann*, hence the new wheels on their trailer.

I asked about freezing the boat in instead. For their first six years in Cambridge Bay, that was what Tom and Gerry had done. And every winter the ice tried to tear the boat apart. If pressure built up in the pack, which it often did in Cambridge Bay, it could squeeze the pulp out of the ship. If you were lucky, the ship might pop up out of the ice and end up on its side, which was far better than being driven *under* the ice. One year, the ice had gotten hold of the *Bergmann*'s prop and tried to twist it off the ship, and in the process tore open a waterproof seal where the prop shaft goes through the hull. Water poured in through the gash and froze solid in the bilge.

And if the vandals and the ice didn't get me, the cold would. Months of −50 degrees Fahrenheit would destroy any electronic device, shatter LCD screens, explode batteries, burst water lines, and maybe, if I was unlucky, crack the engine block. And, of course, there would be no place to get parts come spring because this was, almost literally, the middle of nowhere.

"Or I could give you twenty-five thousand dollars for the boat right now," said Gerry, "and you walk away."

|||

"How'd it go?" asked Ben when I got back to the boat.

"I'm sailing out of here, even if I have to go alone," I replied.

"If that's the call, I'm with you," he said without waiting to hear what had changed my mind.

"Really?" I replied. "But what about getting home to your family?"

"I'm not letting you do this alone," he said. "I'll do one more leg with you to Tuktoyaktuk and then we can reassess there, okay?"

We didn't finish the up-rig until late evening, and then we walked Rudy to the hotel, helping him with his bags. After a quick hug, we stood there staring at one another for a minute. I hoped that he would break the silence by saying that he had changed his mind and decided to continue on with us to Tuk. But instead he grabbed his bag and headed up the steps into the lodge.

We cast off the next morning at six thirty a.m. and backed away from the dock, skirting past the bow of the *Martin Bergmann* into the cold jade-colored water of Cambridge Bay. Not a soul was up and about, and in the early stillness, I almost couldn't believe that less than twenty-four hours since our falling-out, Ben and I were setting off into the unknown once again—only now as a team of two. For most of the voyage, we'd had a crew of at least four. Being double-handed meant that for the next six days, our watch system would be four on, four off. Sleep, I assumed, might be something I'd do when—and if—we got to Tuk.

A mild, cloudless day was dawning, and when we turned the corner into Coronation Gulf, named by John Franklin in 1821 on his ill-fated Coppermine River expedition, we hoisted the Whomper to catch a fair southeast wind. As *Polar Sun* romped westward in light chop and warm sunshine poured into the cockpit, Ben turned to me and said, "So, are *we* good?"

In answer, I reached out and grabbed his hand, and I squeezed it tight while looking into his eyes and chuckling at the absurdity of where our long friendship had brought us.

"Yeah, dude," I said, "we're good. Not great. But good."

Ben smiled. Because right then, good felt good enough.

CHAPTER 14

Tuk-Toyota-Truck

Date: September 4, 2022
Position: Coronation Gulf

Polar Sun's dark blue bow sliced through the sleeping ocean as that gentle southeasterly filled the Whomper like a bright yellow hot-air balloon. It was my shift, but it was so warm and pleasant outside that Ben stretched out beside me in the cockpit, soaking up the sunshine. A lot of harsh things had been said between us the day before, and perhaps even more had been left unspoken. But as I looked over toward my first mate, who was rereading *The Lord of the Rings*, I realized that, at least for right now, the tension between us was gone.

In the afternoon, I watched a distinct line of vapory clouds drift in from the south, and when I filled in the logbook at eight p.m., I noted that the glass in the barometer had fallen precipitously. With the drop in pressure came an increase in the wind, necessitating that we swap the Whomper out for the jib for what promised to be a long night ahead. When I came off watch at midnight, the sky was black as coal tar, and I felt unsettled as I headed for my bunk. I barely slept, and when I showed up for duty at four a.m., I found Ben at the helm wearing his Westport Whaler and a red wool hat. His black

Sony earphones hung around his neck, emitting a faint whisper of opera. "I'm not used to it being so dark for so long," he said. "It kind of wigged me out." We hadn't seen as much as an ice cube since exiting James Ross Strait, but we both knew that we were far from in the clear. Each day, we gained another eight minutes of darkness, and since pack ice doesn't show on radar, it felt a bit like we were playing a game of Russian roulette—with a new bullet added to the gun each day.

By dawn, the warm sunshine of the day before was a distant memory. Hard rain strafed the cockpit enclosure as we surfed down steep, closely spaced waves that seemed far too large for the amount of fetch in the narrow strait. Through the dim light that clung to the water, I saw Murray Island, the smallest of a set of craggy outcrops that hovered like ghost ships in the murk. On the edge of the landmass, a sheer cliff of swirled gray limestone rose out of the ocean. The chart showed deep water up to its edge, so I sailed in close. As *Polar Sun* swept beneath these ramparts, waves boomed against the moss- and guano-streaked walls, sending plumes of spray fifty feet into the air, where cawing seabirds circled in the salty mist.

By late morning, we'd made our turn north out of Coronation Gulf into Dolphin and Union Strait. The wind, however, had shifted into the northwest, which happened to be our course for the next three hundred miles to Amundsen Gulf—the last section of inside water before we'd emerge from the Northwest Passage into the Beaufort Sea. We tended to average about six knots for boat speed, which in twenty-four hours adds up to about 145 nautical miles. (A nautical mile is 1.15 statute miles and is also equal to one minute of latitude. One knot is the speed it takes a boat to cover one nautical mile in an hour.) At this rate, it would take us a couple days to transit Dolphin and Union Strait, and the forecast called for heavy weather overnight. Neither of us relished the idea of bashing into headwinds in the dark, so just before sunset, we found our way into a cove tucked inside the southeast shore of a small island in the middle of the strait.

The horseshoe-shaped bay was about a half mile deep and a quarter mile across, but the chart didn't show a single sounding. We nosed in warily, keeping a close eye on the depth sounder and the forward-looking sonar. Limestone crags reminiscent of Beechey Island guarded both sides of the entrance, and in the back a U-shaped bowl scribed with the telltale horizontal banding we'd seen throughout the Arctic rose about a hundred fifty feet to a windswept plateau. We anchored close to shore in fifty feet of water, and before shutting down the engine, I closed up all the hatches and ran the truck heater on high for about fifteen minutes.

By the time I dipped below deck, the thermometer at the nav station registered an interior temperature of 81 degrees Fahrenheit. I stripped off my two sweaters, poured a drink, and plopped onto the starboard settee. Ben sat in his normal spot on the port side, quietly reading about Bilbo Baggins, Gollum, and Gandalf, which seemed weirdly apropos of our current surroundings.

We'd been sailing hard for months, and as I looked around, I realized that by this point *Polar Sun* felt as much like home as my house back in New Hampshire. The cabin smelled faintly of diesel, and the only sound came from the wavelets that slapped gently against the hull. The net hanging above Ben's head, in which we stored perishables that didn't need refrigeration, was almost empty. We still had hundreds of cans and enough rice, beans, and oats to survive for a year, but not a single fruit or vegetable. The ship felt empty with just the two of us, but I relished all the extra elbow room and having the stern cabin to myself for the first time since leaving Maine. I looked over at Ben and watched him quietly as his eyes flicked back and forth across the page. And in that warm moment, I could see that our differences were a matter of stress and too much time spent in a very small space, and that my own need for validation and affirmation ran against Ben's black-and-white view of the world—and our respective places in it. In the stillness, he didn't notice that I was staring at him or that the silence, while profound, felt easy.

We both knew that we had a long, long way still to go and that time was against us. The next option for a safe harbor was in Tuk, which still lay a few days away, but we'd learned that it had even fewer facilities than Cambridge Bay, and there was no crane or boat trailer in the town. I'd heard that Northwest Passage–bound boats sometimes made their way up the Mackenzie River to overwinter in Inuvik, which lies about eighty-five miles south of Tuk, but unfortunately, *Polar Sun* drew a little too much for that to be possible. This meant that our only option for overwintering was to let the boat freeze into Tuk's harbor—which was still my absolute-worst-case scenario. Sailing onward across the North Slope of Alaska was likely to be our greatest challenge yet, and despite our rekindled bond of friendship, it was something that Ben and I both knew we didn't want to tackle as a crew of two. I had reached out to a few people, but it was a hard sell to get someone to drop everything in their life with little notice and fly to the Northwest Territories to crew on one of the most daunting sailing passages on the planet—at the wrong time of year. I still didn't have a firm commitment from anyone, and I now had only twenty-four hours to find someone and get them on their way.

Back in Gjoa Haven, I had reached out to a veteran high-latitude sailor named David Thoreson. "DT," as his friends call him, is renowned in the sailing world for being the only American to have sailed the Northwest Passage twice: westbound in 2007, then eastbound in 2009. On the second expedition, he continued on to complete a twenty-four-thousand-mile circumnavigation of the Americas via Cape Horn. DT told me that he'd been carefully following our boat tracker, and he was adamant that the weather window for completing the passage was closing quickly. He recommended that we find a place to overwinter. Pronto. "Trust me," he said. "You don't want to be out there when the Bering Sea autumn storm cycle kicks into gear."

To drive home his point, he recounted the story of weathering a superstorm in late September 2007 off the coast of Alaska. For three days, *Cloud Nine*, a fifty-nine-foot ketch, surfed downwind in hurricane-force winds. At one point, the boat accelerated to twenty-two knots as it careened down the face of a wave nearly as tall as the ship. DT steered for hours during that storm, fighting to keep the boat from pitchpoling, a horrifying and deadly scenario in which a ship falls forward, ass over teakettle, off the face of a vertical wave.

DT's advice to call it a season and come back the next year to finish the voyage made perfect sense, but after Ben and I pushed on from Cambridge Bay, it struck me that what we now faced was not unlike a big-summit day, when sometimes you have to throw down all your chips and go for broke. Call it intuition or whatever, but I felt a strangely compelling conviction that we were meant to press on and that if we did, we'd somehow find a way to make it to Nome.

Earlier, I had texted DT and asked him if he would come join us for the sail of our lives. And when I opened my iPad that evening as I sat at the nav station nursing my vodka, I got his response. His son and daughter-in-law had just arrived that day at his home on Lake Okoboji in Iowa for a weeklong visit. "I've been away a lot of my life," he wrote, "missed a lot and lost a lot." It was a nice way of saying, *Thanks but no thanks.*

The only option that remained was an old friend from college named Ben Spiess. Ben was a lifelong waterman and mountaineer who'd grown up sailing on the north shore of Massachusetts. He was now a lawyer in Anchorage, where he was fighting a messy divorce. I'd written him too.

> Hey, amigo. Remember when you said that you wanted in on this mission? Well, here's your chance.

That night, I slept better than I had in weeks; I woke around eight a.m. to the clicking of the igniter on the stovetop as Ben put on the morning kettle. I lay still for a minute and listened for the hum of that northwesterly in the rigging as my mind reminded itself where I was. To my surprise, I heard nothing. I knew that it was blowing out in the strait, which meant that our little nook was providing a perfect refuge from the wind and waves that were bashing the windward side of the island. Hoping against hope that DT had changed his mind or that Ben was able to escape Anchorage and join us, I checked the sat texter. The inbox was empty.

It was another dour gray-bird day and we started slowly, assuming that we'd be taking a rest day to let the northwesterly blow through. Over coffee, I downloaded the forecast. It showed that it was indeed blowing twenty knots out there, but from the northeast, not the northwest. And this 90-degree difference was just enough to give us a sailing angle. The forecasting tool we used, PredictWind, had six weather models, each of which is a GRIB (General Regularly-distributed Information in Binary form); all of them are based on different algorithms. The two most reliable GRIBs, the GFS and the ECMWF, are known as the "American" and the "European" models, respectively. Normally, these were the only two I referenced, but on this day, for some reason, I downloaded all six. All of them showed the wind currently out of the northeast, and all except one, called SPIRE, predicted a shift into the northwest in the next twelve hours. It was obvious that SPIRE was the outlier, and we knew all too well how it would go if we got headed by that northwesterly out in the badlands of Dolphin and Union Strait. But Nome wasn't getting any closer while we were sitting on our asses, so I downed my coffee and suited up to weigh anchor as Ben fired up the engine.

After motoring out of the bay, we caught that northeasterly and sailed all day as close as we could get to the wind with a single reefed main and staysail. By evening the waves had built to majestic proportions, as big as any we'd seen in Davis Strait. But instead of washing over us from behind, they slammed our starboard bow over and over, occasionally knocking *Polar Sun* on her ass and causing the whole boat to tremble as if she'd been given an electrical shock.

As darkness fell, the wind shifted northwest, and we lost our sailing angle. The sea came at us with a flurry of one-two jabs and stiff upper cuts that no boat, let alone its passengers, should ever have to endure. Then a wave slammed into the bow so hard that the boat skidded to a stop. *Fuck this,* I thought as I climbed below to tell Ben that it was time to heave to.

Ben suited up and went forward, while I steered us off the wind to ease the strain on the mainsail. When he'd finished tucking in a second reef, I tacked. But instead of completing the maneuver by releasing the weather sheet on the staysail, I let it backwind and then turned back the way we'd just come. I'm sure this is difficult for a non-sailor to picture, but the gist of it is that the rudder and sails now worked against one another, causing the boat to stall like an airplane that's lost its lift. When a sailboat is perfectly hove to in storm-force conditions, the hydrodynamics of the keel dragging backward through the water creates little vortices inside the incoming waves that can prevent them from breaking and possibly capsizing a vessel.

For the next several hours, *Polar Sun* veered and staggered through the dark night. Every few minutes, a roller would catch the ship just right and send everything not bolted down flying across the salon. Ben was in his cabin, lashed into his bunk, and I was alone on watch. With my legs braced against the leeward side of the cockpit, I stared out into the void. Lacking any visual reference, my mind focused on the sounds around me: the screech of the wind in the rigging, the hiss and fizz of the waves as they swept beneath the hull, and the deep moaning of the ship itself as the rig and the hull twisted against each other in a never-ending battle for equilibrium.

Despite the constant and violent rocking of the boat, I felt myself sliding into a deep and uncontrollable sleep. I tried slapping the side of my face and even holding my eyelids open with my fingers, but it was no use. I couldn't stop myself from shutting down, and every time I passed out, my limp form was tossed around the cockpit. After conking my head hard on the coaming, I knew that I had no choice but to pull the plug and recharge my batteries, at least for a bit. I climbed down the companionway, leaving no one on watch, and set the alarm on my phone for fifteen minutes. The second my head hit that moldy throw pillow, I passed out cold.

To avoid sleeping through my alarm, I had chosen the harshest tone I could find on my phone, which sounded like a siren used at a nuclear power plant. When it went off at full volume, I woke in a panic and lurched up into the cockpit, where I checked the instruments and scanned the black ocean, making sure we weren't about to be run down by an iceberg or a random ship. I then reversed my path back to the settee, reset the alarm, and passed out again. And thus my watch passed in fifteen-minute increments. Each time I lay down, I was instantly carried away from that stormy night into a surreal dreamworld of spies and secret cameras that retracted into the headliner above my bunk right before I woke up.

When Ben finally took over, I headed below with a full four hours off duty. But, of course, I couldn't sleep. I tossed and turned in some gray in-between world until, hours later, I was jolted wide-awake by a shrill engine alarm. The next thing I knew, Ben was in my cabin.

"We overheated," he said. "The engine hit two hundred ten degrees." Our normal operating temp was 175 to 180. This was not good.

The wind and waves had driven us into poorly charted waters near the shore of Victoria Island, and Ben had hoped to use the engine to claw our way north inside a shallow bay where the land might provide enough of a lee to knock down the waves a bit. Now, with a dead engine, the swell was sweeping us onto a lee shore marked on the chart as "outlying shoals." This was serious.

Like automobile engines, most marine diesels use a closed-loop system in which coolant flows through an internal jacket, absorbing heat from the engine's moving parts. In a car, the coolant is air-cooled in a radiator before it's pumped back into the engine. Most marine diesels operate similarly, except that the cooling is done with seawater that is sucked into the engine from a valve below the waterline and then pumped out through the exhaust.

A quick check showed that the coolant was topped off and the expansion tank filled to the proper level, which meant that the problem had to be with the raw water. After letting the engine cool for a bit, I fired it up again, and Ben confirmed that there was no seawater discharging through the exhaust. This meant that either the raw-water pump wasn't working or there was some kind of blockage in the system. During *Polar Sun*'s refit, I had replaced almost the entire exhaust system, including the heat exchanger, so my gut told me the problem was more likely to be the pump. I remembered how that fisherman in Newfoundland had warned me that brash ice could chew up the rubber impeller in the raw-water pump—and we had passed through an awful lot of the stuff since then.

To check the impeller, I had to remove the pump's cover plate, which was about the size of a silver dollar and held in place with six tiny brass screws. Lying on my side as huge rollers swept under the ship, I carefully unscrewed the plate, holding one hand underneath in case a screw tried to drop into the bilge. Under the plate, I saw that the seal was good and the impeller looked brand-new. "Fuuuuck," I moaned, rolling onto my back as *Polar Sun* skidded sideways across another colossal wave. *If it's not the impeller, what the hell could it be?* I wondered. *I'm a hack engine mechanic at best. What if I can't figure this out?*

The raw-water pump on the Ford Lehman is self-priming, which means it's not supposed to vapor-lock when air gets into the system. But with the way the waves were slapping against the hull, it wasn't hard to imagine that one of them might have pushed a bubble of air into the black rubber hose that connects the thru hull to the raw-water pump. To help the pump with its prime, I opened the strainer, which catches debris that gets sucked into the intake, and I poured in fresh water from the sink until all the hoses leading to and from it were overflowing. I called for Ben to try the engine again, and thirty seconds later, he yelled down from above, "We've got water in the exhaust. Whatever you did, it worked!"

After sliding into the nav station, I logged into the satellite texter, and there were two new messages. DT had changed his mind. It looks like I'll be seeing you in Tuk, he wrote. And Alaska Ben had also managed to get the time off of work and arranged for his ex to watch their six-year-old so he could do some sailing in the Arctic with his old college buddy. Elated, I gripped the nav table with both hands and closed my eyes as another roller swept beneath the ship. Now all we had to do was get to Tuk and *Polar Sun* would be back to a crew of four.

We spent the rest of the night motor-sailing north under the southwest shore of Victoria Island. It was a tough slog, but by midmorning we'd reached the mouth of Amundsen Gulf, which lies between the Canadian mainland and Banks Island. Banks, which is about the size of Maine, is the westernmost island in the Canadian Arctic Archipelago. While Inuit have lived on Banks Island for at least thirty-five hundred years, the first European to lay eyes on it was William Edward Parry, who spotted it while overwintering at Melville Island in 1819.

| | |

I'd read a lot about Banks Island as it figures prominently in the history of Arctic exploration. It was here that the expedition aboard HMS *Investigator*, led by Robert McClure, sailed into the Northwest Passage from the west in search of the Franklin expedition—and the £20,000 reward—in 1850. After rounding Cape Horn and stopping in Hawaii, they crossed the Arctic Circle on July 28, proceeded up and over Point Barrow (now Nuvuk), and then east through a narrow strip of water between the North Slope and ten-tenths ice that extended all the way to the North Pole. In early September, they crossed Amundsen Gulf to the south shore of Banks Island. A week later, they entered Prince of Wales Strait, which runs between Banks Island and Victoria Island. They fought their way up this channel to the northeast for about a hundred miles through heavy pack ice before getting frozen in for the winter.

A few weeks later, McClure climbed a six-hundred-foot peak on Banks Island and from the rounded summit he saw that Prince of Wales Strait led directly into the Parry Channel, where Melville Island shimmered on the horizon. The Franklin expedition, of course, was nowhere to be found, but it was a heady moment for McClure and his men as they had just proven for the first time the existence of a Northwest Passage connecting the Atlantic and Pacific Oceans.

Alexander Armstrong, the ship's surgeon, would later write that "thus was established the greatest Maritime Discovery of the age, which for centuries had baffled the skill, enterprise, and energy of the civilized world . . . the solution of this great enigma leaving nothing undone to confirm Great Britain's Queen—Empress of the Sea."

All that was left for McClure was to sail another fifty miles or so into the Parry Channel and complete his historic transit of the passage. But the ice, cruel mistress that she is, had other plans for *Investigator.* Blocked by heavy pack to the north the next summer, McClure and his crew were forced to spend three more winters trapped in the ice until they were eventually located by a crew member of HMS *Resolute*, who had been sent off from Melville Island in search of the missing expedition.

When he finally made it back to England in the fall of 1854, McClure was court-martialed for abandoning *Investigator*, promptly pardoned, and then awarded £10,000 for his completion of the Northwest Passage and the first circumnavigation of the Americas. His voyage, though, had spanned four years and been completed partially on foot and aboard three different ships—two of which had been abandoned in the process.

ı|ı

As the storm subsided, Ben and I sailed from Nunavut into the Northwest Territories and then up and over Cape Bathurst, one of the only peninsulas on the North American continent that crosses above 70 degrees north latitude. As we weathered the cape, we encountered the first ship we'd seen since leaving Cambridge Bay, a shallow-draft tug called *Henry Christoffersen*, which was towing three barges piled high with shipping containers. When we got within a few miles of each other, I hailed them on the radio and asked the skipper for an ice report. According to the latest ice chart, the floe edge lay about forty miles north of us, but it also showed a rogue finger of pack ice several miles long that lay much closer. With all the northerlies we'd been experiencing, it was possible that we might encounter ice in the vicinity of the cape, and we'd been warned by Alan and Annina, who were about two days ahead of us and with whom I'd been communicating regularly via email, to stay vigilant.

In a heavy Scandinavian accent, the skipper responded that the route to Tuk was clear and that we should have no issues. When I asked where he was going, he said that he was en route to a gold mine in Bathurst Inlet (or Hannigayoka in Inuktitut), a deep bay that incises the south coast of Coronation Gulf. This mine, I came to learn, was first established in the early 1980s. Back then, the gold had been carried out via an ice road to Yellowknife in the Northwest Territories. But as the Arctic warmed, the tundra no longer froze hard enough to support heavy trucking. So the mining company constructed a new port facility, airstrip, tank farm, work camp, and 125-mile road connecting the mine to Bathurst Inlet, where the precious metal could be shipped out through the Northwest Passage.

Before long, we were sailing along the Tuktoyaktuk Peninsula. This finger of land, which extends for a hundred miles to the northeast from the mouth of the Mackenzie River Delta, is covered in hundreds of thermokarst lakes that are formed when permafrost thaws and fresh water fills the depressions. The peninsula is home to vast herds of reindeer and caribou and bounded to the north by the Arctic Ocean and to the south by an otherworldly archipelago known as the Husky Lakes. The entrance to the lakes, which is actually an ocean bay, is a maze of long, narrow, parallel inlets separated by fingers of land that mirror one another so perfectly, they look man-made on the chart.

As sea ice has retreated along the southern shore of the Arctic Ocean in recent decades, oil and gas exploration has increased in kind, especially along the Tuktoyaktuk Peninsula, which is easily accessible from the Mackenzie River. The chart showed a series of artificial islands, unmarked shoals left over from abandoned drilling platforms, breakwaters extending from random bays, and more navigational aids than we'd seen yet in the Northwest Passage, most of which were marked as "PRIV" (private). But lost in the perpetual fogbank that is notorious for hugging this coastline, we couldn't see any of it.

The mist didn't thin until we made our turn into the Tuk channel and we caught sight of a row of cone-shaped hills that studded the land like miniature volcanoes. Pingos, as they are called, are essentially giant frost heaves that thrust upward from the permafrost. They can grow at the rate of an inch a year for centuries, and the tallest of them rise two hundred feet above the surrounding tundra. On the surface, they're blanketed with soil, vegetation, and rocks, but on the inside, their cores are formed of solid ice. Pingos form all over the Arctic, from Siberia to Norway, but nowhere are these unique formations more prevalent than in the area around Tuktoyaktuk, where more than a thousand of them rim the shoreline of the Beaufort Sea.

Fighting a twenty-five-knot headwind, we pounded, close-hauled, into a gray-green sea with only a few feet of water below the keel. The sky grew dark as we squeezed through a narrow opening between a gravel island and a marshy headland into Tuk's inner harbor, where dozens of surprisingly colorful red, yellow, blue, and green single-story homes lined the edge of a crumbling peninsula. As we approached the town, I spied two figures waving from the edge of a wharf that appeared to be collapsing into the sea. It was DT and Alaska Ben, both bundled up in foul-weather gear. As I made my first pass by the wharf to scope things out, I made eye contact with DT, who was pumping his fist in the air like an excited teenager, belying his stature as a veteran mariner in his early sixties. Lean and of middle height, he had sharp features and intense blue eyes. Alaska

Ben could have been DT's brother—tall and thin, clean-shaven with a craggy jaw and wire-rimmed glasses that perched atop a patrician nose.

As we motored past, DT called out: "I don't think it's deep enough for you guys to make it in. You should probably anchor."

I looked over at Ben, who stood on deck just outside the enclosure. "I really don't feel like anchoring," he said. "But it's your call." I didn't either, so I spun *Polar Sun* around and lined up for our approach. When the bow got within a few feet of the wharf, Ben tossed our new crew members a line, and I put the boat in reverse to bring in the stern. The spinning prop created little whirlpools in the turbid chocolate-colored water—but the ship moved no closer to the dock.

"I think you're aground," said DT, chuckling, as he tied off our bow line.

Sailors consider it a point of pride not to run aground, and I'd rather not admit how many times it's happened to me over the years. But as the old saying goes, "There are two types of sailors: those that have run aground—and those that lie." The boat appeared secure enough, so Ben leapt off the bow for the wharf, and I followed suit. Soon we were shaking hands, hugging, and backslapping with our new recruits. "Welcome to Tuk-Toyota-truck," said DT.

As we headed into town to find a drink and some dinner, I, for one, had an extra spring in my step, knowing that, for a few hours at least, my beloved ship was docked in a secure harbor—even if some of her was stuck on the bottom.

CHAPTER 15

Merbok

Date: September 10, 2022
Position: Tuktoyaktuk, Northwest Territories

DT, Alaska Ben, Ben, and I sat in a circle in the living room of a small apartment a stone's throw from the wharf. The room had a worn linoleum floor and was sparsely furnished and lit by a multiheaded hydra of a wall lamp that sat in the corner. Cardboard take-out boxes with cheeseburgers and fries were balanced on our laps. I poured glasses of vodka for DT and me, while the two Bens enthusiastically raised their water bottles in a toast to "Tuk-Toyota-truck." DT had come up with the nickname back in the day to entertain school kids who came to see his talk about the 2009 around-the-Americas expedition. And the moniker had stuck.

The fact that we were all here and prepared to finish the Northwest Passage together was the result of a long sequence of small miracles and sacrifices. After seven flights, two days of travel, and one night spent on the floor of the Edmonton Airport, DT and Alaska Ben had arrived in Inuvik, the largest town in the Mackenzie River Delta. From there, they caught a taxi ride with a Syrian refugee on the new highway that now connects Inuvik to the Arctic Ocean.

Up until its completion in 2017, you could only drive to Tuk in the winter via a frozen thoroughfare that you might have seen featured on the reality TV show *Ice Road Truckers*.

DT and Alaska Ben got dropped off at the Northern Store, where the managers, a Lithuanian couple, offered up the keys to their apartment so that Ben and I could take showers when *Polar Sun* landed in Tuk. As we ate, DT described how the burger joint, Grandma's Kitchen, was guarded by a giant junkyard dog that patrolled back and forth in front of the order window. When DT tried to get by, the dog pounced—and started humping him. When he finally got loose and made it to the window, he complimented Grandma on her "very friendly dog."

"Oh, that's not my dog," she replied.

The next morning, *Polar Sun* refloated on a rising tide and we pulled her the rest of the way in to the dock. A local named Brad stopped by and shared that we had run aground because the strong easterly winds had blown all the water out of the bay. Apparently, this was a common occurrence in Tuk. After I changed the oil and filters on the engine, Brad gave us a ride to the gas station in his truck to fill our empty jerry cans with diesel.

We were ready to depart by early afternoon, but before we pushed off, Alaska Ben and I took a walk out to the end of a narrow peninsula that juts into the Beaufort Sea. Years ago, the ocean that surrounds the peninsula was ice-free for only a few months every summer, and back then the pack was rarely far from shore. "The ice used to protect us," Brad had told me. "When I was a kid, freeze-up used to be in September. Nowadays, we're lucky if it starts before mid-November."

Reduced ice might make it easier to sail through the Northwest Passage, but it also makes life significantly more difficult for those who live on Alaska's North Slope. According to Brad, less ice means there is more space for waves to grow, and bigger waves eat away at the permafrost on which towns like Tuktoyaktuk sit. Shorter winters also mean less snow cover, which allows the layer of earth insulating the permafrost to absorb more heat, leading to more melt. And so the cycle continues. By some estimates, the Earth's permafrost holds up to fifteen hundred gigatons of carbon—more than the total amount already in our atmosphere. As permafrost thaws, this carbon is released in the form of carbon dioxide and methane, further reinforcing a doom-cycle feedback loop that scientists call "Arctic amplification."

When the ocean is frozen, its white surface has a high albedo effect, which bounces incoming solar radiation and heat directly back into the atmosphere. But as warmer temperatures cause the polar ice cap to melt and recede, the open water left in its place is much darker, and as a result, it absorbs rather than reflects all the radiation. This raises the water temperature, which in turn melts more ice, which turbocharges the doom cycle.

For thousands of years, a deep-frozen Arctic has acted as a global thermostat, regulating our planet's weather systems. Less ice in the polar regions translates into more heat waves and wildfires in subarctic zones like Alaska and Siberia. It also means more intense winters in certain parts of the northern hemisphere as the polar jet stream wobbles and carries bitter, crop-destroying cold snaps into crucial food-producing areas like the breadbasket of the United States. Data suggests that a melting Arctic is also strengthening hurricanes and allowing them to push farther inland where they can dump torrential rain, as happened during Hurricane Ida in 2021, when eleven people drowned in basement apartments in New York City.

And as we were about to find out, Arctic amplification was also allowing Pacific typhoons to push farther north than ever before.

Alaska Ben and I strolled past a handful of boxy prefabricated houses that line the sheltered eastern shore of the peninsula. But on the other side, all that was left of the homes that once stood here were a few abandoned pilings and the riprap and black textile cloth laid down by the town to slow the erosion of the shoreline. In 2020, when waves started rocking the houses on their foundations, the locals loaded the homes onto skidders and dragged them all a few miles farther inland.

Alaska Ben picked up a stone and threw it into the churning ocean. "These guys in Tuk actually have it pretty good," he said. "In Shishmaref the entire town is being swept into the sea." Ben knows more than most about what's happening on the North Slope, as he represents several native corporations in his work as a real estate lawyer, and in a past life he was a reporter for the *Anchorage Daily News*.

At the end of the peninsula, we came upon a huge blue sign that read "Arctic Ocean" and below this "Nunaryuam Qaangani Tariuq"—Inuvialuit for "ocean at the top of the world." A few picnic tables fringed the riprap where waves crashed, filling the air around us with salty spume. As we took it all in, a mud-splattered Nissan SUV with a red gas can tied to its roof rack pulled up and a young man and woman hopped out. Since the new road was built, Tuk is now connected to the Canadian-American road system, and it's possible to drive from anywhere in the lower forty-eight right to the shore of the Arctic Ocean. I couldn't read their license plate because it was covered in grime, but they quickly bounded over the rocks to the edge of the sea, where they took off their shoes and squealed as they dipped their toes into the frigid water.

In a way, I envied them their brief encounter with the Beaufort Sea. We still had twelve hundred miles to go, and my gut told me that the "ocean at the top of the world" wasn't quite done with us yet.

We pushed off from Tuk under a rippled gray sky and eased back out into the Artic Ocean. As soon as we exited the Tuk channel and made our turn west, the wind settled in directly over the stern. *Polar Sun* can be made happy at almost any wind angle, but her least favorite point of sail is dead downwind. Sailors call it "running," and the name seemed apropos of what we were doing.

Twenty-four hours later, we crossed into American waters and celebrated our reentry by hoisting a large blue Alaska flag that we flew from the spreaders. As a twenty-five-knot easterly hurled us forward, the eight gold stars on the flag (representing the Big Dipper and Polaris) stuck straight out from the mast. The latest ice chart showed a strip of open water about fifty miles wide running between the polar ice cap and Alaska's North Slope. Nome still lay more than a thousand ocean miles away, which we could reel off in less than a week if we kept flying downwind like we had been since leaving Tuk. But our escape from the Arctic depended on the wind continuing to blow from the east or south, which would help keep this channel open. If it shifted north or northwest, the pack ice would be driven in toward shore, blocking our only way out.

As we sailed deeper into Alaska, the wind slowly built to near gale force and the seas grew until we found ourselves surfing down double overhead waves. We doused the main and used a spar called a whisker pole that we guyed with several control lines off the side of the ship to help the jib hold its shape. *Polar Sun* liked it and was giving us a hell of a ride, but when our speed started surging over twelve knots in the troughs, I knew it was time to pull in on the reins. We took one, then two reefs in the poled-out jib, but on September 13, even after pumping the brakes, we still recorded a twenty-four-hour run of 221 miles—a new record for me as the skipper of *Polar Sun* and undoubtedly one of the biggest days she'd ever had at sea.

What I remember most from that wild sled ride across the shallow waters of the continental shelf was how *Polar Sun* seemed to understand that we were crossing through a sort of purgatory, and if there were ever a time to sprout wings and fly, this was it. For days, we hardly touched a line or the helm as our self-steering wind vane piloted us due west through a limbo in which the physical world seemed to have melted away. The image of those days that's stored in my mind is of that little red scrap of double-reefed jib fluttering between a blurry sky and a heaped-up sea streaked with long, lacy strips of foam.

We followed a watch schedule of three hours on and three hours off, with an hour overlap on each end. This way, we were on watch alone for only the middle hour, and we each enjoyed the luxury of a full five hours off. But as we surfed down the twelve-foot waves for days on end, the motion below was so violent and my preoccupation with the well-being of the boat so all-consuming that I remained wide-awake through most of my time off watch. Desperately craving sleep, I rigged a camp towel over the hatch above my bunk to block out the light during daylight hours. Most of the time, the towel just flapped in the wind, creating a sort of strobe effect that only made things worse. On the rare occasion when I did doze off, it always seemed to be within minutes of my next watch. The sleep deprivation, the bizarre dreams, the stroboscopic lighting, and the nuclear-power-plant siren of my alarm clock—all set against a background of going weightless every few seconds as we crested the massive waves—made me feel like I was stuck on a deranged carnival ride from a Stephen King movie.

One night, as Mr. Toad's Wild Ride tossed me around in my bunk, I heard a loud *BANG* that shook the ship and sounded like a massive guitar string breaking. My first thought was that we'd blown one of the wires holding up the mast, which meant that at any second we might lose our rig. I flew up to the cockpit in my long underwear to find DT alone at the helm looking all wide-eyed and a bit freaked out. It was pitch-black with thick fog and a thirty-knot northeasterly coming from dead astern. I threw on my foul-weather gear and a life jacket, clipped into the jack line running down the side deck, and carefully worked my way forward. When I got to the mast, I shone my headlamp into the gloom and saw that the whisker pole was swinging wildly back and forth across the foredeck like a wrecking ball. Every time it swung aft, it collided violently with the lower shroud. The problem, I quickly realized, was that one of the control lines guying down the pole had come loose and it was now bullwhipping the forward half of the boat. The jib had collapsed, and without any sails to drive us, *Polar Sun* was rounding up into the wind. And this, if not checked quickly, could lay us broadside to the waves that were already rolling us like one of those pirate ship rides at the county fair.

I inched forward in a low crouch to avoid the wildly swinging pole, craning my neck to track its swings. As the boat rocked from side to side, I waited for the wind to blow the loose line over the deck. When it did, I snagged it and scooted forward, where I looped it over one of the bow cleats and snubbed it down tight. DT got us back on course, and as the sail caught the wind, I climbed down to my bunk, knowing that sleep would be out of the question for the rest of the night.

We rounded Nuvuk (aka Point Barrow) at three forty p.m. on our third day out from Tuk. After we made our long-awaited turn to the south, the wind eased and the swell, now blocked by land to the east, subsided to three or four feet. Soon a huddle of a dozen walrus frolicked around the boat, surfing in our wake and thrusting their huge white tusks out of the water whenever they broke the surface with their gleaming, earth-toned bodies. A few of them swam alongside, including a mother with a small baby on her back. As we sailed past, Mom stuck her head out of the water to make eye contact with me, and then barked, spewing her warm breath into the air as if to say, *Who the hell are you?*

The motion below deck eased, and after days without sleep, I passed out hard on the starboard settee in the salon. I woke several hours later and climbed up into the cockpit, where the Bens lounged in warm midday sunshine. The fog was gone, and to port, the willow- and dwarf-birch-covered slopes of Cape Lisburne blazed fire-engine red like a stand of maple trees in New Hampshire during peak fall foliage. A few hours earlier, Alaska Ben had been so taken by the breathtaking beauty of the landscape that he tried to wake me up so that I could see it too. Thankfully, though, Ben had stopped him. "It's okay, man. Let him sleep," he told our fresh-faced new crew member. "He's seen a lot of capes."

It was the kind of day that sailors dream about—sunshine, flat water, a fair breeze, wildlife in the water—as we slid along the edge of a landscape that very few people, especially outsiders, have the privilege of witnessing. In hindsight, I wished I'd sat with that moment a little longer before I checked my messages, because when I did, the buoyancy we all felt evaporated in an instant. I read a text from Hampton aloud to the crew: Have you heard about typhoon Merbok? she wrote. You're not anywhere near a place called Point Hope, are you?

Point Hope was our next waypoint, only forty miles to the south. The National Weather Service, I soon learned, was calling Merbok "the strongest storm in over a decade." It had formed a few days earlier to the east of Japan, in an area where the ocean isn't normally warm enough to support the formation of a tropical cyclone. But that summer, oceanographers recorded the highest temperatures in more than a hundred years of tracking the North Pacific. Even still, the typhoon should have died out by the time it got to the Bering Sea. But the waters there were unprecedently warm too—so Merbok just kept on trucking. By the time we learned about the storm, it had already reached the southern Bering Sea, where it was spawning forty-foot waves and winds up to ninety miles an hour—the equivalent of a strong category 1 hurricane. And thanks to good old Murphy's Law, it was now moving northeast directly toward Point Hope, where meteorologists were predicting it would park itself and pound the area for days with tropical-storm-force winds and waves.

Obviously, we needed a place to hide, but I didn't even bother to look at the chart because I already knew that there wasn't a single decent harbor for hundreds of miles in any direction. As messed up as it was, we didn't have any real option but to ride out the storm at Point Hope itself. The settlement is on a cape that runs east to west and projects about twelve miles out into the Chukchi Sea. The northern shore of the cape is a ten-mile-long barrier island composed of a sand-dune-like permafrost bluff that rises about fifteen feet above the sea. The village of Point Hope sits on the western tip and is one of the oldest continuously occupied communities in the entire Arctic. The inlet on the inside would have been an ideal hurricane hole, but its controlling depth was five feet—one foot too shallow for *Polar Sun*. The storm was predicted to come in from the southeast, so when we got to Point Hope, we nosed in under that crumbling grassy bluff and anchored about two hundred feet from shore in the open ocean.

That evening aboard *Polar Sun* was one of the most pleasant of the entire trip. The wind blew softly from the north-northeast, and the sunset lit the world aglow for hours and gave no indication of the violent weather that lay over the horizon to the south. As the clouds lowered and thickened, a vibrant rainbow arced across the sky, and where it touched down to the north, it exploded laterally in a series of pastel stripes that made me think of lines on a piece of litmus paper. If I hadn't seen it with my own eyes and captured it with my camera, I would doubt my memory of such a spectacular sight.

Long, wispy mare's tail clouds greeted us in the morning, and by afternoon, it was blowing a near gale. We busied ourselves by battening down the hatches, securing the running rigging, and presetting a third reef in the mainsail so that we'd be prepared to heave to if we found ourselves getting blown out to sea. Every few minutes, my satellite texter would buzz with a new message about the destruction being wreaked along the Alaskan coast: "Water levels rising 11 feet above normal tides," "41-foot seas recorded," and "The strongest storm in the Bering Sea in the past 70 years" were just some of the highlights. Eventually, I switched the device off.

As the storm built, I went topside every few hours to check on our ground tackle. Each time, I let out a bit of line on the snubber to spread the wear. To do this, I had to hang over the bow, with DT holding on to my legs as *Polar Sun* hobbyhorsed violently in the chop. At one point, the bow plunged so deep that my head went completely underwater.

Down below, though, the motion was more or less tolerable, and we kept a cozy fire going in the woodstove. Alaska Ben cooked us his specialty, Spam sandwiches with melted American cheese and hot sauce on English muffins, and Ben made popcorn. While I sat at the nav station taking notes and studying the GRIBs, which showed a dark red vortex swirling over the Bering and Chukchi Seas for the next several days, the crew played endless games of Scrabble. Ben usually won.

Some people appeared on the beach a few hundred feet away. They parked their four-wheeler and stared out at us. From the deck, I yelled, "Radio," waving the handheld VHF over my head. But either they didn't hear me or they didn't have a radio. Without a way to communicate, all we could do was wonder about one another's stories. The village of Point Hope, home to around six hundred fifty people, sits ten feet above sea level, which happened to be the height of the forecasted storm surge. We didn't know it at the time, but some folks had already gone to the school, the highest place in town, for shelter. The good news for the village's residents is that it was moved a few miles in from the end of the spit in the early 1970s, and unlike a lot of other communities along this coast, it sits on a thick bed of gravel. So far, this configuration has prevented the underlying permafrost from melting.

I learned over email that Alan and Annina had anchored in a similarly exposed spot outside Shishmaref, which lies just north of the Bering Strait on the Seward Peninsula. Shishmaref, as Alaska Ben had told us earlier, is a tiny place, but it's well-known in the climate change world because it's under imminent threat of being swallowed by the sea. Ice used to protect the village, which sits atop permafrost on a tiny barrier island that is only a quarter mile wide. As the sea ice has disappeared, waves have been violently chewing away at the town, sometimes swallowing up to fifteen feet of shoreline in a single storm. In 2016, the community voted to move the village inland, and Barack Obama signed an executive order pledging financial support. But Donald Trump rescinded that order and, so far, Congress has been unwilling to foot the bill for the move, which may cost as much as $180 million.

|||

As Merbok screamed in the rigging and *Polar Sun* danced the funky chicken in the waves, I passed the hours below deck rereading a book by David Woodman entitled *Strangers Among Us*. I hadn't been able, quite yet, to let go of the eternal question about what had really happened to Franklin and his men, and Woodman's hypothesis—that a small group of Franklin survivors, perhaps as few as three or four men, might have almost made it out—fired my imagination. I kept thinking that if I could just arrange the clues in the right order, the whole Franklin puzzle would finally fall into place.

As Woodman details in *Strangers Among Us*, Charles Francis Hall collected Inuit testimony about men who wore rifles and left strange boot prints in the snow on the Melville Peninsula in the mid-1850s. This is a solid five years after the last Franklin survivors were seen on King William Island and hundreds of miles to the east of Starvation Cove. Elsewhere on the Melville Peninsula, Inuit reported finding campfires with red-painted tins and feces that looked nothing like those of the native people. This small party, which was reportedly led by a "great officer," might have split off from the group that met Tetqataq, Ukuararsuuk, and Mangaq in Washington Bay, or perhaps they were stragglers who had pushed on from *Erebus* after it was abandoned in the ice at Oot-joo-lik. Of course, there are dozens of Franklin sailors who have never been accounted for, so it could also have been an entirely different group that went in its own direction early on, perhaps heading down the east coast of King William Island instead of the west coast. A few searchers have looked for evidence of this, and to date, none has been found. But nothing has disproved it either. What we do know is that as the end for the Franklin party drew closer, they split into smaller and smaller parties, which is perhaps what one might expect.

According to Hall, the last Franklin survivors were seen by Inuit on King William Island in the summer of 1849. So if a small group was still alive in the Arctic in 1855, where had they been and what had they been doing for the past six years?

With a group of thirty or forty men (let alone the 105 who left together when they initially abandoned the ships), it was almost impossible, once the rations ran out, for them to hunt enough game to meet their caloric needs, especially when you consider the loads they must have been carrying and the long days they spent trudging across the tundra and ice. Any Inuit these large groups encountered might have wanted to help, but there were simply too many mouths, and, of course, their primary focus would have been to feed their own families. But a smaller group of just a few men would have changed this calculus, and if they'd met Inuit who weren't experiencing food insecurity of their own, perhaps they might have been invited to spend the winter in one of the igloo villages. If Inuit accounts of seeing "strangers" out on the Melville Peninsula as late as 1855 are true and these men were indeed from the Franklin expedition, this seems to be the only explanation for how they could have survived for so long.

One of the more intriguing stories about these strangers comes from an American whaler named Thomas Barry, who spent two winters frozen in at Repulse Bay from 1871 to 1873 aboard a ship called *Glacier*. Repulse Bay, which both Rae and Hall used as a base for their expeditions, is located on the southeast shore of the Melville Peninsula, to the north of Hudson Bay. Inuit call it Iwillik. In the late-nineteenth century, it was well-known as an Inuit gathering place and it was also the location that Aglooka mentioned as his destination when he met Tetqataq, Ukuararsuuk, and Mangaq in Washington Bay.

In the *Journal of the American Geographic Society*, Barry wrote that two Inuk men visited his ship that winter and reported having met a group of white people led by a chief with three stripes on his sleeve who had placed some books in a cairn lying somewhere off to the north. Barry pulled out a chart and "they traced the coast along Melville peninsula northward, as far as the projection of Cape Englefield. Then they looked for an island on the chart to the N.W., in the gulf of boothia, which they could not find as it was not on the chart. They said that it was *'connie tuck-a-lu'* (not far off or close to that), used the word *kig-a-tunn*, Which is their word in Esquimaux for island."

Four years later, while overwintering off Marble Island in Hudson Bay, Barry had a similar encounter with a different Inuit group who told him a nearly identical story. The white men, they said, had died of hunger and cold on a small island in the Gulf of Boothia. They too spoke of a cairn in which books, similar to the ones they had seen in Barry's cabin, had been cached, and they also pointed to a blank spot on Barry's chart and said, *"No-taca-noona"* ("Where is the island?"). They produced several silver spoons they had acquired from these men. Barry traded for one of them, and it was later determined that the spoon bore Franklin's crest.

The charts aboard *Glacier* didn't show the island where Inuit placed the cairn because it hadn't been discovered yet by American or European explorers. Today, this place is known as Crown Prince Frederik Island. In 2008, Jim Savelle, an anthropology professor from McGill University, and a geologist named Arthur Dyke spent a week surveying the south, east, and northeast parts of the island aboard four-wheelers. No cairns or artifacts were found, but they skipped over the north-central and northwest parts of the island because they were low and marshy, so it's conceivable that they missed something. To date, this mysterious cairn, which may possibly hold journals and logbooks that could shed light on the Franklin mystery, has never been located. To my knowledge, Savelle and Dyke are the only modern researchers who have ever bothered to look for it.

Around the same time that Barry's story was circulating in the American press, an old Inuk hunter presented a sword to a trader at an HBC outpost, reporting that it had been gifted to him in 1857 by a "great officer" of the Franklin expedition in appreciation for taking care of his men and him over the winter. The blade bears the insignia "W IV," which stands for "King William IV," namesake of King William Island, who ruled England from 1830 to 1837. As these particular swords were given to military officers upon their receiving a lieutenant's commission, it should have been easy to cross-reference which member of the Franklin crew had been promoted during this time frame. But, intriguingly, no one had. To add to the mystery, the sword is a kind used by the British infantry, not the navy. To date, this relic, which resides in the HBC archives in Winnipeg, Manitoba, has never been positively identified.

Inuit referred to the great officer seen on the Melville Peninsula, in several instances, as "Aglooka." As I've mentioned already, it's a somewhat generic name that Inuit gave to white men. But it was famously bestowed upon Francis Crozier in 1821, and when he showed back up on King William Island twenty-five years later, some of the older Inuit remembered him from Igloolik and referred to him by his old nickname.

There is some room to speculate that Crozier was *the* Aglooka that the Inuit met in Washington Bay. And because he was one of the most experienced Arctic hands on the expedition, it's not hard to imagine that the officer who might have lived with the Inuit well into the 1850s, surviving under their care, could have been Crozier. In December 1864, Inuit in Pelly Bay told Hall that the "same man, Crozier, who was at Igloo-lik when Parry and [George] Lyon were there, was Esh-e-muta [meaning "captain" in this case, the literal "chief"] of the two ships lost in the ice at Neitchille [King William Island]" and that "Crozier was the only man that would not eat any of the meat of the Koblunas as the others all did."

Whoever he was, by the time Aglooka got to Washington Bay, there would have been no hiding the fact that this group had zero chance of surviving the coming winter without assistance and support from the Inuit. If Aglooka was Crozier, he would have known from his time in Igloolik that many Inuit families overwintered in Repulse Bay and that there were likely Inuit camps on the way there. When the possible fate of this group is seen in this light, it makes sense that Aglooka told Mangaq that he was trying to get to Repulse Bay.

Of course, without those missing logbooks, we have no way to know what Crozier did or thought after he scribbled in the top corner of the Victory Point Record, "And start on tomorrow 26th for Backs Fish River." But as Woodman writes in *Strangers Among Us*, "the tradition that Crozier himself led his men and was one of the last to die was widespread."

Not long before he left the Arctic in the summer of 1869, Hall collected a last piece of testimony from a man named Joseph Fisher, captain of *Ansel Gibbs*—the same ship on which Peter Bayne and the other whalers sailed back to New Bedford after Hall murdered Patrick Coleman. Fisher told Hall that his Inuit contacts from the area near Chesterfield Inlet, which lies on the western shore of Hudson Bay, told him that "a white man arrived among the Kin-na-pa-toos many years ago & that he was finally murdered . . . & *this whaite* [*sic*] *man,* Capt. Fisher says, *was Crozier.*"

As I lay in my bunk waiting for Merbok to blow through, I thought about Aglooka. If all these stories are true, whoever he was, he must have been one hell of a survivor. And to have come so close to making it out, only to fall just short in the end . . . had to be the saddest ending imaginable.

|||

When the storm peaked on our third night off Point Hope, our Windex recorded gusts over thirty-five knots. But I later learned that the anemometer at the top of our mast had lost two of its wind cups. The highest winds recorded in the village just a few miles away were in excess of sixty knots. By the next morning, the governor of Alaska had declared a state of emergency and Merbok was on the front page of the *New York Times*. Bits of news filtering into us via text reported widespread flooding across thirteen hundred miles of the Alaskan coastline. Roads and houses had been washed away and thousands of people had lost boats, four-wheelers, snowmobiles, hunting camps, and years' worth of food supplies. Thanks to the fact that many people had evacuated to higher ground, there were no fatalities as of yet, although there had been some dramatic rescues, including one of an old man who'd been carried from his home in the bucket of a front-end loader.

One of the hardest-hit areas was the tiny community of Golovin, seventy miles east of Nome in Norton Sound. Here, waves jumped a twenty-foot-high berm and washed into town, blowing through the front doors of people's homes and sending community members running for higher ground. When the floodwaters receded, the village was covered in three feet of sand and muck mixed with leaking fuel tanks, destroyed vehicles, massive piles of driftwood, and the personal contents of the destroyed homes.

In comparison, Point Hope made it through relatively unscathed, with only minor flooding on the airport road. But we were not yet in the clear. Overnight, the wind had shifted into the southwest, robbing us of protection from the bluffs on the point. The waves came from one direction and the wind from another, which set the boat rocking violently from side to side. The motion was wearisome in the extreme, especially since we all knew that relief was nowhere in sight. Our plan was to press on for the Bering Strait once the storm blew through, but as predicted, Merbok had stalled out somewhere off to southwest over the Chukchi Sea, and now another low-pressure system was forming in the central North Pacific. If we waited any longer for conditions to moderate, we'd likely be setting ourselves up to get hit by the next storm when we got to the Bering Strait. This scenario, where low-pressure systems tag-team their way across the Bering Sea for months on end, is exactly why you have to get out of the Arctic before the autumn storm cycle kicks in.

We discussed the situation and unanimously decided that if we were going to be uncomfortable, we might as well be uncomfortable while making headway toward home. Decision made, I suited up into my foul-weather gear and headed topside to reel in the anchor. When I stepped out from the enclosure and peered toward Point Hope through the fog, a cold rain, driven by a twenty-five-knot gust, stung my face. We'd sailed through a lot of shit since leaving Maine, but intentionally setting off into the remnants of an Arctic typhoon set a new nadir for the expedition. Strangely, though, the vibe aboard wasn't as gloomy as I might have expected. DT, in particular, exuded confidence and good cheer.

"It's gonna be ugly out there," he said when I got back to the cockpit. "But don't worry. We'll get it done." To make sure I got the point, he gripped my shoulder with his hand and looked into my eyes. DT and I had already bonded to an uncommon degree. I kind of saw him as the older brother I had never had, and we frequently seemed to read each other's minds when it came to keeping *Polar Sun* happy.

As we sailed out from behind the point with a double-reefed main and staysail, the seas, now wide-open to the full fetch of the Chukchi, where Merbok still swirled, grew into the largest and angriest waves that I'd ever encountered. The wind, of course, was coming from exactly the direction we needed to go, and a ripping northerly current carried us back toward Nuvuk. We pointed as high on the wind as we could manage, but in effect, we were heading toward northeastern Siberia. Sailing is a funny, unpredictable thing, and sometimes you get surprised when you tack and find that one direction is a lot better than another, even though on paper they are similar. This wasn't one of those times. As soon as we tacked back toward Alaska, *Polar Sun* responded by doing her best impression of a submarine and stuffing her bow into the churning, confused seas. Green water rushed across the deck and burst in from under the dodger, flooding the cockpit and pouring down the companionway into the cabin. Our sanctuary inside the enclosure, which had always stayed relatively dry, now sloshed with water, and there was nowhere to sit without getting a wet ass. We were sailing too close on the wind, and we needed to fall off. But if we did, we wouldn't clear Point Hope, so we just kept bashing into it. For the next twenty-four hours, moving anywhere aboard, even from one part of the cockpit to another, was nothing less than full combat. At four a.m., Ben described the sea state in the logbook as "crashy."

By the next morning, the barometer showed a slight uptick for the first time in a week. The seas moderated and the wind shifted slightly west, allowing us to point toward the eastern side of the Bering Strait. We crossed the Arctic Circle at 66 degrees 33 minutes north at three p.m., and according to the Scott Polar Research Institute, in that moment we officially completed our transit of the Northwest Passage. I tried to rally the crew to celebrate, but no one had any appetite for doing so.

Ben headed off watch, DT downloaded the latest forecast, and Alaska Ben tended to a small brown bird that had landed on deck and hopped right into his palm. It appeared to be an old sparrow with tufts of ruffled, downy feathers sticking out from its chest. It must have been caught by the winds of Merbok and carried out to sea. Before heading below, Ben had declared that the bird had exhausted itself in the storm and was not long for this world. He was right. It died a few minutes later, and Alaska Ben tossed it into the ocean.

It took us all night working against a nasty current to transit the Bering Strait between the Diomede Islands and the twinkling lights of Wales, the westernmost town on the continent of North America. When the sun rose, we were greeted by a bright and placid day, with light southwesterly winds and spectacular views into the interior of the Seward Peninsula, where rolling tundra butted up against a range of sharply etched three-thousand-foot mountains. While the rest of the crew prepped the ship for our arrival in Nome, I took an afternoon siesta. Lying quietly in my bunk, I listened to the burble of the water as it sloshed against the hull, and I tried to decide how I felt about the voyage actually coming to an end.

I'd been yearning for weeks now for the trip to be over. I wanted to go home; to hug Hampton, Tommy, and my other children; to sleep in a bed that didn't move; to hang out with friends, tell stories, drink beer, and eat home-cooked food that didn't come out of a rusty can. I wanted to walk on land until my knees ached and to stand in the shower until I ran out of hot water. But now that all that was nearly within my grasp, I realized there was another part of me that wanted to keep sailing, to stay immersed in this Arctic wonderland—and to remain in the thrall of not knowing if we would actually make it out. Yes, I was tired, perhaps more so than I'd ever been in my life. But I'd managed to latch on to something irresistible in the Northwest Passage, something I'd been seeking for a long time. And that was how I found myself in the confusing place of wanting it all to be over while at the same time feeling totally content right where I lay.

As the sun slid toward the horizon and the city of Nome appeared off to the east, we passed a smattering of weather-beaten shacks and an abandoned yellow dump truck sinking into a beach on the outskirts of town. Before making our turn between the massive breakwaters guarding the harbor entrance, we untangled our giant Alaska flag from one of the shrouds, and DT took a picture of Ben and me in the golden light, arm in arm on the foredeck.

In the photo, we're both wearing our Icelandic sweaters and smiling broadly. I have my hat and reading glasses in my hand. The breeze ruffles my mop of unruly hair, which is noticeably more gray than it had been when I left Maine. In the lenses of Ben's mirrored shades, the sun sets over the sparkling Bering Sea. Looking at that image now, I remember the moment precisely and how good it felt to share it with Ben. I've said a lot about my old friend, not all of it flattering, but the bottom line is that there may not be a single other soul on this planet who would have shown up for me the way he did when he did. And he stuck with me to the bitter end, just like he always said he would.

That's what I see in the photo: two friends who see the world through radically different lenses and yet somehow found enough common ground, solidarity, and goodwill to succeed at the hardest thing that either of them had ever set out to do.

We motored past the breakwater, which was stacked four stories tall with red, yellow, and green shipping containers, and made our turn into the inner harbor. Every square inch of dock space was packed with an assortment of vessels, including at least a dozen gold dredgers, most of which looked like they'd been thrown together in a junkyard. Three days earlier, Merbok's ten-foot storm surge had washed through here, traveling a quarter mile inland, flooding buildings, destroying roads, and lifting a house off its foundation before it floated down the Snake River and got jammed under a bridge. A popular restaurant had burned to the ground, and Front Street, which runs parallel to the seashore, was awash in a thick layer of mud, sand, and debris. From sea level, though, no obvious signs of damage were immediately visible, and the only indications of trouble were the military helicopters flying back and forth overhead.

Deep in the back of the harbor, I spied *Taya* tied to the end of a floating-finger dock. Alan and Annina stood on deck, waving us over to raft up as we'd done in Pasley Bay. A few minutes later, DT and Ben tossed them our dock lines, I dropped the transmission into reverse, and we slid to a gentle stop.

On September 20, 2022, after 112 days and 6,736 miles, *Polar Sun*'s east-to-west transit of the Northwest Passage came to an end.

A previous owner of *Polar Sun*, Kelly Overman, had called ahead to the liquor store in town and bought us three bottles of champagne. I swung in right before closing, and that night we celebrated aboard *Taya*. After we finished the first bottle, Alan invited me to step up into the cockpit to join him for a cigarette. "Ahh, thanks, man," I said, "but I don't smoke." Then I thought about it for a second and realized that a cigarette sounded pretty good. And it was.

DT and I took the two Bens to the airport in the morning. Alaska Ben had to get back to lawyering, his looming divorce, and caring for his six-year-old, and Ben had that birthday party to attend; he made it with hours to spare. DT had his own work and family to get home to, but he knew what I was up against with *Polar Sun*'s decommissioning, and so he decided to stay on for a few more days.

Twenty-four hours later, the two of us stood on the deck of *Polar Sun* in a cold drizzle as it lurched up the town boat ramp on a homemade trailer owned by a bald bear of a man named Rolland Trowbridge. Rolland ran an automotive shop in town, and he was no stranger to the Northwest Passage, having completed it himself in 2009 with his wife and two young children aboard a thirty-two-foot gaff-rigged cutter called *Precipice*, which he had built in Michigan in his front yard. Nome wasn't supposed to be the end point of their voyage, but while waiting there for a new dinghy (they'd lost theirs while caught in the ice in James Ross Strait), his wife fell in love with the place. The way Rolland tells it, without even checking with him, she got a job, enrolled the kids in school, and found a place to live.

Rolland wasn't completely confident that if *Polar Sun* would fit on his trailer, and sure enough as soon as she began to lose buoyancy, she tipped over, nearly tossing DT and me overboard. The boat slid unceremoniously back into the water, almost as if she were saying, *Uh-uh*. This happened two more times, with *Polar Sun* leaving chunks of fiberglass on the trailer skids before Rolland, cussing like the sailor he is, dove into the freezing, muddy harbor water with a handful of ratchet straps. He was blue and shivering when he crawled back onto land, but he'd somehow managed to secure *Polar Sun* to the trailer.

Pulled by a John Deere backhoe belching clouds of black smoke, *Polar Sun* finally emerged from the water and shook herself off, and Rolland set us down a few hundred feet away in a muddy paddock. But when he released the tongue of the trailer from the hitch, the whole contraption tilted forward by about 20 degrees.

It rained nonstop for the next few days as DT and I took the boat apart and winterized her systems. We slept aboard on steeply canted mattresses, and because the boat was so far off its trim, rainwater pooled in the cockpit and poured over the sill into the ceiling panels, where it dripped through cracks onto my forehead. I didn't know what came next for *Polar Sun*, only that these were my last few nights aboard until I'd have to leave her for the better part of a year. One way or another, I still had to tackle the Bering Sea, but in the Aleutian Islands, I'd have a big decision to make. If I followed one of Mercator's great circle lines due south across the Pacific, I'd hit Hawaii in about twenty-five hundred miles. Or I could cut north up to Kodiak Island and then across the Gulf of Alaska into the rainforest archipelago of the Inside Passage. Only time would tell where the winds of fate would next lead my beloved boat.

DT left before me, and on the morning of my flight back to Boston, I climbed down the aluminum ladder leaning against *Polar Sun*'s rail with a bulging duffel bag on my back. Before hopping into the beat-up old Suburban that Rolland had loaned me, I stood in the ankle-deep mud for a minute, staring at *Polar Sun*. As rain pooled on the shoulder patches of my jacket, the tarp I'd rigged over the cockpit to keep the rain out fluttered in the breeze, bringing me back to those wild days on the North Slope.

When my plane took off and banked over town, I pressed my face to the window and caught a brief glimpse of *Polar Sun* before we disappeared into the clouds. And long after we'd turned south and the window had fogged with my breath and condensation, I kept staring down—as if I could somehow hold on.

Epilogue

On every side of us are men who hunt perpetually for their personal northwest passage, too often sacrificing health, strength, and life itself to the search; and who shall say they are not happier in their vain but hopeful quest than wiser, duller folks who sit at home, venturing nothing and, with sour laughs, deriding the seekers for that fabled thoroughfare?

—KENNETH ROBERTS

Ben and I didn't communicate at all in those early weeks after getting home. When I finally emailed to tell him about *Polar Sun*'s haul out and the decommissioning in Nome, he responded right away.

"It's mind boggling to come back from such an awesome and epic trip and realize that almost nobody realizes what the Northwest Passage even is," he wrote. "Or where, or what we were doing all summer, or how huge the coast of Alaska is. They're like, 'Oh yeah, how was it?' as though I could tell them in six words."

I could relate. I live in a small village of only one thousand people where word gets around. Everyone seemed to know that I'd been away for a long time and had done something big—even if they weren't quite sure what it was. But when I tried to explain my time in the Northwest Passage, the words felt empty and meaningless. It just didn't seem possible to distill the experience into a pithy sound bite. Back at home, I was spending hours at my desk, staring at a blank document that was supposed to be the first pages of this book. But even with an unlimited supply of words, I still wasn't sure what to say.

What I did know was that my experience in the Arctic had real heft, perhaps more so than anything I'd ever done; and the buzz from it was always there, resonating just below the surface. When I was walking in the woods, sitting by the woodstove at night with Tommy, or just staring at that blank computer screen, moments would suddenly percolate to the surface of my awareness unprovoked: the oceanic power and majesty of the gray-green waves that carried us through the gale in Davis Strait; the way the evening light etched the cliffs of Devon Island as we ghosted through the milky plumes of glacial runoff; the never-ending, time-expanding sunset at Point Hope as we awaited the arrival of Typhoon Merbok; and so many other "sharp glimpses of eternity," as William Finnegan once observed, that will forever be etched inside my psyche. These were the things I carried home.

One thing I did not bring back with me, though, were answers to the mystery that had initially drawn me to the Northwest Passage. While I worked on this book, Tom Gross returned to King William Island three more times, all without finding Franklin's tomb. Last I heard, he's planning another expedition, his forty-fourth. "I know we must be close," he wrote to me recently—words that I've heard him say before. Tom hasn't invited me back, perhaps because he knows that I'm done with Franklin. My search for Franklin's tomb was a lot like the other obsessions I've had over the years: Even if inherently pointless, it gave brief structure to my peripatetic life. And maybe that's the nut of it all: In pursuit of the absurd, we find the freedom to discover ourselves, to shed the pressures of productivity and social status and simply *explore*. I know this is why Tom will never let this go, and it's why, for his sake, a part of me hopes he never finds it.

It was Jimmy Pauloosie, the director of the Guardians in Gjoa Haven, who pointed out the illogic that underlies the very notion of the Northwest Passage. "The term 'Northwest Passage' is something that we might have to rethink," he told me. Even the name itself, he explained, is a construct we've inherited from European explorers who needed a catchy sobriquet to achieve glory and conquest—a Holy Grail–type quest that we can now see in hindsight had no practical value whatsoever.

The incredible irony, which John Ross himself came the closest to understanding through his interactions with Ikmallik the Hydrographer, is that Inuit always held the keys to this kingdom. Not only did they live and raise their families in what the British considered terra incognita, but they had long ago explored every inlet, strait, and island in this Arctic maze. All the explorers had to do was ask, and Inuit would have filled in the blanks on their maps for them. For the lack of that knowledge, already long possessed and passed down through centuries of oral tradition, Franklin and every single one of his men paid with their lives. And whether you believe that their restless spirits are still out there trying to find their way home, many Inuit wish that Franklin and those who followed him—including us—had never set out in search of that shortcut to the Pacific. One very uncomfortable truth about our trip is that we were never fully welcomed anywhere in the Arctic. And I guess I understand that if Franklin's voyage was ultimately pointless, misguided, and foolhardy, then perhaps so too was our own.

I know, of course, that the very conditions that made it possible for us to sail a fiberglass boat through the Northwest Passage are the manifestation of deadly forces—set in motion in our southern, industrialized world—over which Inuit have no control. If temperatures in the Arctic continue to rise as they have in recent years, scientists predict that a "Blue Ocean Event"—the point at which *all* Arctic ice melts over the summer—will occur for the first time before the end of this century. And as we witnessed ourselves, the Northwest Passage is already transforming into a major hub for commercial shipping, resource extraction, and tourism. Ross, Franklin, and McClintock probably never would have believed it, but the reality of an Open Polar Sea—a myth that drove the exploration and exploitation of the Arctic for centuries—is near.

The final chapters of the history of the Northwest Passage have yet to be written, and it's hard for me not to feel conflicted about my tiny part in the story. I'm proud of having added *Polar Sun* to the list of private yachts that have completed the Everest of sailing, and my time spent among Inuit in the land of the midnight sun makes me feel like I have touched something divine.

ı|ı

A little more than a year after setting off from Biddeford, Maine, on our voyage into the Arctic, I flew back to Nome with Mr. Dirt in tow. We met up with DT and an old friend and climbing partner named Jared Ogden, and the four of us sailed *Polar Sun* through the Bering Sea, then turned the corner in the Aleutian Islands and headed east across the Gulf of Alaska into the Inside Passage. There, a day's sail south of Glacier Bay, we pulled into a quirky, boardwalk-lined village called Elfin Cove.

As we tied up at the town dock, a well-tanned woman in her mid-forties with long brown hair came by to say hello. Her name was Nicole, and we now sat alongside her sailboat, a fifty-one-foot Beneteau Oceanis that she and her husband, Jacques, a retired bush pilot, had sailed from Elfin Cove to Fiji and back. Before she had even introduced herself, Nicole announced that she had spent her morning listening to a National Geographic podcast called *Overheard* about some guy who sailed his boat from Maine through the Northwest Passage. "It's an incredible story," she said. "It was so good, I actually listened to it twice. I think the boat was called *Polar Sun*."

I'd been unsure where we would end that summer's voyage and lay up for another winter, and here, falling into my lap, was yet another example of the miraculous serendipity that sometimes makes me wonder if maybe there is some method to the madness after all.

A lot of uncertainties lie ahead, but as I write these words, I'm sitting at the galley table aboard *Polar Sun* as Hampton, Tommy, and I work our way across Frederick Sound in the Inside Passage on a bluebird July day.

Before flying back to Elfin Cove, Hampton and I put our house in New Hampshire up for rent, sold our cars, and pulled Tommy out of school. And with any luck, by the time you're reading this, we'll be on our way to the South Pacific, just as we plotted so many years ago on that warm summer night in Cape Porpoise. And if you're curious how it's all going, you can check our boat tracker, the link to which I've included in the endnotes.

Maybe you're wondering, like I am, if sailing with Hampton and Tommy will be enough to satisfy my quest for adventure. It has, after all, proven to be rather insatiable. I guess it's too soon to tell, but at least this time, I'm taking a cue from my old friend Ben's playbook—and setting off into the wild with my family by my side.

Mark Synnott
SV Polar Sun
Frederick Sound, Alaska
July 8, 2024

NOTES ON SOURCES

When I left my home in New Hampshire in June 2024 to continue voyaging on *Polar Sun*, I was still working on these notes. As such, I had some hard decisions to make in terms of which titles to take with me and which to leave behind. The e-books were a no-brainer, of course, but when it came to hard copies, there was only so much room in my duffel bags, so I had to prioritize. The first two books I grabbed off my desk were *Unravelling the Franklin Mystery: Inuit Testimony* and *Strangers Among Us,* both by David Woodman. And it's fair to say that were it not for Woodman and the decades he has dedicated to solving this mystery, I probably never would have added my own small chapter to this epic saga. The few tidbits that I mined from Woodman's research and insights into the Franklin mystery merely brush the surface of a subject that is deep enough to consume a lifetime. If you've gotten this far into my story and you feel yourself being drawn deeper into this vortex, Woodman's work is where I recommend you turn next.

My to-go book pile also included *The Arctic Grail*, which is where my obsession with the far north first started when I was in college. That book has now drawn me to the Arctic (and near Arctic) eight times. Pierre Berton was no fan of Franklin, but his overview of Arctic exploration history is simply a great read, and he has done more than perhaps any other writer to revive interest in the Franklin mystery. *May We Be Spared to Meet on Earth: Letters of the Lost Franklin Arctic Expedition*, edited by Russell Potter and others, is described in its blurb as "a privileged glimpse into the private correspondence of the [Franklin] officers and sailors," and indeed, there is nothing that brings these men more to life than reading some of the last letters they wrote before they disappeared. By the same token, two of the ships that figure prominently in this story came to life and became characters in their own right through Michael Palin's *Erebus: The Story of a Ship* and *The Voyage of the 'Fox' in the Arctic Seas* by Sir Francis Leopold McClintock.

The following, in no particular order, is a bibliography of the books that informed my writing of *Into the Ice*. I've also included several titles that I read during this process purely for inspiration—and to help me find my voice as a sailing writer. And, yes, if you have a deep and abiding interest in this subject matter, as I do, I would recommend every book listed below.

The Arctic Grail: The Quest for the North West Passage and the North Pole, 1818–1909 by Pierre Berton

Unravelling the Franklin Mystery: Inuit Testimony by David C. Woodman

Strangers Among Us by David C. Woodman

Finding Franklin: The Untold Story of a 165-Year Search by Russell A. Potter

Frozen in Time: The Fate of the Franklin Expedition by Owen Beattie and John Geiger

Erebus: The Story of a Ship by Michael Palin

May We Be Spared to Meet on Earth, edited by Russell Potter, Regina Koellner, Peter Carney, and Mary Williamson

Narrative of the Second Arctic Expedition Made by Charles F. Hall, edited by J. E. Nourse

Sir John Franklin's Last Arctic Expedition by Richard J. Cyriax

The Life of Sir John Franklin by H. D. Traill

The Voyage of the 'Fox' in the Arctic Seas by Sir Francis Leopold McClintock

Narrative of a Second Voyage in Search of a North-west Passage by Sir John Ross

The Search for Franklin: A Narrative of the American Expedition Under Lieutenant Schwatka, 1878 to 1880 by Frederick Schwatka

Schwatka's Search: Sledging in the Arctic in Quest of the Franklin Records by William H. Gilder

The Terror by Dan Simmons

The Spectral Arctic: A History of Ghosts and Dreams in Polar Exploration by Shane McCorristine

The Last Winter: The Scientists, Adventurers, Journeymen, and Mavericks Trying to Save the World by Porter Fox

Leviathan: The History of Whaling in America by Eric Jay Dolin

The Last Viking: The Life of Roald Amundsen by Stephen R. Brown

James Fitzjames: The Mystery Man of the Franklin Expedition by William Battersby

Ice Ghosts: The Epic Hunt for the Lost Franklin Expedition by Paul Watson

Fatal Passage: The Story of John Rae, the Arctic Hero Time Forgot by Ken McGoogan

John Rae, Arctic Explorer: The Unfinished Autobiography, edited by William Barr

Weird and Tragic Shores: The Story of Charles Francis Hall, Explorer by Chauncey Loomis

Fury Beach: The Four-Year Odyssey of Captain John Ross and the Victory by Ray Edinger

The Last Voyage of Capt. Sir John Ross, R.N. to the Arctic

Regions: For the Discovery of a North West Passage by William Light

North with Franklin: The Lost Journals of James Fitzjames by John Wilson

Stray Leaves from an Arctic Journal by Lt. Sherard Osborn

Arctic Passage: A Unique Small-Boat Voyage in the Great Northern Waterway by John Bockstoce

Ice Blink: The Tragic Fate of Sir John Franklin's Lost Polar Expedition by Scott Cookman

Loss and Cultural Remains in Performance: The Ghosts of the Franklin Expedition by Heather Davis-Fisch

Prisoners of the North by Pierre Berton

The New Northwest Passage: A Voyage to the Front Line of Climate Change by Cameron Dueck

The Norse Atlantic Saga: Being the Norse Voyages of Discovery and Settlement to Iceland, Greenland, and North America, edited by Gwyn Jones

Unknown Shore: The Lost History of England's Arctic Colony by Robert Ruby

North-West Passage by Willy de Roos

Eskimo Diary by Thomas Frederiksen

The Wager: A Tale of Shipwreck, Mutiny and Murder by David Grann

And for inspiration:

Sailing a Serious Ocean: Sailboats, Storms, Stories, and Lessons Learned from 30 Years at Sea by John Kretschmer

North to the Night: A Year in the Arctic Ice by Alvah Simon

In the Kingdom of Ice: The Grand and Terrible Polar Voyage of the USS Jeanette and *The Wide Wide Sea: Imperial Ambition, First Contact and the Fateful Final Voyage of Captain James Cook* by Hampton Sides

Ice with Everything by H. W. Tilman
Arctic Dreams by Barry Lopez
The Memoirs of Stockholm Sven by Nathaniel Ian Miller
Over the Horizon: Exploring the Edges of a Changing Planet by David Thoreson
One Island, One Ocean: Around the Americas Aboard Ocean Watch by Herb McCormick
The Figure 8 Voyage: Five Oceans, Three Continents, One Year, Solo by Randall Reeves
To the Ice and Beyond: Kiwi Yachtman's Epic Solo Circumnavigation via the Arctic Northwest Passage by Graeme Kendall
Across the Western Ocean by William Snaith
The Long Way by Bernard Moitessier
A Voyage for Madmen by Peter Nichols
Dove by Robin Lee Graham with Derek L. T. Gill
Tinkerbelle by Robert Manry
Trustee from the Toolroom by Nevil Shute
The Old Man and the Sea by Ernest Hemingway

Perhaps of equal importance to these books were a few key websites, first and foremost amongst which is Visions of the North by Russell Potter—https://visionsnorth.blogspot.com. Potter's website is encyclopedic (as he is himself) when it comes to the Franklin mystery, and often, if I was looking for a tidbit of information, I'd pop it into the search function of this website and nine times out of ten, I got a hit. Often, not only did I find what I was looking for, but a dozen other things I hadn't even thought of or even remotely conceived.

At the time of this writing, David Woodman is working on a website of his own: Aglooka—Long Strider (aglooka.ca). It is already a treasure trove of information, but I'm sure it will only grow in size and importance as Woodman transfers his prodigious knowledge on this subject to the digital realm.

Peter Carney's website/blog, Erebus & Terror Files (http://erebusandterrorfiles.blogspot.com), follows a format similar to the one used in Visions of the North and covers a lot of new ground. It has been hugely useful for my research, especially when it comes to some of the more arcane corners of the Franklin mystery. For example, it was on Carney's website that I pieced together the details surrounding the John Ross and James Clark Ross chart of Poet's Bay (today's Rasmussen's Bay)—the imaginary backwater (and cartographic error) that likely doomed the Franklin expedition.

I also really enjoyed my time with Fabiënne Tetteroo's website at www.jamesfitzjames.com. As mentioned previously, Tetteroo's interest in Fitzjames began after she watched the AMC miniseries *The Terror* (highly recommended), which in turn led her to William Battersby's excellent biography *James Fitzjames: The Mystery Man of the Franklin Expedition*. Her website includes, among other things, copies of Fitzjames's drawings, some done during the expedition itself (and sent home from the Whale Fish Islands); it opens with a line Fitzjames wrote to John Barrow Jr. in 1844: "I think I should make up in perseverance what I might want in sense."

And I'd be remiss if I didn't mention Russell Taichman, the dean of the University of Alabama at Birmingham School of Dentistry and a Franklinite extraordinaire. Taichman has published many scholarly articles on the Franklin mystery and is currently working on a book about Charles Francis Hall entitled *Dancing with Sŭ-pung-er: The Witness Tales*, with a foreword by David Woodman. He also spent a lot of time on the ground on King William Island, including several expeditions with Tom Gross; and he was in the back seat of the plane the day that Tom saw the stone house somewhere out on Cape Felix. When I contacted Taichman with questions about the Bayne story, not only did he answer them, but he sent me all of his research material, which saved me from days of digging through historical archives. Most of the sources referenced in the Bayne section of this book came directly from Taichman.

Much of the dialogue in *Into the Ice* is quoted verbatim. National Geographic Television transcribed hours of the vérité footage and all the interviews, and was kind enough to supply me with the transcripts, which I printed and had bound in a book that ran to more than two hundred pages. I also had access to outtakes of the video footage shot by Renan, Rudy, and Matt Irving. In cases where conversations were not filmed, I made digital recordings or took notes on my phone either contemporaneously or as soon as possible afterward. If I wasn't sure that I had recorded an exchange accurately, I checked with those quoted. Additionally, most of the main characters in the book had the opportunity to read the manuscript before it went to press. David Woodman, who knows more about the Franklin mystery than just about anyone alive, read a galley of the manuscript and provided me with critically important factual corrections and feedback. Woodman has been remarkably supportive and thoughtful of my speculations throughout the process of writing this book, even though we do not always agree. Case in point: While I lean toward the possibility that Aglooka was Crozier, he does not. "Your speculation about Aglooka being Crozier is, like all speculation, feasible," he commented. "I don't believe it, however, since the Washington Bay Aglooka couldn't speak Inuktitut very well and passed on the communication duties to another. Crozier would have been relatively fluent."

Another important source of information for the sailing portion of the story comes from our ship's log, which includes Date, Time, Position, Course, Speed, Barometric Pressure, Sea State, Visibility, Wind Speed / Direction, Sail Combination, Engine Power, and

Remarks—in four-hour increments starting on June 2, 2022, and ending on September 20, 2022. Here's a sample remark from June 4, as *Polar Sun* crossed the Gulf of Maine early in the voyage: "Antares in scorpious visible in the south sky 11:35pm, strong bioluminescence from stern and hull."

In terms of reconciliation and inclusivity, I have endeavored to refer to the indigenous inhabitants of the Arctic in the ways they prefer. Since the word "Inuit" means "the people" in Inuktitut, I've tried wherever possible to avoid using the definite article "the" in front of Inuit because literally translated this means "the the people." A careful reader will notice, though, that there are a few places where I have written "the Inuit," such as instances where I'm talking about the Inuit in general and in places where a sentence otherwise didn't read properly. Inuit is already plural, so you won't find the word "Inuits" in this story, and a singular person is referred to as an Inuk, which is always capitalized. I have also tried to use Inuit names for geographical locations in the Arctic, with the caveat that in many places English names continue to supersede Inuktitut names that were bestowed on these places long before the first European explorers arrived. For any omissions or errors I have made regarding proper respect shown to Inuit—a group of people whom I could not admire more—I sincerely apologize.

You'll also find that I haven't used "the" before *Erebus*, *Terror*, *Polar Sun* (except in this book's subtitle), or many of the other ships referred to in the text. The convention when talking about ships is to leave off "the" because it is understood. When referring to a given ship for the first time and then not thereafter, I often used HMS, which stands for "His/Her Majesty's ship." Ships are always, officially, either SS (steamship), SV (sailing vessel), FV (fishing vessel), RV (research vessel), or RMS (Royal Mail ship; e.g., RMS *Titanic*), to name a few designations. But this rule doesn't seem to be perfectly consistent (for reasons I still don't understand), and there are some ships for which a definite article is used, such "the *Fox*" and "the *Titanic*." When I asked David Woodman for advice on this subject, he pointed out that the *Fox* was technically a motor yacht (MY), but no one referred to it thus because of how confusing it would be to call her MY *Fox*. It's all rather befuddling, but I tried to be as consistent as possible while also endeavoring to write about these ships in a way that wouldn't draw undue attention to the language.

Introduction

xxviii **can be traced back to a map:** All original copies and early facsimiles of the Mercator map have been lost. Reproductions currently reside at the Bibliothèque Nationale de France, the University of Basel, and the Maritiem Museum in Rotterdam. A digital version is available for free download on Wikimedia Commons. Peter Mercator, CC BY-SA 3.0.

xxix **Mercator turned to some sketchy sources:** Cara Giaimo, "The Mysteries of the First-Ever Map of the North Pole," Atlas Obscura (February 27, 2017), https://www.atlasobscura.com/articles/north-pole-map-mercator.

xxix **written by an English friar:** A translation of the Latin text from Mercator's map can be found in *Hydrographics Review* 9, no. 2 (1932): 7–45. "In 1360, an English minor friar of Oxford, who was a mathematician, reached these isles and then, having departed therefrom and having pushed on further by magical arts, he had described all and measured the whole by means of an astrolabe. . . . He averred that the waters of these 4 arms of the sea were drawn towards the abyss with such violence that no wind is strong enough to bring vessels back again once they have entered; the wind there is, however, never sufficient to turn the arms of a corn mill. Exactly similar matters are related by Giraldus Cambrensis in his book on the marvels of Ireland . . . : 'Not far from the isles [Hebrides, Iceland, etc.] towards the North there is a monstrous gulf in the sea towards which from all sides the billows of the sea coming from remote parts converge and run together as though brought there by a conduit; pouring into these mysterious abysses of nature, they are as though devoured thereby and, should it happen that a vessel pass there, it is seized and drawn away with such powerful violence of the waves that this hungry force immediately swallows it up never to appear again.'" For more on Mercator, see also M. A. Zuber, "The Armchair Discovery of the Unknown Southern Continent: Gerardus Mercator, Philosophical Pretensions and a Competitive Trade," *Early Science and Medicine* 16, no. 6 (2011): 505–41, https://doi.org/10.1163/157338211X607772; Dirk Imhof, Iris Kockelbergh, Ad

Meskens, and Jan Parmentier, *Mercator: Exploring New Horizons* (Antwerp: Plantin-Moretus Museum, 2012).

xxx **appointed him to run the Admiralty:** Fergus Fleming, *Barrow's Boys: A Stirring Story of Daring, Fortitude, and Outright Lunacy* (New York: Grove Press, 2001), 5; "Sir John Barrow 1764–1848," the Ulverston Town Council, https://ulverstoncouncil.org.uk/education/sir-john-barrow-1764-1848/; "Sir John Barrow 1764–1848," Princeton University Library, https://library.princeton.edu/visual_materials/maps/websites/africa/barrow/barrow.html; "Scientist of the Day—John Barrow," Linda Hall Library (November 23, 2022), https://www.lindahall.org/about/news/scientist-of-the-day/john-barrow/; Pierre Berton, *The Arctic Grail: The Quest for the North West Passage and the North Pole, 1818–1909* (Toronto: Anchor Canada, 1988), 19–20; Michael Palin, *Erebus: The Story of a Ship* (London: Arrow Books, 2018), 20.

xxxi **in the name of exploration:** Russell Potter, *Finding Franklin: The Untold Story of a 165-Year Search* (Montreal: McGill–Queen's University Press, 2016), 6.

xxxi **duty to spread civilization:** David Grann, *The Wager: A Tale of Shipwreck, Mutiny and Murder* (New York: Doubleday, 2023), 12–13.

xxxii **offering a prize of £20,000:** Parliament, House of Commons, *Statutes at Large* 18 Geo. II c. 17 (1745).

xxxii **"laughed at by all the world":** Berton, *The Arctic Grail*, 143.

CHAPTER 1: Peanut Butter and Jelly

3 **"the Everest of sailing":** "Cruising Guru Jimmy Cornell Transits the North West Passage in Aventura IV," *Yachting World*, February 23, 2016, https://www.yachtingworld.com/features/jimmy-cornell-transits-the-north-west-passage-70357.

3 **list maintained by the Scott Polar Research Institute:** https://www.spri.cam.ac.uk/resources/infosheets/northwestpassage.pdf.

5 **Most notable of these:** Herb McCormick, *"Fortitudine Vincimus," Cruising World*, July 9, 2012, https://www.cruisingworld.com/fortitudine-vincimus-0/.

6 **According to data collected:** David McGee and Elizabeth Gribkoff, "Permafrost," Massachusetts Institute of Technology Climate Portal, August 4, 2022, https://climate.mit.edu/explainers/permafrost.

7 **declined by 95 percent:** R. D. C. Mallett, J. C. Stroeve, S. B. Cornish, et al., "Record Winter Winds in 2020/21 Drove Exceptional Arctic Sea Ice Transport," *Communications Earth & Environment* 2, no. 149 (2021), https://doi.org/10.1038/s43247-021-00221-8; Gloria Dickie, "Researchers Express Alarm as Arctic Multiyear Sea Ice Hits Record Low," Mongabay, September 17, 2021, https://news.mongabay.com/2021/09/researchers-express-alarm-as-arctic-multiyear-sea-ice-hits-record-low/; Porter Fox, *The Last Winter: The Scientists,*

Adventurers, Journeymen, and Mavericks Trying to Save the World (New York: Little, Brown and Company, 2021), 71.

CHAPTER 2: Pretty Damn Western

26 **@SmellyBagOfDirt Instagram account:** He has since changed his handle to @Mister.Dirt.

CHAPTER 3: Sir John

My primary sources for Franklin's biography in chapter three include *The Life of Sir John Franklin* by Henry Duff Traill as well as Franklin's *Narrative of a Journey to the Shores of the Polar Sea*. Much of Traill's source material for Franklin's biography came from a packet of material, including letters and personal correspondence of the Franklin family, that was collected by his niece Sophia Cracroft after his disappearance. She wanted to write her own biography of her uncle but never managed to do so, in part because her vision became impaired in her later years.

31 **"I'll get a ladder":** H. D. Traill, *The Life of Sir John Franklin* (London: John Murray, 1896), 433.

31 **"the boy returned home":** Traill, *The Life of Sir John Franklin*, 23.

32 **"Mr. Tycho Brahe":** Traill, *The Life of Sir John Franklin*, 19.

32 **later praised Franklin:** Traill, *The Life of Sir John Franklin*, 22–23.

33 **"Franklin usually felt great loyalty":** Janice Cavell, "Franklin, Sir John," *Dictionary of Canadian Biography*, vol. 7, University of Toronto / Université Laval, http://www.biographi.ca/en/bio/franklin_john_7E.html.

35 **"now tolled so continuously":** Frederick William Beechey, *A Voyage of Discovery Towards the North Pole, Performed in His Majesty's Ships Dorothea and Trent, Under the Command of Captain David Buchan, R.N.; 1818* (London: Richard Bentley, 1843), 92; Traill, *The Life of Sir John Franklin*, 63.

36 **Samuel Hearne had reached:** Samuel Hearne, *A Journey to the Northern Ocean: The Adventures of Samuel Hearne* (Surrey, British Columbia: TouchWood Editions, 2007).

38 **"great penetration and shrewdness":** John Franklin, *Narrative of a Journey to the Shores of the Polar Sea*, vol. 2 (London: John Murray, 1823), 21.

40 **"the whole party ate":** Franklin, *Narrative of a Journey to the Shores of the Polar Sea*, 297.

40 **"convinced of the necessity":** C. Stuart Houston, ed., *Arctic Ordeal: The Journal of John Richardson, Surgeon-Naturalist with Franklin, 1820–1822* (Montreal:

McGill–Queen's University Press, 1994); P. L. Simmonds, *The Arctic Regions and Polar Discoveries During the Nineteenth Century* (London: George Routledge & Sons, 1875), 47.

41 **"Tea is indispensable":** Stephen Brown, *The Company: The Rise and Fall of the Hudson's Bay Empire* (Toronto: Doubleday Canada, 2020), 324.

41 **"That's the man who ate his shoes!":** Traill, *The Life of Sir John Franklin*, 435.

41 **"in singing to baby":** Traill, *The Life of Sir John Franklin*, 116.

42 **"embrace both of you":** Traill, *The Life of Sir John Franklin*, 121.

43 **"Franklin's Paradise":** Traill, *The Life of Sir John Franklin*, 194.

45 **"he will die of disappointment":** James Alex Browne, *The North-West Passage and the Fate of Sir John Franklin* (Woolwich, UK: W. P. Jackson, 1860), 32.

47 **told that he was the first person:** Woodman later told me that this was a mistake. Hall's biographer Chauncey Loomis had gone through Hall's papers in the 1960s.

48 **"I started to get the sense":** Daphne Bramham, "Accidental Historian's Work Was Key to Finding Franklin's Lost Ship," *Vancouver Sun*, August 22, 2016, https://vancouversun.com/news/local-news/accidental-historians-work-was-key-to-finding-franklins-lost-ship.

49 **where the ships had been abandoned:** According to Woodman, *Erebus* was found in a "rather unexpected place," thirteen miles south of the spot indicated by Inuit testimony and about three miles east of the eastern limit of both his and Parks Canada's surveys.

50 **Supunger claimed that he:** Potter, *Finding Franklin*, 161–64.

53 **a program called "Buried in Ice":** https://www.youtube.com/watch?v=TQ2LTV32hjM.

CHAPTER 4: Povl

64 **enjoys a relatively mild climate:** Petter Dannevig, "Climate in Greenland," Norwegian Meteorological Institute, January 26, 2019, https://snl.no/Klima_p%C3%A5_Gr%C3%B8nland.

65 **had stalls for a hundred sixty cows:** Jared Diamond, *Collapse: How Societies Choose to Fail or Succeed* (New York: Viking, 2005), 272.

65 **from abandoned Dorset camps:** Robert W. Park, "The Dorset-Thule Succession in Arctic North America: Assessing Claims for Culture Contact," *American Antiquity* 58, no. 2 (1993): 203, https://doi.org/10.2307/281966.

66 **chronology of Arctic cultures:** Arctic Chronology, Avataq Cultural Institute, 2015, http://www.avataq.qc.ca/en/Institute/Departments/Archaeology/Discovering-Archaeology/Arctic-Chronology.

66 **from a Saqqaq man:** Andrea Thompson, "Frozen Hair Yields First Ancient Human Genome," Live Science, February 10, 2010, https://www.livescience.com

/6098-frozen-hair-yields-ancient-human-genome.html; Nicholas Wade, "Ancient Man in Greenland Has Genome Decoded," *New York Times*, February 10, 2010, https://www.nytimes.com/2010/02/11/science/11genome.html.

66 **did not interbreed:** Anatolijs Venovcevs, "North America's First Contact: Norse-Inuit Relations," Medievalists.net, November 2013, https://www.medievalists.net/2013/11/north-americas-first-contact-norse-inuit-relations/; http://www.fitp.ca/articles/FITPXX/First_Contact.pdf.

66 **"The *skraelings* assaulted":** Diamond, *Collapse*, 35.

67 **the Little Ice Age:** Gwyn Jones, ed., *The Norse Atlantic Saga: Being the Norse Voyages of Discovery and Settlement to Iceland, Greenland, and North America* (Oxford: Oxford University Press, 1986), 90.

70 **more than a dozen expeditions:** A list of all the various Franklin search expeditions up until McClintock's discovery of the Victory Point Record in 1859 can be found in J. E. Nourse, ed., *Narrative of the Second Arctic Expedition Made by Charles F. Hall* (Washington, DC: Government Printing Office, 1879), xxix–xxxii.

70 **the Victory Point Record:** The original copy of the Victory Point Record is stored at the United Kingdom Hydrographic Office in Taunton, Somerset, England. An excellent analysis of the document can be found on Russell Potter's blog, *Visions of the North*, at https://visionsnorth.blogspot.com/2009/04/it-is-perhaps-most-evocative-document.html; a digital facsimile and transcript of the record can be found here: https://www.historymuseum.ca/blog/a-very-special-piece-of-paper/. Further analysis by Dave Woodman can be found here: https://www.aglooka.ca/two-words-the-victory-point-record/.

71 **In the early 1860s:** Francis Leopold McClintock, *The Voyage of the 'Fox' in the Arctic Seas* (London: John Murray, 1859), 6.

CHAPTER 5: The Whale Fish Islands

83 **aboriginal whaling in Greenland:** "Management and Utilisation of Large Whales in Greenland," International Whaling Commission, https://iwc.int/management-and-conservation/whaling/aboriginal/greenland; for more background on whaling in Greenland, see: Peter Matthiessen, "Survival of the Hunter," *The New Yorker*, April 24, 1995, 67–77; Regin Winther Poulsen, "Greenland Votes to Move Whaling Away from Tourists' Eyes," *Hakai Magazine*, May 20, 2021, https://hakaimagazine.com/news/greenland-votes-to-move-whaling-away-from-tourists-eyes/.

84 **used to frame houses:** Peter C. Dawson and Richard M. Levy, "A Three-Dimensional Model of a Thule Inuit Whale Bone House," *Journal of Field Archaeology* 30, no. 4 (Winter 2005): 443–55.

85 **The Basques dominated whaling:** Alex Aguilar, "A Review of Old Basque Whaling and Its Effect on the Right Whales of the North Atlantic," in *Right Whales:*

Past and Present Status, ed. by R. L. Brownell Jr., P. B. Best, and J. H. Prescott (Cambridge, UK: International Whaling Commission, 1986), 191–99, https://www.researchgate.net/publication/235407504_A_review_of_old_Basque_whaling_and_its_effect_on_the_right_whales_of_the_North_Atlantic.

86 **"simple American fisherman":** Eric Jay Dolin, *Leviathan: The History of Whaling in America* (New York: W. W. Norton & Company, 2007), 214.

88 **began to gain ice:** Carol Rasmussen, "Cold Water Currently Slowing Fastest Greenland Glacier," NASA Jet Propulsion Laboratory, California Institute of Technology, March 25, 2019, https://www.jpl.nasa.gov/news/cold-water-currently-slowing-fastest-greenland-glacier; Ala Khazendar et al., "Interruption of Two Decades of Jakobshavn Isbrae Acceleration and Thinning as Regional Ocean Cools," *Nature Geoscience* 12 (March 2019): 277–83, https://www.nature.com/articles/s41561-019-0329-3.

94 **"The waves of ice, 30 feet high":** John McCarron, "HMS Terror and Lough Swilly," Ships of the Swilly, July 15, 2020, https://slavaryghost.wordpress.com; Lincoln Paine, *Ships of Discovery and Exploration* (Boston: Houghton Mifflin, 2000), 139–40.

95 **London and Croydon Railway:** Peter Carney, "Archimedes and Croydon: The Engines of Erebus and Terror," Erebus & Terror Files, July 30, 2010, http://erebusandterrorfiles.blogspot.com/2010/07/archimedes-and-croydon-engines-of.html.

95 **At the Woolwich Dockyard:** Russell Potter et al., eds., *May We Be Spared to Meet on Earth: Letters of the Lost Franklin Arctic Expedition* (Montreal: McGill–Queen's University Press, 2022), 78.

95 **thirty-two-foot-long iron shafts:** Palin, *Erebus*, 195.

96 **to reduce drag:** William Battersby and Peter Carney, "Equipping HM Ships *Erebus* and *Terror*, 1845," *International Journal for the History of Engineering & Technology* 81, no. 2 (July 2011): 192–211, https://doi.org/10.1179/175812111X13033852943147.

96 **on the ships' manifests:** Peter Tyson and Russell Potter, "Franklin's Provisions," NOVA: Arctic Passage, https://www.pbs.org/wgbh/nova/arctic/provisions.html.

96 **brass oil lamps:** Palin, *Erebus*, 203.

96 **had died on the way:** Potter, *May We Be Spared to Meet on Earth*, 425.

97 **quit shortly before:** Traill, *The Life of Sir John Franklin*, 435.

99 **"exceedingly good old chap":** Potter, *May We Be Spared to Meet on Earth*, 240.

99 **"beloved by us all":** Potter, *May We Be Spared to Meet on Earth*, 13.

99 **circumnavigated the Americas:** William Battersby, *James Fitzjames: The Mystery Man of the Franklin Expedition* (Toronto: Dundurn Press, 2010), 155–56.

99 **He was born in Brazil:** Battersby, *James Fitzjames*, 26.

100 **truth about his origins:** Battersby, *James Fitzjames*, 24.

100 **as a national hero:** "Liverpool 1835," James Fitzjames, https://jamesfitzjames.com/liverpool-1835/.

100 **held hostage by a sheikh:** Battersby, *James Fitzjames*, 103.

101 **began calling Crozier:** David Woodman, *Unravelling the Franklin Mystery: Inuit Testimony* (Montreal: McGill–Queen's University Press, 1991), 195.

103 **he was "quite reconciled":** Olga Kimmins, "The Heart of a Broken Timeline," The Thousandth Part, May 27, 2022, https://www.thethousandthpart.com/notes/the-heart-of-a-broken-timeline (this article does a superb job of analyzing the relationship between Crozier and Cracroft; see its sources); Sophia Cracroft's journal (MS 248/239), June 1844, Scott Polar Research Institute, Cambridge, UK; Francis Rawdon Moira Crozier, letter to James Clark Ross (MS 248/364/17), September 9, 1844, Scott Polar Research Institute, Cambridge, UK; Francis Rawdon Moira Crozier, letter to John Henderson (AGC/C/5/1), July 4, 1845, National Maritime Museum, Greenwich, London, https://www.rmg.co.uk/collections/archive/rmgc-object-1112518.

103 **was daunted by the responsibility:** Michael Smith, *Captain Francis Crozier: Last Man Standing?* (Gloucestershire, UK: Stroud, 2016), 126.

104 **"has not good judgement":** Berton, *The Arctic Grail*, 146.

104 **"Except to kick up a row":** Potter, *May We Be Spared to Meet on Earth*, 298.

106 **they ever went AWOL:** Ralph Lloyd-Jones, "The Men Who Sailed with Franklin," *Polar Record* 41, no. 219 (2005): 311–18, https://www.cambridge.org/core/services/aop-cambridge-core/content/view/6D2310DBE0F8AF801001A2F476528247/S0032247405004651a.pdf/the-men-who-sailed-with-franklin.pdf.

106 **"perfectly useless":** Potter, *May We Be Spared to Meet on Earth*, 299.

107 **oldest noncommissioned member:** Lloyd-Jones, "The Men Who Sailed with Franklin," 311–17.

108 **"Failure in no way diminishes":** Lloyd-Jones, "The Men Who Sailed with Franklin," 318.

108 **a tightly spaced fourteen-page letter:** Potter, *May We Be Spared to Meet on Earth*, 268–79.

CHAPTER 6: Qikiqtaaluk (Baffin Island)

114 **corresponded to "eggs":** For a detailed explanation on how to interpret ice charts and the egg code, see: https://www.canada.ca/en/environment-climate-change/services/ice-forecasts-observations/publications/interpreting-charts/chapter-1.html.

116 **McClintock was considered:** Berton, *The Arctic Grail*, 186–91; Clements Markham, *Life of Admiral Sir Leopold McClintock* (London: John Murray, 1909); William

Barr, "Francis Leopold McClintock (1819–1907), *Arctic* 40, no. 4 (1987): 352–53, https://pubs.aina.ucalgary.ca/arctic/Arctic40-4-352.pdf.

117 **"ever betokened an anxious mind":** Markham, *Life of Admiral Sir Leopold McClintock*, 261; Edward Parry, *Memorials of Commander Charles Parry, Commander Royal Navy* (London: Strahan & Co., 1870), 219.

119 **"hair had turned grey":** McClintock, *The Voyage of the 'Fox' in the Arctic Seas*, 109.

127 **"cultural genocide":** Ian Austen, "Canada Settles $2 Billion Suit over 'Cultural Genocide' at Residential Schools," *New York Times*, January 21, 2023.

129 **780 million metric tons:** Ian Austen, "ArcelorMittal Moves Ahead with Bid for Arctic Miner," *New York Times*, January 25, 2011.

130 **story about the protest:** Dustin Patar, "Mary River Mine at a Standstill as Blockades Enter 5th Day," *Nunatsiaq News*, February 8, 2021, https://nunatsiaq.com/stories/article/mary-river-mine-comes-to-standstill-as-blockades-enters-5th-day/; Dustin Patar, "Hunters Block Mary River Mine Airstrip, Road to Protest Baffinland Expansion," *Nunatsiaq News*, February 5, 2021, https://nunatsiaq.com/stories/article/hunters-block-mary-river-mine-airstrip-road-to-protest-baffinland-expansion/.

CHAPTER 7: Consider Your Ways

137 **Dannett later recorded:** Parliament, House of Commons, *Arctic Expedition. Instructions to Captain Sir J. Franklin, in Reference to Arctic Expedition of 1845; Correspondence in Reference to Arctic Expedition, 1845–48*, HC 1847–48, vol. 264 (London: House of Commons, 1847); Richard Cyriax, *Sir John Franklin's Last Arctic Expedition* (London: Methuen & Co., 1939), 67–68.

138 **the sun set:** https://www.timeanddate.com/sun/@5897411?month=11&year=1845; https://www.timeanddate.com/sun/@5897411?month=2&year=1846.

138 **passed away on April 3:** The fourth grave on the beach is that of Thomas Morgan, a Franklin searcher who passed away aboard HMS *North Star* on May 22, 1854.

138 **"in a bag with holes":** The idea that Franklin might have included the verse from Haggai 1:7 in the Old Testament was proposed in Lloyd-Jones, "The Men Who Sailed with Franklin," 316.

139 **pouring hot water:** Owen Beattie and John Geiger, *Frozen in Time: The Fate of the Franklin Expedition* (Vancouver: Greystone Books, 1987), 184.

140 **"Lords will may we be spared":** Potter, *May We Be Spared to Meet on Earth*, 304.

141 **theory that lead poisoning:** Russell Potter, "Sir John Franklin: Fact and Fiction (Part 1 of 2)," Visions of the North, February 21, 2009, https://visionsnorth.blogspot.com/2009/02/sir-john-franklin-fact-and-fiction-part.html.

142 **discovered by American and British:** Sherard Osborn, *Stray Leaves from an Arctic Journal* (London: Longman, Brown, Green, and Longmans, 1852), 71–77; Berton, *The Arctic Grail*, 179.

142 **"dark and frowning cliffs":** Sherard Osborn, *The Career, Last Voyage, and Fate of Sir John Franklin* (London: Bradbury and Evans, 1860), 51.

143 **a faint heart embroidered:** Osborn, *Stray Leaves from an Arctic Journal*, 74; the gloves, along with many other Franklin artifacts, are housed at the National Maritime Museum, Greenwich, London, and they can be seen here: https://www.rmg.co.uk/collections/objects/rmgc-object-2034.

145 **probably "patent fuel":** Palin, *Erebus*, 196.

145 **both sides of the hulls:** Palin, *Erebus*, 194.

CHAPTER 8: The Hydrographer

Sources for the section on the sinking of SV *Anahita* in Bellot Strait include an article in the daily Spanish-language newspaper *El País*, various Facebook posts made by Pablo David Saad, blog posts on the website Arctic Northwest Passage (www.arcticnorthwestpassage.blogspot.com), as well as an interview I conducted with Saad via WhatsApp. At the time, he was sailing the new *Anahita* along the coast of Somalia, en route from Kenya to Djibouti.

150 **"mountains of crystal":** John Ross, *Narrative of a Second Voyage in Search of a North-west Passage* (Brussels: A. Wahlen, 1835), e-book 2328–34.

150 **crack like a nut:** Henry Larsen, *The North-West Passage, 1940–1942 and 1944: The Famous Voyages of the Royal Canadian Mounted Police Schooner "St. Roch"* (Ottawa: Queen's Printer, 1969), https://gutenberg.ca/ebooks/larsenh-northwestpassage/larsenh-northwestpassage-00-h-dir/larsenh-northwestpassage-00-h.html.

153 ***"Estamos muertos"*:** Federico Bianchini, "The Day a Polar Bear Could Have Eaten Us in the Arctic," *El País*, February 19, 2021, https://elpais.com/eps/2021-02-21/el-dia-que-nos-pudo-comer-un-oso-polar-en-el-artico.html.

156 **"joyful beach":** Woodman, *Unravelling the Franklin Mystery*, 17.

156 **"the dreariest prospect":** Ross, *Narrative of a Second Voyage in Search of a North-west Passage*, 3416–22.

156 **lost his lower leg:** Ray Edinger, *Fury Beach: The Four-Year Odyssey of Captain John Ross and the Victory* (New York: Berkley Books, 2003), 69; Ross, *Narrative of a Second Voyage in Search of a North-west Passage*, 3580.

157 **"chased Agliluktoq out":** Woodman, *Unravelling the Franklin Mystery*, 13.

158 **"the opulence of luxury":** Ross, *Narrative of a Second Voyage in Search of a North-west Passage*, e-book 3509–16.

158 **"man of unusual power":** Ross, *Narrative of a Second Voyage in Search of a North-west Passage*, e-book 3659–61.

159 **"downright cruelty and inhumanity":** William Light, *The Last Voyage of Capt. Sir John Ross, R.N. to the Arctic Regions: For the Discovery of a North West Passage* (London: John Saunders, 1835), 243.

159 **masses of ice heaped:** Ross, *Narrative of a Second Voyage in Search of a North-west Passage*, e-book 5615.

160 **after English poets:** Peter Carney, "Return to Poctes Bay," Erebus & Terror Files, August 1, 2014, http://erebusandterrorfiles.blogspot.com/2014/08/return-to-poctes-bay.html.

161 **movement over time measured:** Subsequent expeditions, including those of Franklin and Roald Amundsen, did their own magnetic observations, which over time helped to establish that through the centuries the magnetic north pole has slowly been moving north-northwest. In Ross's day, it moved about ten miles a year, but today it shifts about thirty-four miles a year. Alan Buis, "Flip Flop: Why Variations in Earth's Magnetic Field Aren't Causing Today's Climate Change," NASA Jet Propulsion Laboratory, California Institute of Technology, August 3, 2021, https://science.nasa.gov/science-research/earth-science/flip-flop-why-variations-in-earths-magnetic-field-arent-causing-todays-climate-change/.

161 **at a random spot:** Ross recorded the coordinates as 70 degrees 5 minutes 17 seconds north, 96 degrees 46 minutes 45 seconds west.

161 **"a pyramid of more importance":** Ross, *Narrative of a Second Voyage in Search of a North-west Passage*, e-book 7455–59.

162 **"suffered smartly":** Ross, *Narrative of a Second Voyage in Search of a North-west Passage*, e-book 8667–68.

163 **"blunder headedness of men":** Ross, *Narrative of a Second Voyage in Search of a North-west Passage*, e-book 9540–45.

165 **Barrow called the *Victory*:** Berton, *The Arctic Grail*, 119–21; Russell Potter, "The Library of the Erebus and Terror," Visions of the North, April 26, 2009, https://visionsnorth.blogspot.com/search?q=library+of+the+erebus+and+terror.

168 **"ere we return":** Potter, *May We Be Spared to Meet on Earth*, 297.

CHAPTER 9: Franklin's Tomb

Much of the source material on the Peter Bayne story came from Russell Taichman, a friend of Tom Gross and a noted Franklin and Arctic historian. He kindly provided me with a digital folder full of old newspaper clippings, photos, reports, and book excerpts that saved me hours of research.

173 **36,563 islands:** "The Canadian Arctic: Canada's Largest Ocean Area," Government of Canada, https://www.dfo-mpo.gc.ca/about-notre-sujet/publications/infographics-infographies/soto-rceo-arctic-arctique/page-01-eng.html.

175 **thinking about an essay:** John Harries, "Enough with the Northwest Passage, Already," Attainable Adventure Cruising, October 9, 2014, https://www.morganscloud.com/2014/10/09/enough-with-the-northwest-passage-already/.

178 **"the finest little harbor":** James Martin Miller, ed., *Discovery of the North Pole: Dr. Frederick A. Cook's Own Story* (New York: Polar Publishing Co., 1909), 310.

184 **the dubious honor:** Woodman, *Unravelling the Franklin Mystery*, 98.

185 **to have killed:** Jonathan Lamb, *Scurvy: The Disease of Discovery* (Princeton, NJ: Princeton University Press, 2017); Philip K. Allan, "Finding the Cure for Scurvy," *Naval History* 35, no. 1 (February 2021): 6, https://www.usni.org/magazines/naval-history-magazine/2021/february/finding-cure-scurvy.

186 **"an unwholesome stench":** Grann, *The Wager*, 75.

186 **James Lind discovered:** U. Tröhler, "Lind and Scurvy: 1747 to 1795," *Journal of the Royal Society of Medicine* 98, no. 11 (November 2005): 519–22.

186 **"never been consolidated":** George Anson, *A Voyage Around the World* (London: John and Paul Knapton, 1748), 145.

187 **"the most dreadful terrors":** Anson, *A Voyage Around the World*, 139.

187 **"fastens upon the breast":** Emily Wilkinson, "Only the Lonely," *Washington Examiner*, February 4, 2012, https://www.washingtonexaminer.com/magazine/2042187/only-the-lonely-2/; Susan J. Matt, "Home, Sweet Home," *New York Times*, April 19, 2012, https://archive.nytimes.com/opinionator.blogs.nytimes.com/2012/04/19/home-sweet-home/.

189 **According to Bayne:** L. T. Burwash, *Canada's Western Arctic: Report on Investigations in 1925–26, 1928–29, and 1930* (Ottawa: F. A. Acland, 1931), appendix C 112–16; Woodman, *Unravelling the Franklin Mystery*, 229–32.

189 **Bayne's account of the incident:** "Hall's Arctic Career: How He Shot Pat Coleman in the Region of the Heart—An Appeal to the British Government," *Louisville Courier-Journal*, April 18, 1870, accessed via ProQuest Historical Newspapers.

191 **exhumed Hall's body:** Chauncey Loomis, *Weird and Tragic Shores: The Story of Charles Francis Hall, Explorer* (New York: Modern Library, 2000), 305–20.

191 **sculptor named Vinnie Ream:** Russell Potter, "A Motive for the Murder of Charles Francis Hall," Visions of the North, July 6, 2015, https://visionsnorth.blogspot.com/2015/07/a-motive-for-murder-of-charles-francis.html.

192 **was beached in Nome:** "Wreck of the Power Schooner DUXBURY on the Beach at Nome, Alaska, 1925," University Libraries, University of Washington, https://digitalcollections.lib.washington.edu/digital/collection/transportation/id/505.

192 **personally witnessed the burial:** "Franklin's Grave: Clues to Location; Expedition to Search," *Evening Post*, February 10, 1930.

198 **at least eleven individuals:** Anne Keenleyside, Margaret Bertulli, and Henry C. Fricke, "The Final Days of the Franklin Expedition: New Skeletal Evidence," *Arctic* 50, no. 1 (March 1997): 36–46.

199 **human flesh had been boiled:** Woodman, *Unravelling the Franklin Mystery*, 186–88.

199 **remains of two men:** McClintock, *The Voyage of the 'Fox' in the Arctic Seas*, 235.

200 ***"returning to the ships"*****:** McClintock, *The Voyage of the 'Fox' in the Arctic Seas*, 237–38.

201 **"Awarded to John Irving":** Potter, *Finding Franklin*, 132. An excerpt of Schwatka's narrative about finding Irving's grave can be accessed here: https://w3.ric.edu/faculty/rpotter/temp/Schwatka_and_Supunger.pdf.

202 **in the *Journal of Archaeological Science*:** Douglas R. Stenton, Stephen Fratpietro, and Robert W. Park, "Identification of a Senior Officer from Sir John Franklin's Northwest Passage Expedition," *Journal of Archaeological Science: Reports* 59 (November 2024), https://doi.org/10.1016/j.jasrep.2024.104748.

203 **thanks to a Dutch academic:** Alexander Nazaryan, "Researchers Find Cannibalized Victim of 19th-Century Arctic Voyage," *New York Times*, October 4, 2024, https://www.nytimes.com/2024/10/04/science/franklin-expedition-cannibalism.html.

203 **"neither rank nor status":** Media Relations, "Another Franklin Expedition Crew Member Has Been Identified," *Waterloo News*, September 24, 2024, https://uwaterloo.ca/news/media/another-franklin-expedition-crew-member-has-been-identified.

CHAPTER 10: Icebound

227 **check the level:** I borrowed this idea from Bernard Moitessier, *The Long Way* (Dobbs Ferry, NY: Sheridan House, 1995).

CHAPTER 11: Gjoa (*Jo-Uh*), Come Again

245 **Per Oo-na-lee's recounting:** McClintock, *The Voyage of the 'Fox' in the Arctic Seas*, 210–12.

245 **determined that it lay:** Woodman, *Unravelling the Franklin Mystery*, 248; Nourse ed., *Narrative of the Second Arctic Expedition Made by Charles F. Hall*, 403–4.

246 **"four were still unopened":** Frederick Schwatka, *The Search for Franklin: A Narrative of the American Expedition Under Lieutenant Schwatka, 1878 to 1880* (New York: American Geographical Society, 1882), 30–31; William H. Gilder, *Schwatka's Search: Sledging in the Arctic in Quest of the Franklin Records* (New York: Charles Scribner's Son, 1881), 78–79.

246 **"potential inaccuracies":** Potter, *Finding Franklin*, 112.

248 **ran aground in 1996:** "Marine Investigation Report M96H0016: Grounding—Passenger Vessel 'HANSEATIC,' Simpson Strait, Northwest Territories," Transportation Safety Board of Canada, August 29, 1996, https://www.tsb.gc.ca/eng/rapports-reports/marine/1996/m96h0016/m96h0016.html.

CHAPTER 12: *Terror*

251 **an Inuk man named Kok-lee-arng-nun:** The convention of using hyphens to separate the syllables in hard-to-pronounce Inuit names was introduced by Hall in his journals. Woodman followed Hall's lead, and I have done the same.

251 **"He was a very cheerful man":** Woodman, *Unravelling the Franklin Mystery*, 198; Nourse, ed., *Narrative of the Second Arctic Expedition Made by Charles F. Hall*, 255–56.

252 **"were carried down with her":** Nourse, ed., *Narrative of the Second Arctic Expedition Made by Charles F. Hall*, 257.

252 **"a great many men—black men":** Woodman, *Unravelling the Franklin Mystery*, 204–6.

253 **"flaring torches and savage yells":** Potter, *Finding Franklin*, 178–79.

255 **granddaughter of "Patsy" Klengenberg:** "Klengenberg, Patsy, d. 1946," Dartmouth Libraries, Archives and Manuscripts, https://archives-manuscripts.dartmouth.edu/agents/people/1363; "Patsy Klengenberg," Angulalik: Inuinnaq Fur Trader, Kitikmeot Heritage Society, https://www.kitikmeotheritage.ca/angulalik-patsy-klengenberg.

256 **carefully replaced the stones:** Burwash, *Canada's Western Arctic*, 94.

256 **an HBC publication:** William Gibson, "Sir John Franklin's Last Voyage," *The Beaver: A Journal of Progress* 268, no. 1 (June 1937): 44–75.

258 **drawn him a sketch:** McClintock, *The Voyage of the 'Fox' in the Arctic*, 199.

258 **and two large canvas tents:** Nourse, ed., *Narrative of the Second Arctic Expedition Made by Charles F. Hall*, 407.

259 **"as though he was only asleep":** Nourse, ed., *Narrative of the Second Arctic Expedition Made by Charles F. Hall*, 595; Woodman, *Unravelling the Franklin Mystery*, 156–57.

259 **Hall later identified the spot:** Woodman, *Unravelling the Franklin Mystery*, 156; Douglas R. Stenton, "Finding the Dead: Bodies, Bones and Burials from the 1845 Franklin Northwest Passage Expedition," *Polar Record* 54, no. 3 (May 2018): 202, https://doi.org/10.1017/S0032247418000359.

262 **Jimmy and I ended up talking:** With Jimmy's permission, our conversation was recorded and everything in quotes is verbatim from the transcript.

262 **"If any bodies are found":** Dean Beeby, "Tragedies in Arctic Hamlet Sparking Talk of a Franklin 'Curse,'" CBC News, December 18, 2018, https://www.cbc.ca/news

/politics/franklin-gjoa-haven-nunavut-wrecks-erebus-terror-keanik-mckenna-blessing-curse-1.4946976.

266 **little punji sticks:** I didn't know about it at the time, but the noted Franklin archaeologist Doug Stenton had visited Fitzjames Island in 2012. He believes that the rectangular moss bed I found is the remains of a sealskin curing operation—not the site of the Terror Bay "tent place." Russell Potter, however, left this door open, writing to me in an email that it is "always possible that there are earlier features here, perhaps obscured by subsequent use, that could point back to Franklin!"

CHAPTER 13: Not Great but Good Enough

271 **at thirty-five different locations:** Stenton, "Finding the Dead," 197–212.

272 **Tetqataq, Ukuararsuuk, and Mangaq:** The varied translations and spellings of Inuit names recorded by Hall and subsequent historians have been a source of great confusion over the years. Tetqataq was also known as Terqetsaq/Teekeeta, and Mangaq was sometimes referred to as Tooshooarthariu. According to David Woodman, this confused Hall, who was told that Tooshooarthariu had met with the men in Washington Bay and that another Tooshooarthariu had nursed four survivors through a winter. Later, when Hall interviewed Mangaq/Tooshooarthariu, Mangaq said he knew nothing about nursing any survivors through a winter. Hall, not realizing that there were two men named Tooshooarthariu, got frustrated and "concluded that everything he was told was untrustworthy." To my knowledge, no one has made more of an effort to sort through these names and figure out who's who than Woodman, so in this book, I have chosen to go with the spellings/translations used in *Unravelling the Franklin Mystery.*

272 **"shook in the wind":** Woodman, *Unravelling the Franklin Mystery*, 124–33.

273 **"Innuits were in a hurry":** Woodman, *Unravelling the Franklin Mystery*, 130.

274 **what appeared to be German:** Russell Potter, "The Mystery of the 'Peglar' Papers," Visions of the North, March 31, 2009, https://visionsnorth.blogspot.com/2009/03/mystery-of-peglar-papers.html.

274 **Thomas Armitage and William Gibson:** Potter, *Finding Franklin*, 41.

274 **who took the wallet:** Woodman, *Unravelling the Franklin Mystery*, 153.

274 **housed in the Caird Library:** https://www.rmg.co.uk/collections/objects/rmgc-object-2113.

275 **whose 1939 magnum opus:** Cyriax, *Sir John Franklin's Last Arctic Expedition*, 170–71.

276 **most of the main party:** Potter, *Finding Franklin*, 179–80.

277 **marked with the initials "S.C.":** William Barr, ed., *John Rae, Arctic Explorer: The Unfinished Biography* (Edmonton, Alberta: Polynya Press, 2019), 353–54; for

more background on Rae, see: Ken McGoogan, *Fatal Passage: The Story of John Rae, the Arctic Hero Time Forgot* (New York: Carroll & Graf, 2001).

277 **trek across the south coast:** R. I. Murchison, "Address to the Royal Geographical Society of London," May 24, 1852 (London: W. Clowes and Sons, 1852), 4; a pdf of Murchison's address can be found on Wikimedia.

279 **"to the last dread alternative":** A transcript of Rae's September 1, 1854, report can be found here: https://canadianmysteries.ca/sites/franklin/archive/text/RaeProceedings1854_en.htm; Berton, *The Arctic Grail*, 266; Woodman, *Unravelling the Franklin Mystery*, 261, 298–99; Potter, *Finding Franklin*, 22.

280 **"would be acting a falsehood":** Heather Davis-Fisch, *Loss and Cultural Remains in Performance: The Ghosts of the Franklin Expedition* (New York: Palgrave Macmillan, 2012), 1; for more background on Lady Jane and the various expeditions she launched in search of her husband and crew, see: Pierre Berton, *Prisoners of the North* (Toronto: Doubleday Canada, 2004).

280 **"domesticity of blood and blubber":** Charles Dickens, "The Lost Arctic Voyagers," *Household Words*, December 2, 1854, 362–65, and December 9, 1854, 387–93, https://victorianweb.org/authors/dickens/arctic/pva342.html; Berton, *The Arctic Grail*, 268.

281 **"Mother Nature could probably not produce":** Heinrich Klutschak, *Overland to Starvation Cove: With the Inuit in Search of Franklin 1878–1880*, trans. and ed. William Barr (Toronto: University of Toronto Press, 1987), 133; Garth Walpole, *Relics of the Franklin Expedition: Discovering Artifacts from the Doomed Arctic Voyage of 1845*, ed. Russell Potter (Jefferson, NC: McFarland & Co., 2017), 120.

286 **announced plans for a new deepwater port:** Petra Dolata, "A New Canada in the Arctic? Arctic Policies Under Harper," *Canadian Studies* 78 (June 2015): 131–54, https://doi.org/10.4000/eccs.521; "Prime Minister Stephen Harper Announces New Arctic Offshore Patrol Ships," Government of Canada, July 9, 2007, https://www.canada.ca/en/news/archive/2007/07/prime-minister-stephen-harper-announces-new-arctic-offshore-patrol-ships.html.

286 **he formed a nonprofit:** Amy Dempsey, "Franklin Discovery Is Canada's Moon Shot: The Back Room Story of the History-Making Franklin Expedition Search," *Toronto Star*, September 13, 2014.

287 **"Canada's North broke Franklin":** Beattie and Geiger, *Frozen in Time*, 7.

287 **Speech from the Throne:** "Full Text: Throne Speech 2013," Global News, October 16, 2013, https://globalnews.ca/news/906578/full-text-throne-speech-2013/.

289 **best dive of his life:** "Finding HMS *Erebus*," Government of Canada, https://parks.canada.ca/lhn-nhs/nu/epaveswrecks/culture/archeologie-archeology/decouvertes-discoveries/erebus#; this site includes a nearly six-minute video with underwater footage of the wreck.

290 **"I have a really important story":** The Sammy Kogvik / HMS *Terror* discovery story was told to me by Adrian Schimnowski. All of the quotes, except where otherwise noted, come from the transcript of that interview. See also: Paul Watson, *Ice Ghosts: The Epic Hunt for the Lost Franklin Expedition* (New York: W. W. Norton & Company, 2017), 325–32; Steve Ducharme, "Ship Director and Inuk Ranger Tell the Tale of Terror," *Nunatsiaq News*, September 23, 2016.

293 **triggering its quick-release mechanism:** Watson, *Ice Ghosts*, 330.

294 **still attached to the rigging:** John Roobol, "The Condition of HMS *Terror* (New Data)," John Franklin Forum 119 (July 11, 2021). This website has since been taken down. The analysis of the *Terror* wreck comes from David Woodman. Here is an excerpt from the now defunct page: "Terror lies in 24m of water, upright on a hard substrate. Her gunwale stands 6m proud of the bottom . . . images show that, as with almost all ocean-going vessels, her freeboard (waterline to deck) measurement is significantly less than her draft (keel to waterline). We know from Whalefish letters that she was considered 'very deep' at 17-foot draft, so generously we could give her a keel to deck height of twice that (10m). These measurements show, as was verified by the Parks video that very little of her hull has penetrated the bottom, most of it visible (surprisingly there is no damage). The masts all lie off the starboard side, broken off leaving stumps (tallest is 5m) and still largely attached to rigging points on the ship. This strongly indicates that the ship did heel far to starboard when sinking and that the rigging could no longer support the masts, rather than having been broken by ice motion."

294 **Footage captured on a later dive:** "HMS *Terror*: New Video from Inside Arctic Wreck Reveals Artifacts Frozen in Time," Global News, https://www.youtube.com/watch?v=im2DtfgmMTc.

295 ***Terror*'s propellor was engaged:** Megan Gannon, "Divers Get an Eerie First Look Inside the Arctic Shipwreck of the HMS *Terror*," *Smithsonian*, August 29, 2019.

295 **Watson's exclusive on the story:** Paul Watson, "Ship Found in Arctic 168 Years After Doomed Northwest Passage Attempt," *The Guardian*, September 12, 2016.

298 **His role in the enterprise:** I shared the manuscript with Ben prior to publication and he shared a number of corrections, including this about the falling out we had in Cambridge Bay: "I didn't storm off angrily to find breakfast—I tried to get you guys to come with me, but you were too bent out of shape. I have my own version of our scrap in Cambridge Bay, but again, this is your story to tell. You should know, though, that I never wanted to be top dog nor intentionally to display superiority in front of others. I was there to do my bit as best I could, since our lives depended on everyone going all in. I did my best to sail the boat as best I knew; to maneuver it safely, to stand my watches on time, to make sure we ate as well as possible—not to be superior in any way, but to carry my share of the burden and do what I could to ensure success. But I never desired to be in charge—it was your boat and your trip.

I just wanted to be sure I was giving good value. That's why it astonished me so much that you were feeling so much animosity that you'd skip breakfast."

CHAPTER 14: Tuk-Toyota-Truck

308 **sailed the Northwest Passage twice:** David Thoreson, *Over the Horizon: Exploring the Edges of a Changing Planet* (Platteville, WI: Mark Hirsch Publishing, 2016); Herb McCormick, *One Island, One Ocean: Around the Americas Aboard Ocean Watch* (San Rafael, CA: Weldon Owen, 2011).

316 **"Empress of the Sea":** Alexander Armstrong, *A Personal Narrative of the Discovery of the North-West Passage* (London: Hurst & Blackett, 1857), 279; Berton, *The Arctic Grail*, 211–27.

CHAPTER 15: Merbok

321 **new highway that now connects:** Peter Kujawinski, "The Road to the Top of the World," *New York Times*, February 11, 2016, https://www.nytimes.com/2016/02/14/travel/canada-tuk-northwest-territory.html.

323 **fifteen hundred gigatons of carbon:** McGee and Gribkoff, "Permafrost."

323 **turbocharges the doom cycle:** Fox, *The Last Winter*, 71.

323 **is also strengthening hurricanes:** Chris Remington, "Researcher Examines Impact of Melting Polar Ice on Hurricanes," WLRN, April 30, 2019, https://www.wlrn.org/show/sundial/2019-04-30/researcher-examines-impact-of-melting-polar-ice-on-hurricanes; C. L. Parker, P. A. Mooney, M. A. Webster, and L. N. Boisvert, "The Influence of Recent and Future Climate Change on Spring Arctic Cyclones," *Nature Communications* 13, no. 6514 (November 2022), https://doi.org/10.1038/s41467-022-34126-7.

324 **waves started rocking the houses:** Nick Faris, "Eroding Tuktoyaktuk: Every Year Homes in This Northern Hamlet Are Getting Closer to the Sea," *National Post*, February 28, 2019, https://nationalpost.com/news/canada/eroding-tuktoyaktuk-every-day-homes-in-this-northern-hamlet-are-getting-closer-to-the-sea.

328 **"strongest storm in over a decade":** Chuck Johnston and Aya Elamroussi, "Alaska's Western Coast Is Expecting High Winds and Flooding in Powerful Storm This Weekend," CNN, September 16, 2022, https://www.cnn.com/2022/09/16/weather/storm-merbok-alaska-friday/index.html.

331 **Trump rescinded that order:** Amy Martin, "An Alaskan Village Is Falling into the Sea. Washington Is Looking the Other Way," *The World*, October 19, 2018, https://theworld.org/stories/2018/10/19/shishmaref-island-alaska-ground-zero-climate-change.

332 **red-painted tins and feces:** Nourse, ed., *Narrative of the Second Arctic Expedition Made by Charles F. Hall*, 601–2; David Woodman, *Strangers Among Us* (Montreal: McGill–Queen's University Press, 1995), 78–79.

332 **led by a "great officer":** Potter, *Finding Franklin*, 182.

333 **Barry wrote that two Inuk:** Barry's statement and analysis from Joe Eberling (an Inuk companion of Charles Francis Hall), McClintock, and Rae can be found in the *Journal of the American Geographical Society of New York* 12 (1880): 275–92, https://www.jstor.org/stable/196515.

334 **an anthropology professor from McGill:** The information regarding the 2008 expedition to Crown Prince Frederik Island comes from an email from Jim Savelle to Russell Potter. Here is that excerpt:

> *Re: Crown Prince Frederik Island . . . Arthur Dyke (the geologist I work with) and I surveyed the entire south, east, and northeast part of the island from sea level to the highest elevations on ATVs during a seven day period in 2008, recording all archaeological and paleontological sites. No cairns or any sites/material that might relate to Franklin/other explorers were noted. The north-central and northwest part of the island is very low, wet marshes; a Franklin-related camp(s) on the west coast is possible, but unlikely. Also that year, we spent seven days conducting similar surveys at Cape Chapman (northern tip of Simpson Peninsula) with similar results.*

334 **gifted to him in 1857:** Potter, *Finding Franklin*,182.

335 **"Crozier was the only man":** Nourse, ed., *Narrative of the Second Arctic Expedition Made by Charles F. Hall*, 589.

335 **"last to die was widespread":** Woodman, *Strangers Among Us*, 96.

336 **"& *this whaite* [*sic*] *man*":** Woodman, *Strangers Among Us*, 128.

336 **front page of the *New York Times*:** Derrick Bryson Taylor, "Western Alaska Lashed by Strongest Storm in Years," *New York Times*, September 17, 2022, https://www.nytimes.com/2022/09/16/us/alaska-storm-typhoon-merbok.html.

337 **When the floodwaters receded:** Emily Schwing and Mike Baker, "Storm Surge in Alaska Pulls Homes from Their Foundations," *New York Times*, September 17, 2022.

Epilogue

349 **check our boat tracker:** https://forecast.predictwind.com/tracking/display/SV_PolarSun/.

ACKNOWLEDGMENTS

I've already written about how this story was born, but I'd like to acknowledge once more that were it not for the unconditional love and support of my wife, Hampton, the epic adventure that forms the backbone of this narrative would never have taken place. When I first pitched Hampton on the idea of sailing *Polar Sun* through the Northwest Passage, she reacted as she always does when I come forth with my latest mad scheme: with unwavering belief in me and the conviction that with enough planning, preparation, hard work, grit, and vision, any of us can accomplish whatever it is we set our minds to. Life partners like this are not easy to find, and my only regret at this point is that Hampton, Tommy, and my other children, Will, Matt, and Lilla, didn't participate in more of *Polar Sun*'s voyage through the Arctic.

It's also true that I could not—and would not—have undertaken such an ambitious voyage without someone like Ben Zartman by my side. We had our differences along the way, but we emerged from the Northwest Passage not only as better friends than we were when we started, but with a bond that has now been tempered by the unforgiving elements that one can only find in the high latitudes.

I owe a similar debt of gratitude to Renan Ozturk and Rudy Lehfeldt-Ehlinger, who sailed aboard *Polar Sun* uncomplainingly for thousands of miles, during which they not only took care of the boat and its crew but somehow found the motivation and energy to document everything we did for National Geographic. How these two managed to do double duty throughout the expedition is something I am still trying to understand. I'm also further indebted to Renan for supplying the cover photo, as well as the bulk of the images in the photo insert. I hope one day I can find a way to repay his incredible generosity. And I'm most grateful to Rudy for the portrait of Renan that you'll find in the insert.

All told, a dozen people served aboard *Polar Sun* on this mission, and together we formed a team that was so much stronger than the sum of our individual parts. My sincere thanks go out to the rest of my crew, which included Erik Howes (aka Mr. Dirt), Stephen Morrow (my former editor at Dutton who sailed with us from Maine to Newfoundland), Taylor Rees, Jacob Keanik, David Thoreson, and Ben Spiess. Thoreson also provided sage council during the planning and preparation stage of the expedition, and contributed two photos to the insert.

The crew on our overland four-wheeler expedition on King William Island enjoyed similar solidarity and goodwill thanks to the leadership of the inveterate Franklin searcher Tom Gross and his old friend Jacob Keanik. Many thanks to the other members of this crew, which included Pam Gross, Devon Oniak, Eileen Gross, and Matthew Irving. It was especially gratifying and rewarding to spend time with Jacob in his element out on the land. And were it not for Tom and his tireless, multi-decade effort to find Franklin's tomb, I never would have landed National Geographic as a sponsor and media partner. Tom embraced me from the moment I first reached out to him, and in years since, he has always been there for me as a historian, researcher, collaborator, sounding board, and friend.

At National Geographic, show runner and editor Drew Pulley bit down hard on the idea behind this expedition from the moment I first pitched it to him; and his early enthusiasm for the project put wind in my sails just when I needed it most. There were many others at National Geographic who contributed to our success but I'd especially like to thank Bengt Anderson and John Block. The television special we created for *Explorer* is called "Lost in the Arctic" and can be found on Disney+. It was released alongside a cover story in *National Geographic* magazine that appeared in the August 2023 issue. Thanks to Nathan Lump, Peter Gwin, Sadie Quarrier, and Anne Farrar for having the faith that we could pull off such an ambitious assignment.

Additional support for the expedition was provided by my longtime sponsor the North Face. James Kelly, in particular, made sure that the entire team was well-equipped to face the elements in the Arctic. Sony, Yeti, Gill, North Atlantic Inflatables, and Good To-Go provided additional support. And for anyone looking to learn more about what went on behind the scenes, Sony's film—*Polar Passage Arctic Adventure: Field Notes with Renan Ozturk*—can be found on YouTube. Massive thanks also to North Sails for providing *Polar Sun* with new sails for the expedition.

They say it takes a village, and in the case of *Polar Sun*, I jokingly refer to this community of people as the *Polar Sun* Mechanical Consultancy Group. Some of these people even answer the phone accordingly when they see my name on the caller ID. No one has contributed more to my understanding of *Polar Sun*'s myriad systems than Kelly Overman, a previous owner who loves this boat as much as I do. I've called Kelly countless times for advice and direction, and I'm not sure if he has ever failed to pick up. And it was he and his wife, BJ, who called ahead to Nome to make sure there would be champagne waiting for us at the liquor store upon our arrival. Loric Weymouth, of Weymouth Yacht Rigging, and my old friend and sailing mentor, Thom Perkins, have similarly had my back as I've worked through endless issues aboard, and both will drop anything to pick up whenever I call for help. The folks at Rumery's Boat Yard in Biddeford, Maine, took excellent care of both me and *Polar Sun* over the course of a long winter I spent refitting in their yard. For anyone looking for a boatyard in Maine, you will not find better. I'd also like to mention Alan Creaser and Sandy MacMillan, two lifelong sailors from Nova Scotia who would, I'm sure, give any sailor in need the shirt off their backs.

No list of supporters would be complete without mentioning John Kretschmer, a hero and mentor whose enthusiasm for my bluewater dreams seems to know no bounds. JK, as we call him, is one of the most experienced deep-ocean sailors alive in the world today, and I've found endless inspiration from his books and through his friendship and stories. It was fitting, then, that we crossed paths with him and his beloved sailboat *Quetzal* at our first stop in Lunenburg, Nova Scotia, where we shared a dock and a wild night of storytelling.

So many people generously shared their stories with me, but Alan Cresswell and Annina Barandun, who we were trapped alongside us in Pasley Bay, also gave us their friendship during a time of intense emotional strain when I needed it more than they will ever know. Much like sharing a rope on a serious climb, our short but consequential time together forged a bond that has linked us irrevocably. In the years since our experience in the Northwest Passage, *Taya* rounded South America (again) and took on 1,300 pounds of fair-trade coffee beans in Brazil, which Alan and Annina then delivered to a speciality coffee roaster in Lausanne, Switzerland. *Taya* now sits on the hard in southern France awaiting her next adventure.

Others whose stories and direction helped inform both our voyage and this work of nonfiction include David Woodman (see Notes on Sources for more on David's contribution), Sam Lowry, Herb McCormick, Darcy King, Dave Gardner (a fisherman who befriended us in Newfoundland and took us to L'Anse aux Meadows), Jens Kjeldsen, Povl Linnet, Jimmy Pauloosie, Tom Surian, Gerry Chidley, Adrian Schimnowski, and Pablo David Saad. I'd also like to offer a special thanks to Nigel Buckley at Oxford's Balliol College, who was always there for me (as he was for my previous book, *The*

Third Pole) whenever I ran into a dead-end with my historical research or sourcing.

Adrienne White and Amanda Prysizney at the Canadian Ice Service went above and beyond in helping me to interpret the ice charts and the weather, even texting with me at critical junctures to make sure I had all the information I needed. I owe thanks to Danielle Zartman for giving her husband her blessing to join me on this epic mission, and she also spent many hours downsizing ice charts and emailing them to *Polar Sun*. Marc de Keyser and Fritz Buyl at weather4expeditions provided us with professional weather forecasts early in the expedition. Andrew Cassels, the skipper of SV *Draco*, preceded us through the Northwest Passage and shared intel on the ice, weather, and routing. He and his crew squeaked through the James Ross Strait a couple days before us and then sailed on without incident while we found ourselves trapped in Pasley Bay.

No one followed our voyage more closely than Hampton's father, Alan Kew, aka the YouTube sensation "Captain Q," who is another of my long-time sailing mentors. Alan communicated with me more than anyone (including Hampton) during the voyage, offering endless support and advice about the the ice, weather, and other piloting decisions. In fact, so closely did he follow our tracker from his easy chair in southern Maine (just a few miles from where we originally set off in Biddeford) that on several occasions he texted me that it was time to tack. Usually, he was right.

For images in the photo insert, I'm indebted to Alamy, the New York Public Library, Wikimedia Commons, and *The Illustrated London News*. Thanks and gratitude as well to Daniel Huffman (www.somethingaboutmaps.com) for their brilliant work on the map, which I hope you referenced frequently while reading. I met Daniel thanks to Soren Walljasper, a cartographer at National Geographic who created the brilliant, award-winning map that ran as part of the spread in *National Geographic* magazine.

ACKNOWLEDGMENTS

When I returned home to New Hampshire from the Northwest Passage in September 2022, the first person I called was my old friend and collaborator Ben Ayers. I needed someone to help me wrap my head around how to tell this story, and I'm not sure if I know anyone who understands the storytelling craft better than Ben. If you enjoyed this book, it's due in large part to the many hours Ben spent consulting, reading, analyzing, editing, rereading, criticizing (and sometimes praising) my work, while always maintaining an uncanny ability to zoom out and see the big picture. "What is this book supposed to be about?" he would often ask. Thank you, Ben, for asking the tough questions and for always pushing me to dig deeper. I couldn't have done this without you.

Thanks also to Madeline Woda, Matt Samet, Ben Spiess, and Alan Kew, who provided additional editorial support, and to John Climaco and Jeff Chapman who read first drafts of the manuscript.

One thing I hadn't expected is that I'd be switching editors partway through the writing of this book. I lost my longtime editor Stephen Morrow, whom I worked with on my previous two books, when he left Dutton in the fall of 2023. I was sad to see him go as he's a good friend and he had helped me to launch and outline this story. But as is usually the case in life, there was an unseen benefit because I soon found myself in the capable hands of veteran editor Jill Schwartzman. Most of the book had already been written by the time I submitted the first pages to Jill, and I'll admit that I suffered through some sleepless nights worrying that she might not appreciate the direction I had taken with the narrative. It was a huge relief, then, when I learned that she loved the story. Thankfully, though, she tempered her enthusiasm with some critical feedback on the book's structure, particularly relating to Part One. With her fresh eyes and deft skills as an editor, Jill directed a major restructuring that slowly turned this into the work that I always imagined it could be. Assistant editor Charlotte Peters added further insight and refinement, as did senior production editor LeeAnn Pemberton and associate director of publicity Emily Canders. The rest of the team at Dutton, including John Parsley, Erica Rose, Dora Mak, Laura Corless, Melissa Solis, Clare Shearer, Nicole Jarvis, Amelia Zalcman, and Dominique Jones, worked tirelessly to make sure that this title had every opportunity to shine. Special thanks also to copy editor Frank Walgren and proofreader Kim Lewis, who poured over every word and polished this up incredibly.

None of this, of course, would have come together the way it did were it not for my gifted and tireless agent, Gillian MacKenzie, whose advice and direction I have never failed to heed. She is a faithful friend who believes in me as much as Hampton does. Thanks also to her associates Liz Rudnick, Kirsten Wolf, and to the Marsh agency, which does a stellar job of finding foreign publishers for my books.

But perhaps the greatest gift that this project has given me is the realization that in sailing I have finally found an endeavor through which I can knit my two greatest passions in life: adventure and family. As *Polar Sun* sails on, putting as much distance as she can between her fiberglass hull and those cold and capricious waters of the Northwest Passage, she does so now in the warm embrace of the Synnott family. And if you're curious where we might have ended up—and I hope you will be—you can learn more via our boat tracker, which can be found at: https://forecast.predictwind.com/tracking/display/SV_PolarSun/.

INDEX

ABOUT THE AUTHOR

Mark Synnott is a twenty-eight-year member of the North Face Global Athlete Team, an internationally certified mountain guide, and a trainer for the Pararescuemen of the United States Air Force. A regular contributor to *National Geographic* magazine, he is the author of the *New York Times* bestseller *The Impossible Climb* and *The Third Pole*. When not living on *Polar Sun*, Mark and his family reside in the Mt. Washington Valley of New Hampshire.